SNOWY *to the* SOMME

A Muddy and Bloody Campaign, 1916-1918

BIG SKY PUBLISHING
www.bigskypublishing.com.au

TIMOTHY J. COOK

First published 2014

Big Sky Publishing Pty Ltd
PO Box 303, Newport, NSW 2106, Australia
Phone: 1300 364 611
Fax: (61 2) 9918 2396
Email: info@bigskypublishing.com.au
Web: www.bigskypublishing.com.au

Cover design and typesetting: Think Productions
Printed in China by Asia Pacific Offset Ltd.

See National Library of Australia for Cataloguing-in-Publication entry

ISBN: 978-1-922132-63-5

Front cover image: A line of Australian soldiers making their way along a duckboard track behind Delville Wood. AWM E00138
Back cover image: A mule team bogged in thigh deep mud near Potijze Farm in the Ypres Sector. Note the sandbags and tents in the background. The soldier on the left, smoking a cigarette, is carrying his gas mask. AWM E00962

SNOWY *to the* SOMME

A Muddy and Bloody Campaign, 1916-1918

BIG SKY PUBLISHING
www.bigskypublishing.com.au

TIMOTHY J. COOK

THE AUSTRALIAN ARMY HISTORY COLLECTION

ARMY • HISTORY • UNIT

PROTECTING ARMY HERITAGE
PROMOTING ARMY HISTORY

Winning with Intelligence
Judy Thomas

Duntroon
Darren Moore

The Warrior Poets
Robert Morrison

The History of the Royal Australian Corps of Transport 1973–2000
Albert Palazzo

Defenders of Australia
Albert Palazzo

The Fight Leaders
D. Butler, A. Argent and J. Shelton

Operation Orders
Pat Beale

Little by Little
Michael Tyquin

Red Coats to Cams
Ian Kuring

Bowler of Gallipoli
Frank Glen

Vets at War
Ian M. Parsonson

Only One River to Cross
A.M. Harris

The Fragile Forts
Peter Oppenheim

Hassett: Australian Leader
John Essex-Clark

Persian Expedition
Alan Stewart

The Chiefs of the Australian Army
James Wood

Never Late
Gordon Dickens

To Villers-Bretonneux
Peter Edgar

Madness and the Military
Michael Tyquin

The Battle of Anzac Ridge 25 April 1915
Peter D. Williams

Doves Over the Pacific
Reuben R.E. Bowd

The Lionheart
David Coombes

Battlefield Korea
Maurie Pears

Chemical Warfare in Australia
Geoff Plunkett

A Most Unusual Regiment
M.J. Ryan

Between Victor and Vanquished
Arthur Page

Country Victoria's Own
Neil Leckie

Surgeon and General
Ian Howie-Willis

Willingly into the Fray
Catherine McCullagh

Beyond Adversity
William Park

Crumps and Camouflets
Damien Finlayson

More than Bombs and Bandages
Kirsty Harris

The Last Knight
Robert Lowry

Forgotten Men
Michael Tyquin

Battle Scarred
Craig Deayton

Crossing the Wire
David Coombes

Do Unto Others
Alan H Smith

Fallen Sentinel
Peter Beale

Sir William Glasgow
Peter Edger

Training The Bodes
Terry Smith

Bully Beef and Balderdash
Graham Wilson

Fire Support Bases Vietnam
Bruce Picken

Toowoomba to Torokina
Bob Doneley

A Medical Emergency
Ian Howie-Willis

Dust, Donkeys and Delusions
Graham Wilson

The Backroom Boys
Graeme Sligo

Captains of the Soul
Michael Gladwin

TABLE OF CONTENTS

AWM P06539.001

In memory of
2142 Sergeant William Henry Cook, 55th Battalion, AIF,
killed in action at Polygon Wood, Belgium, on 26 September 1917.

His brother,
7020 Private Arthur Frederick Cook, 17th Battalion, AIF,
wounded in action at Mont St Quentin, France, on 31 August 1918.
Returned to Australia in 1919.

And
the Glorious Dead of the 55th Battalion, AIF.

I was eighteen when I was first sent into battle at Fromelles. The hard thing was to see so many mates drop beside you and to control yourself to face up to the things happening that you couldn't really believe … War is both the silliest and cruellest activity that a man can get mixed up in. It breaks my heart even now to think of it.[1]

Walter Herbert 'Bert' Bishop, ex-55th Battalion,
speaking almost seven decades after the end of the Great War.

Introduction

January 1916: war rages in Europe. A dozen men from Delegate, a small community in the high country near the border between New South Wales and Victoria, begin to make their way to the Australian Imperial Force (AIF) Training Depot in Goulburn. Along the route other men join the walkers, rural workers for the most part, all keen to play their part in the Great War. The 'Men from Snowy River' recruitment march passes through all the major regional centres of the Monaro district. A total of 144 'Snowies' arrive in Goulburn on 28 January, most allotted to the 4th Reinforcements of the 55th Battalion. Eleven months would pass before the 'Snowies' reach the Western Front and join the battalion. By the time they arrive, the unit has been blooded at Fromelles, spent months garrisoning the front line, and is now enduring the cruel conditions of the Somme battlefield in winter. To the reinforcements, it has been a long journey from the Snowy to the Somme.

The 55th Battalion was raised in the deserts of Egypt in early 1916 and moved to the Western Front in the middle of that year. The members of the battalion came from all sections of Australian society. Whatever their background they shared four qualities: they were volunteers, none was a professional soldier, they had enlisted in New South Wales and all believed, to varying degrees, in the righteousness of the cause for which they were fighting.

By the time the 'war to end all wars' was over, every man had paid a price for belonging to the battalion. Some 531 had died and thousands had been wounded. Many of the survivors, tormented by physical and mental wounds, would argue that the living, and not the dead, were the ones who had made the ultimate sacrifice.

In the years that followed the Armistice, most AIF units published accounts of their wartime deeds. At various times, the veterans of the 55th attempted a similar undertaking, but this task was never completed. This book fills that gap in the historical record and gives a voice to the thousands of men who, for three tempestuous years, called the 55th Battalion their home.

Acknowledgements

As a boy in the 1960s, I often visited my grandmother's house in Braidwood, a small community in rural New South Wales. In a hallway hung a mildewed parchment adorned with imagery of God and Empire. It spoke of the death of my great uncle, William, at Polygon Wood in 1917 as a member of the 55th Battalion. In the guest bedroom, the mantelpiece supported a coaster-sized bronze disk, a 'dead man's penny', embossed with William's name and the reassuring epitaph that 'He Died for Freedom and Honour'. These two artefacts were enough to spark a child's interest in his dead relative and the men who fought alongside him.

Over the decades since then, it has been my privilege to tramp alongside the ghosts of the 55th Battalion. I have spent the past few years attempting to capture the humanity of these men in the most inhumane of environments. This book is the progeny of those labours.

This undertaking would not have been possible without the assistance of many people. I'd like to begin by thanking a number of the descendants of men from the battalion for providing me with many of the stories I have used: Pamela Grolsch (still researching the life of her father, Bert Bishop), Daphne Bishop, Richard and Gaewyn Hurst, Darrel Cunnington, Sally-Ann Twardochleb, Alan Cheers, Barbara Brady, Bob Buckingham, Sally Twardochleb, Mandy Keevil, John Pearson, Ondrae Campbell and Desley Woodcock. One of the true delights and surprises of researching and writing this book was the unconditional friendship of these wonderful people who are quietly proud that their relatives wore the colour patch of the 55th.

Professor Peter Dennis kindly assisted me to compile the nominal roll; Professor Bill Gammage helped me navigate the resources of the Mitchell Library. Nick Fletcher and Craig Tibbetts at the Australian War Memorial assisted me by sharing much of their research on the 56th Battalion, and introduced me to the national treasure that is the AWM. Damian Madden's creativity and drive to tell the stories of the AIF to a new generation stimulated me. The 'Families and Friends of the First AIF' helped me connect with others sharing my passion for the AIF. Glenn Mason and Mick Martin from Regimental Books were generous in their advice on the publishing industry, as well as pointing the friends and relatives of the 55th in my direction.

The hospitality extended by Michael Woods and his wife Kate during my many stays in Canberra can never be repaid — thanks mate. Ben Waugh helped me with some of the technology. Natalie Le Hanie showed me the Somme flowers on a bright Sydney winter's day. Jodie Siganto reviewed my early drafts and urged me to persist.

Dr Andrew Richardson and staff of the Army History Unit provided me a grant that supported my visit to the battlefields of France and Belgium. One can read many books, but a true understanding of the manner in which a battle unfolded and the reasons many decisions were made is only possible after 'walking the course'. I appreciate your giving me the opportunity to do so.

My manuscript was polished using the editing advice I received from Kathy Stewart and Cathy McCullagh who offered expert and impartial guidance to the amateur author I am. I appreciate the time you spent assisting me to shape this book.

My son, Andrew, offered his critiques on early drafts of the manuscript — thanks for your pertinent comments and constant attacks on my overuse of the comma and the word 'that'. Amelia, my artistic and talented daughter, spent many hours drawing the maps. Mum and Dad never failed to ask after the progress of my endeavours.

Finally, my wife, Jane-Louise was happy for me to disappear into the study most evenings after dinner, and accompanied me as I traipsed through the muddy fields of Flanders. Thanks for putting up with my 'hobby' and keeping the home fires burning. I could not have completed this book without your love and support.

In spite of the varied assistance I have been offered by many people, I alone am responsible for any errors or omissions in this book. Should these be drawn to my attention, I will make every effort to have them rectified in any future editions.

Timothy J. Cook
March 2014

List of maps

List of figures

List of symbols used in maps

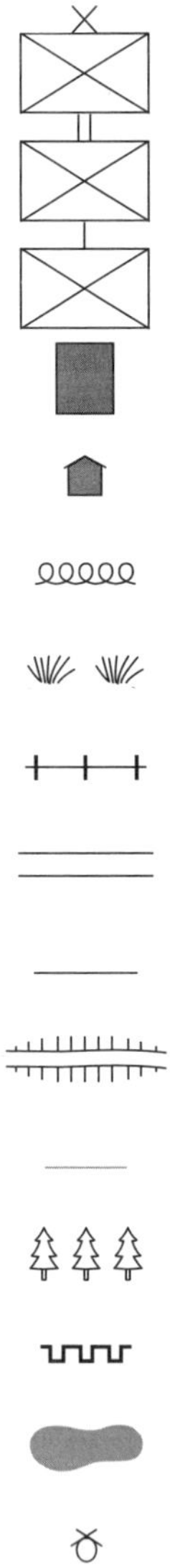

Infantry Brigade
Infantry Battalion
Infantry Company
Built-up Area
Building
Barbed Wire
Marsh or Swamp
Railway Line
Road
Path
Sunken Road
Watercourse
Forest
Entrenchments or Breastworks
Water
Windmill

Organisation of an AIF division

The composition of an AIF division varied somewhat throughout the war, changing to suit the operational circumstances. Nevertheless, these changes did not fundamentally affect the structure or employment of the division as a fighting formation.

Throughout the war, a division contained three infantry brigades and several artillery brigades, as well as engineers, pioneers, signals, supply and medical staff. At full strength, each infantry brigade totalled some 4000 men and comprised four infantry battalions, a machine-gun company and a mortar battery.

Structure of a generic AIF division:

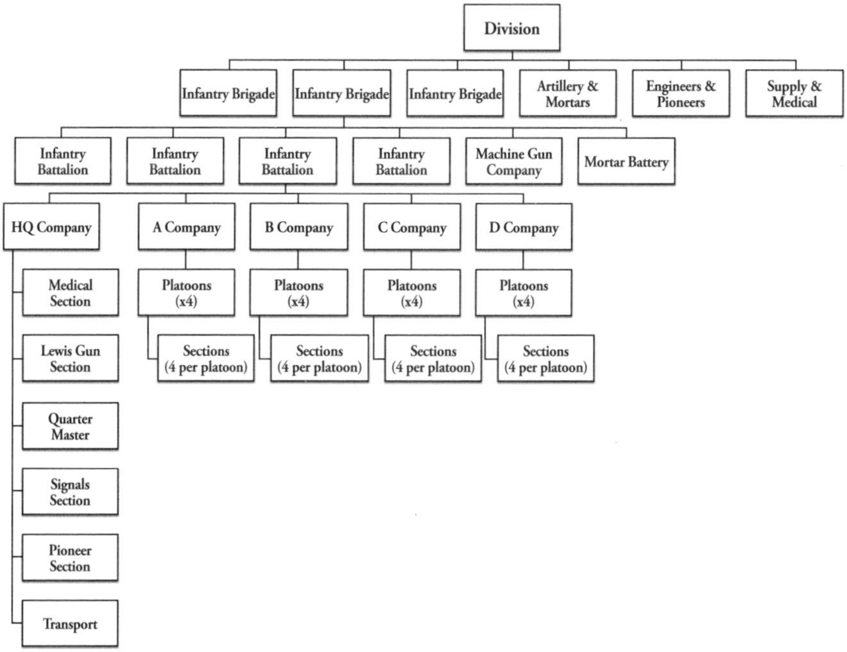

An infantry battalion numbered around 1000 men, most of whom were organised into four infantry companies commanded by a major or captain; each company consisted of four platoons led by a lieutenant or second lieutenant. Platoons were sub-divided into four sections, each consisting of 12 to 15 men under command of a corporal. Those soldiers not in the infantry companies were placed in a headquarters (HQ) company consisting of the battalion leadership (the commanding officer, senior major and adjutant), and specialist sections including medical, Lewis guns, quartermaster (QM), signals, pioneer and transport.[2]

Chapter 1: Egypt – 'too hot for a white man'

At no other time in Australia's history have the flames of nationalism burned as brightly as during the early years of the Great War. Andrew Fisher, Australia's Prime Minister, had committed his country to wage war to the 'last man and the last shilling'.[3] To honour the pledge, thousands of men poured into army recruiting depots around the nation. Troopships crammed with reinforcements steamed to Egypt where their human cargoes were awaited by the 1st and 2nd divisions that had recently withdrawn from the Gallipoli peninsula and were now refitting.

By early 1916 the Australian military authorities in Egypt faced a significant problem: the huge influx of fresh troops was well beyond what was needed to replace the losses suffered by the 1st and 2nd divisions in Turkey. The solution chosen by officialdom was straightforward: increase the AIF from two to five infantry divisions. The means of achieving this expansion involved splitting the existing 1st and 2nd divisions in half, creating two additional infantry divisions in Egypt. The pool of excess manpower was then used to bring all four divisions to full strength. Those enlistees still in Australia were allotted to the last of the planned divisions.

As part of the AIF's expansion, the veteran 3rd Battalion, sited in the dusty desert camp of Tel el Kebir, was reorganised into two 'wings'.[4] The troops in the 'Headquarters Wing' were to reconstitute the 3rd Battalion, while those men in the 'Second Wing' formed the cadre of a new 'daughter' or 'pup' battalion — the 55th.

The 3rd Battalion's superior headquarters, the 1st Brigade, was also to break in two, with one half to become the nucleus of the 14th Brigade. The new brigade, consisting of the 53rd, 54th, 55th and 56th battalions and supporting units, was combined with the 8th and 15th brigades to form the 5th Division. As the 1st Brigade had been raised in New South Wales (NSW), reinforcements for the 14th Brigade would only be drawn from those who had enlisted in NSW.

Brigadier General Godfrey Irving was appointed the 14th Brigade's commander and proved a controversial selection.[5]

To create the 'Headquarters Wing' and 'Second Wing' the 3rd Battalion's Commanding Officer (CO), Major David McConaghy, held a battalion parade and performed the difficult and unpleasant task of subdividing his unit. By the evening of 13 February, the officers and men had been allocated to their respective 'wings'.[6] McConaghy, a 28-year-old decorated Gallipoli veteran, then relinquished command of the 3rd Battalion and took up the mantle as CO of the 55th.[7] His officer cadre from the 3rd Battalion comprised Captains Robert Cowey[8] and Selwyn Holland;[9] Lieutenants Percy Woods,[10] Norman Gibbins,[11] John Marshall,[12] Eric Stutchbury,[13] John Matthews,[14] Sidney Pinkstone,[15] Herbert Palmer,[16] William Denoon and Lionel Sheppard.[17] Second Lieutenants Harry Wilson,[18] James McCarthy,[19] William Giblett,[20] Fred Cotterell,[21] Norman Pinkstone[22] and Percy 'Bob' Chapman[23] completed the list of officers.[24]

At full strength, an AIF battalion consisted of 35 officers and 970 soldiers. To make up numbers in the 55th, 50% of the original 3rd Battalion accompanied Major McConaghy, along with the entire 9th Reinforcements and some members of the 10th and 11th Reinforcements for that battalion. About one-quarter of the men in the new battalion had served at Gallipoli, including every officer McConaghy brought with him. Over the coming months additional reinforcements, many originally slated for the 17th and 19th battalions, would trickle into the 55th.

All units of the AIF wore a unique shoulder patch. In the case of infantry this patch consisted of coloured stripes representing the respective brigade and battalion. The colour of the 1st Brigade patch was green. The 3rd Battalion's patch was brown and worn horizontally above the green patch of the brigade. To maintain affiliation with its 'parent' battalion, the 55th adopted the same green and brown colours but wore the patch vertically, the green to the rear.

The 55th took over the 3rd Battalion's tents at Tel el Kebir. Prior to the arrival of the AIF, Tel el Kebir had comprised a railway line, railway station and a collection of mud huts sprawling among some date palms. By the time the 55th Battalion was formed, the settlement had grown to the size of a large town. Tel el Kebir lay astride the Sweet Water Canal,[25] a freshwater canal running from the Nile River near Cairo eastwards across the desert to Ismailia.[26] The surrounding countryside consisted of tracts of sand interspersed with hard, gravelly hillocks.

The days following the formation of the unit were chaotic as officers and non-commissioned officers (NCOs) struggled to impose order on the disorderly. Every day fresh reinforcements made their way to the camp, adding to the swirling mass of men. Private Herbert Allen[27]described the battalion's early days:

> In the full Egyptian sunlight they [the troops] saw their new bivouac which began and ended in sand ... The camp extended for miles. It lay on the slope of the desert facing a sweet water canal with its strip of vegetation on either side. Ranged along the top of the hill was [sic] different Divisional and Brigade Headquarters. Below them lay the battalions of infantry ... Down the slopes the tents lay in rows. The 55th Battalion occupied five rows. Its four companies each occupied one row, the remaining row was occupied by Headquarters. One tent was allotted to ten men or two officers. The tents and ground covered by the Battalion bivouac were kept scrupulously clean ... Below the tents and at the bottom of the hill were the huts of matting in which the Battalion messed. Beyond these again were the water-taps, showers, latrines and incinerators. Beyond these were the railway lines with its [sic] detraining platforms. Beyond the railway line were the cultivations, the canal, more cultivations and then the desert again.
>
> The Battalion, before it settled down to work and the desert, looked around to see with whom it had cast its lot. All eyes were on the Battalion leader – the CO [McConaghy]. We saw him first as a tall stiff figure on a horse, facing the Battalion as it formed up in companies of platoons each morning. We noted his strong, far seeing look, his firm mouth and chin, his slow sure voice. We decided he was a squatter – a man of the plains, who had fought droughts and won. We learnt later that he was a Sydney businessman.
>
> A Company was in charge of Captain Cowly [Cowey], a tall, reserved man, like his chief. His ideal was thorough competence.
>
> B Company knew its Captain Gibbins as a fearless man who had earned his reputation on the Turkish Peninsula.
>
> C Company was led by Captain Holland, a Sydney solicitor. He was also a veteran and a keen and clever soldier.
>
> D Company was in charge of Lieutenant [Sidney] Pinkstone, an officer of stout bulldog type.
>
> The Adjutant, Captain Woods, [was] a sporting and real Australian officer ... The [Regimental Medical Officer – RMO] was Captain Wyllie[28] [and] it was soon learned that he was no friend of slackers.[29]

Basic training began. Most of the reinforcements had received little in the way of formal military instruction in Australia or during their voyage to the Middle East. The mixture of desert sand and firm ground around Tel el Kebir was the ideal terrain to begin moulding recruits into soldiers, in the process weeding out those unable to withstand the rigours of an infantryman's life. Private Allen, a naïve newcomer, wrote of the training:

> It was impressed upon the Battalion that the serious task, for which it was training, demanded a serious endeavour ... A day's work in the desert was no work for weaklings. Every day the Battalion drilled, marched or manoeuvred under a burning sun over soft sand and rough rubble. The Battalion became thin, lithe and active ... Six mornings of hard drill and six afternoons of lectures made up a typical week's work ... The 8th of March was notable for a lecture by the C.O. on 'Trench Warfare'. We learnt with interest of the sedentary but vigilant life of the trenches. We learnt of the stress and excitement and intensity of an attack, when the trenches ceased to be trenches. We decided that artillery was very useful, that bombs ad lib were essential, that barbed wire was necessary and that bayonets were handy.[30]

For his part, Private Archie Winter noted of this period: 'We were training hard on the desert. It was no joke marching and doing extended order in the sand.'[31] The marches often saw troops carrying heavy packs through soft sand, sinking up to their ankles with each step. The days were warm and officers enforced water discipline, limiting the quantity consumed by the men at any given time. It was common for men to collapse from this combination of physical exhaustion and dehydration.

One of the more colourful characters in the battalion was Sergeant 'Williams'. An Englishman, 'Williams' had originally landed at Gallipoli with the French Foreign Legion but had deserted and made his way to Anzac Cove where he unofficially attached himself to the 3rd Battalion. Eric Wren, the battalion historian, described 'Williams' as 'Tall, lean and sinewy, with a rigid military carriage and soldierly bearing.'[32] 'Williams' spoke little of his past, disclosing only his flight from the Legion and that he had served with the Coldstream Guards. He proved a valuable soldier and remained with the 3rd Battalion until the unit withdrew to Lemnos. Here he was placed under arrest for attempting to stir up trouble and handed over to the French, who promptly sentenced him to be shot for desertion. As he was being led to his execution, he was recognised by an officer of the Coldstreams. This officer was able to negotiate a reprieve conditional on the 3rd Battalion agreeing to retain him. This was achieved and 'Williams' continued to serve with the 3rd until he was transferred to the 55th Battalion. According to Wren:

> It was a splendid sight to see Williams, who was promoted to Sergeant's rank at Tel El Kebir, drilling his platoon. Combining the cold, calculating efficiency of the Coldstream methods with the magnetic fire of the Legion, inspiring those under his command to parade ground manoeuvres that had perhaps no parallel in the AIF.[33]

'Williams', however, had contracted tuberculosis and was struck off the strength of the battalion at the end of April 1916.[34]

Life was not all military work. Unless rostered for duty, the men spent the evenings at leisure, visiting friends, writing letters, going to movies, attending boxing bouts and so on. Private Allen observed that

> The chief delight of the Battalion was beer. After beer came the band. 'Our Band' was first mooted when the C.O. put up the necessary financial backing. All that was wanted were bandsmen. Feverish appeals were made in orders for men who could play cornet, trombone, or bass or even bang a drum. Likely men were buttonholed til at last all modesty and shyness was overcome. The Battalion band became a reality with real instruments, real bandsmen, and real practices every day under a real conductor (Sergeant Cosgrave[35]) … Beer, with which the band was often associated, was retailed by thoughtful authorities in canvas booths known as 'wet canteens'. While the YMCA was the abode of love, wet canteens were the abode of uproar.[36]

Private Fred Farrall[37] also reflected on life at Tel el Kebir, commenting that:

> There was nothing attractive about the camp with its many long lines of tents standing on the sands. Everything at Tel el Kebir called for stamina, for toughness. Sand was in everything, the water and the food. This was primitive as far as cooking was concerned, it was being done in the open, where plenty of sand added to the stew didn't improve its taste … Sooner or later nearly everyone fell victim to severe stomach pains and loss of appetite … Life was made more unpleasant for us by an arrogant officer, Captain J.J. Marshall, under whom we served. He could safely be described as a tyrant.[38]

Marshall was just one of the many officers and NCOs Farrall grew to dislike during his time in the army.

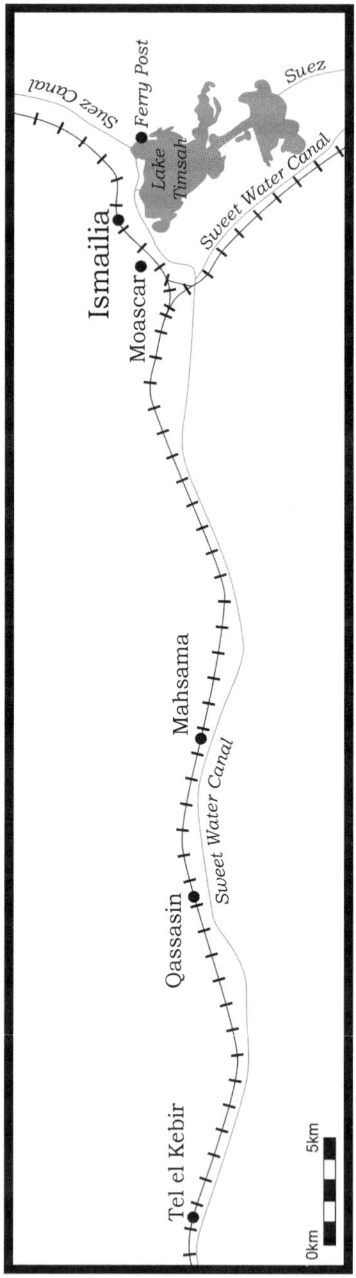

Map 1: Route of desert march.

On 20 March, the 5th Division was instructed to take over a section of the Suez Canal defences 15 kilometres east of Ferry Post. The 8th Brigade was detailed to take over the sector of the front allocated to the division. From Tel el Kebir this brigade travelled by train to Moascar, on the outskirts of Ismailia, and then used the pontoon bridges and ferries spanning the Suez Canal to reach Ferry Post, one in a chain of forts on the eastern side of the canal protecting the waterway from Turkish attack across the Sinai. After two days' rest at Ferry Post, the 8th Brigade made its way to the front line on 27 March.

The rapid move to the front by the 8th Brigade removed any imperative for the other two brigades in the 5th Division to rush to Ferry Post. As a result, Major General James McCay, who had assumed command of the 5th Division on 22 March, decided that the 14th and 15th brigades would march the 57 kilometres from Tel el Kebir to Ferry Post.[39] McCay regarded the activity as an opportunity to test the mettle of his troops. He instructed the 15th Brigade to make its way to Moascar following a route along the hard sand next to the Sweet Water Canal. The 14th Brigade was issued different orders: it would make its way across the desert sands. The men in both brigades were directed to carry a full pack, 120 rounds of ammunition and their rifles.

The 14th Brigade's march was to commence on 27 March with that day's objective Mahsama, a small village and railway siding 23 kilometres away. The following day the men would make their way to Moascar, a distance of 27 kilometres. The last day's journey would comprise the seven kilometres from Moascar to Ferry Post.

The 20 officers and 526 other ranks of the battalion moved off at 7.00 am on 27 March.[40] It was a hot day and the marching troops stirred up a cloud of fine, choking dust. At midday, the battalion halted for lunch near Quassin, a settlement adjacent to the railway line and the Sweet Water Canal.[41] No additional supplies of potable water were made available and guards were placed along the canal to prevent the thirsty men drinking its noisome water.

The march resumed in the early afternoon. As Private Sinclair 'Clair' Hunt recollected later:

> We marched across undulating country, sharp rises and hollows, with a dozen yards of soft sand at the bottom. After a couple of hours going chaps began to drop out, and all day hundreds of them followed the [rest of the Brigade] as best they could. [42]

The battalion finally reached the encampment and white-coloured houses of Mahsama. Parched men jostled around the water tanks, but found there was only sufficient water to refill their water bottles. Many soldiers took matters into their own hands, Private Hunt recording that:

> The men rushed the canal [Sweet Water Canal] and drank the water, which is unfit for human consumption ... We were all dead tired, and set to work to drink as much water as we could. Before morning I had drunk three bottles of water and two dixies of tea.[43]

Twenty-four of the battalion's men had fallen out during the march. By 9.00 pm, 17 of these had rejoined the unit, while ambulances following the infantry column picked up the remaining stragglers. That evening Private George Gill reflected on the day's trek:

> Tired from the hardest walk or march I ever had … With a full pack jammed full of our belongings and 120 rounds of ammunition could hardly have imagined that I could have carried it one mile much less the distance that we did and we have to do another 30. The heat was terrible and all the way they were dropping out, some completely some for a spell. I managed so far to keep well up. [44]

Private Robert Harpley was similarly exhausted:

> [The march] was horribly trying, the heat being considerably over 100 degrees and we had to travel in the heat of the day arriving at the first stage station … minus about 5 or 6 hundred men. And those you would think strongest were fagged out already and were struggling in until about 9 p.m. dead beat.[45]

After a night spent sleeping under the stars, the troops were woken at 5.00 am. For Private Harpley and many of his mates, 28 March was to become 'The day that I will never forget as long as I have breath in my body.' The acute shortage of water meant that many men remained dehydrated, primed for physical collapse.

The battalion set off after consuming a breakfast of bacon, biscuits, jam and a little tea. According to Private Hunt, they 'Marched out at 7 a.m. to music of the band. The day promised to be a scorcher ... We marched in short spells for 40-50 mins.' Each man carried a single, one-litre water bottle.[46] While a number of troops fell out during the morning, as the going underfoot was firm, the column maintained its integrity. Brigadier General Irving halted his brigade for lunch at around 11.30 am. The place he selected was pitiless: no trees, no bushes, just burning sand. Private Harpley wrote later:

> We camped for dinner [lunch] now minus 700-800 men who were already knocked up. We had no extra water or tea for dinner so we had to drink out of our little supply in the bottle. Many of the poor fellows who went a little rash with their water had none left at dinner time. The reason for a lot of this is that our Brigadier General told us we only had 3 more miles to do but we found we had 8 miles to do and the heat was intense at least 110 degrees [Fahrenheit – about 43° Celsius].

After a pause of an hour, the men were roused and the march recommenced. It was now the hottest part of the day and conditions were akin to a furnace, with the sun beating down and the bright sand reflecting the heat. Almost immediately the dehydrated troops struck trouble — dunes of fine drift sand. Private Hunt captured the disintegration that followed:

> We then entered upon yellow sand – seven miles of it. If you can imagine the sand as Sydney dunes as above the beaches and one hill after another of it, with several precipices by way of a change, you will have some idea of the country ... The line began to lengthen, and the men to drop ... At halts we just dropped, gasping. Men cried on all sides for water; many fainted. We struck some salt water, and a few drank it. As we struggled on men collapsed beside us and we could do nothing. Rifles, packs and cartridges were thrown everywhere.

Private Allen recorded similar scenes of suffering:

> Our Brigade ... struggled up and passed over ridges of soft sand. It passed over – but at each ridge there remained a line of black figures, prone upon the white sand. Many suffered from exhaustion. Most suffered from thirst, and from thirst there was no relief. Some swam in the brackish pools that lay in the bottom of sandy dips for the sake of temporary coolness and refreshment. Some drank the salt-water and made their suffering the greater. One or two men went quite mad. A few became dazed and foolish. Men that had started out with well-stuffed packs plus rifle and ammunition arrived in with uniforms only.

Troops foamed at the mouth with exhaustion and thirst. The demented men argued and fought over the dregs remaining in water bottles; some tried to drink rifle oil.

The first groups of men eventually staggered into Moascar. The 14th Brigade had become a rabble. Companies had diminished to handfuls of men; where battalions had started the march, platoons were now arriving.

Soldiers from a nearby New Zealand brigade beheld the pitiful condition of the early arrivals and rushed into the desert to assist the remainder of the column. The Kiwis worked through the night, rescuing men in all stages of debilitation. Some were found unconscious, some naked and raving; others were suffering from sunstroke and were too weak to open their lips and take a sip of water. Private Hunt was one of the many who had fallen out:

> I passed by 'Bac' talking of Mummy and home. He was done ... How I got in I don't remember. I only remember getting up and dropping, and doing it I don't know how many times. When I reached camp I fainted, and was rescued by New Zealand cooks, who treated me to everything they had. I was in at 4 p.m. and could walk at 5 p.m.

For his part, Sergeant Arthur Vogan:

> Dropped out being knocked up with general fatigue and shortage of water ... The rest of the journey was a nightmare. We rested for twenty minutes to half an hour and walked for 10-15 minutes wetting our mouths from water bottles now and again but not sufficient water to have a proper drink. Men fell out right and left ... About three-quarters of a mile [from] camp we were met by New Zealanders coming out with water in bottles, dixies etc. and they gave us a drink and carried our packs in for us. I could not eat more than a slice of bread and jam for tea but drank tea and water galore then vomited it all up.[47]

Those men still able to walk drifted in from the desert over the next few hours. The soldiers were furious at the manner in which they had been treated. Discipline broke down completely and in the darkness officers struggled to contain the near mutinous situation. Gradually military order reasserted itself — the surly troops were simply too exhausted to maintain their fury.

The cover-up of the calamity began that evening. The War Diary, the official record of the unit's activities, contained the following nonsensical entry for this tumultuous period:

> Throughout the day the conditions were much the same as the previous day and Moascar reached at 1830. During the day 4 men fell out, making a total of 11 missing.[48]

The battalion left Moascar Camp for Ferry Post at 8.45 am the next morning, the troops uncharacteristically quiet. As they departed, the soldiers strode past the Prince of Wales, escorted by Brigadier General Irving and a posse of senior officers.[49] Private Harpley later recalled: 'We gave him [the Prince of Wales] three hearty Australian cheers and then we roared "One for our Brigadier General" and he received groans from every man.' Private Winter's account supports Harpley's version of events: 'The Prince of Wales came along accompanied by the great "leader" [Irving] who brought us across the sand – someone called for three cheers for the Prince, but Irving got hoots.' The embarrassment suffered by Irving in the face of such insubordination by his brigade must have been acute.

As they passed through Ismailia the footsore men appreciated the relief offered by the town's formed roads, shady streets and well-kept gardens, but this still did not stop eight of them falling out. The battalion used a pontoon bridge to cross the Suez Canal and made its way to Ferry Post. After an inspection by the RMO, they were given time off. Private Allen wrote:

> All those who could walk made for the Canal, a few hundred yards away, which was soon lined with several battalions of nakedness. In the famous waterway the dust and weariness of our three days pilgrimage in the wilderness was washed away.

The nightmare may have been over but the recriminations were yet to begin.

During the afternoon, Major General McCay wrote a message to all ranks of the 14th Brigade. He was scathing of their conduct during the march, noting particularly the degree of straggling on the Moascar–Ferry Post stretch, writing: 'I am compelled to say plainly that to-day's failure in soldierliness of the 14th Brigade ... has been a great disappointment to me, and that the blame rests largely with regimental officers, as well as NCOs and the men themselves.' Officers and soldiers alike were incensed by McCay's comments. Until now they had solely blamed Brigadier General Irving for their plight, but McCay's injudicious outburst began the slow erosion of trust in his judgement and abilities.

Rumours of fatalities resulting from the march swirled around the camp; there were still 19 men from the battalion whose whereabouts were unknown. Private Harpley listed the brigade's death toll as two officers, two NCOs and seven enlisted men, with some 200 hospitalised. He also asserted that madness had driven one soldier to shoot himself. There is no evidence that anyone died on the march, although journalist W.B. Dalley later suggested that at least one man in the brigade was left so broken that he was repatriated to Australia.[50] The 55th's missing men eventually made their way to Ferry Post.

The CO of the 8th Field Ambulance, Lieutenant Colonel Arthur Shepherd, produced a report describing precisely why the men had struggled during the march. [51] Shepherd concluded that the underlying reasons comprised an overall lack of fitness, worn-out boots, poor water discipline, malingering, and the conduct of the activity during the hottest part of the day. He mentioned the lack of water only in passing. Other sources blamed the after-effects of inoculations administered during the days preceding the march. No-one seems to have questioned the sufficiency of a single water bottle per man. Given the hot conditions, one water bottle was vastly inadequate for a demanding trek by an unencumbered man. By loading the men with a 45-kilogram burden, Major General McCay ensured that dehydration alone would be sufficient to create a catastrophe.

Brigadier General Irving must have realised that he would be held accountable for the debacle and was careful where he apportioned responsibility in his report to Major General McCay. He offered many reasons to explain why a night march was not possible and emphasised that his brigade had followed the route specified by 5th Division Headquarters. These explanations did not save Irving: McCay sacked him on 1 May 1916, appointing Colonel Harold Pope in his place.

In his turn, Major General McCay sought to deflect any possible stain on his military record as a consequence of the march. He was vulnerable to criticism; McCay had ordered the march, stipulated that it be conducted with full packs and set the demanding timetable. Irving, however, became his scapegoat, and McCay's removal of Irving became characteristic of the man's *modus operandi*. Months later in France he would dismiss Pope after another shambles — Fromelles.

The 55th Battalion took over garrison duties the day after arriving at Ferry Post, manning the 'fort' and guarding strategic locations in the surrounding area. The duties of the garrison battalion were not onerous. Men not on guard duties were employed in various fatigues, chiefly unloading stores and hauling heavy punts from one side of the canal to the other. After the rigours of the preceding weeks, Private Harpley appreciated the easy life: 'We are now camped right on the edge of the canal. It is a glorious camp, very little work and tons of time to swim.'

After eight days of rest and garrison duty the battalion moved to a large training precinct outside the 'fort' known as Ferry Post Staging Camp to continue their training. Private Allen wrote of the camp's regime:

> Training was to be conducted under active service conditions ... Route marches on the same old sand were our 'red letter days'. The Battalion became more thin and more wiry than ever … We marched and we manoeuvred – we extended, we closed in, we advanced in platoons, we advanced in line, we twisted and turned this way and that.

The sophistication of training increased and the unit began participating in battalion, brigade and even divisional-level exercises. Additional reinforcements arrived and the battalion soon reached full strength.

Private Herbert Harris, one of the 'elders' of the battalion at the age of 42, chafed at the camp's conditions:

> The best thing about it is the occasional swim. The worst is the food and sand. The men are always hungry and don't get enough to keep a ten-year-old boy growing. Can't even buy anything – canteens are out of things the same day they arrive. Have been here two weeks and could only get one pound of biscuits and one tin of syrup in all that time. Unless we get away from here or better food (that is in quantity) is supplied there is likely to be trouble.[52]

Private Rupert Campbell was another soldier not enjoying himself at Ferry Post Staging Camp: 'It's nothing but sand, heat, flies and thousands of other discomforts here … We are in fairly good condition but trudging for miles through this loose sand with full packs up is breaking our hearts. It's not training men, it's killing them.'[53]

With the approach of summer the hot winds of the khamsin arrived, carrying great quantities of sand and dust. Like everyone else, 2nd Lieutenant Chapman had no choice but to endure these wind storms:

> Of course it would not do to be in Egypt without experiencing all the vicissitudes of the climate. After the wind had done its best to fill our ears, eyes and mouths with sand, a shower of rain tried to occur, it succeeded just long enough to lay the dust, but it was not an enjoyable night! Now … the ground has dried, the wind is trying to blow the tent down ... Sand is everywhere, outside it beats against the tent like sleet. Inside, well it is inside of everything. I grind my teeth and crunch sand. I have swallowed a good deal also, and foolishly had a glass of water which appears to have made mud inside me. Anyway, I feel rather peculiar.[54]

On 25 April, the men marked the anniversary of the Anzac landing with a church service and sports carnival. For Private Winter the unquestionable highlight of the day was a communiqué in which '[Brigadier General] Irving broke the glad news to us that he was leaving us, which was a great relief to all the boys. He would be more at home leading a bullock team instead of men.'

On 14 May, the battalion was informed that it was going to the front line the next day. Rumours of Turkish incursions across the Sinai added a touch of nervous excitement to the coming deployment.

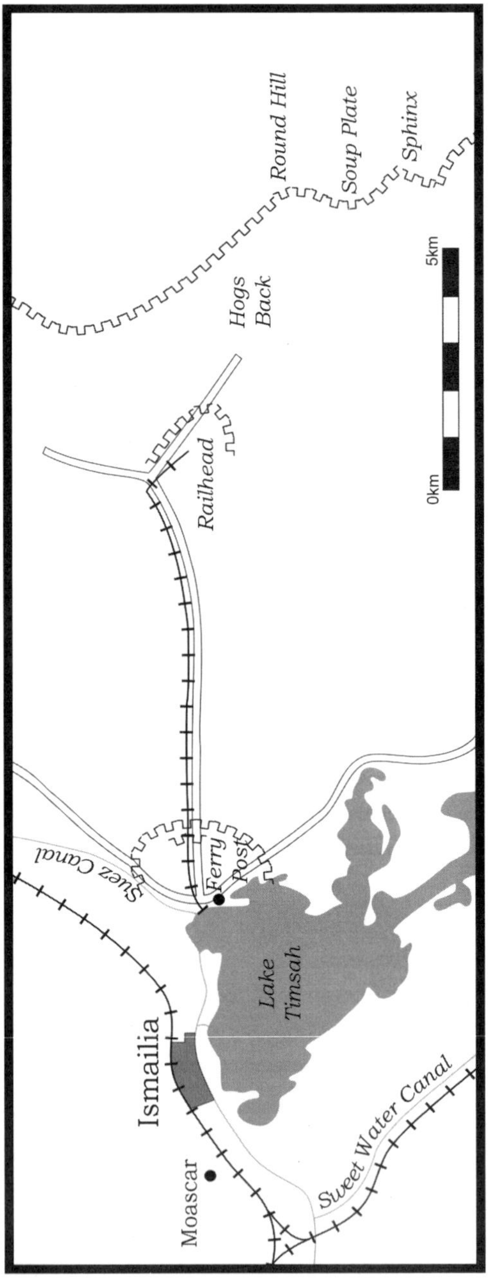

Map 2: Suez sector – May 1916.

The Australian front line was situated in the middle of a 'blank hopeless desert'.[55] To resupply the forward troops a light railway had been constructed from Ferry Post to a terminus creatively named 'Railhead'; adjacent to the railway ran a water pipeline and a road made of crushed stone. Camels were used to move rations, water and other supplies from Railhead to the battalions dispersed along kilometres of front-line trenches.

As directed, the 55th moved to the front line early on 15 May. Private Bishop[56] described the march:

> The morning was searing hot, with full pack, extra-water bottle, ammunition and bombs, each man carried more than eighty pounds. Our boots sank to their tops in the light sandy areas, gravelly patches now and then giving some relief. In less than an hour it was not a march. In my platoon man after man dropped out, flopped with his pack to support him and said: 'I'm dead'. Within two hours our battalion was an untidy shambles of exhausted men strung along more than a mile of sizzling desert. We drank sparingly, those of us with full water bottles helping those with no water.[57]

The heat was intense and the battalion was obliged to rest at Railhead where some huts provided a little shade. Private Winter recorded the temperature inside one of these huts as 134°F (approximately 57°C).

The slog was resumed once the blowtorch heat had abated. Sphinx Post, a flat piece of desert surrounded by high sandhills and the principal defensive position in the battalion's sector, was reached in an hour, the men taking over from the 15th Brigade. The battalion also occupied secondary strongpoints, known as Club Foot, Soup Plate, and Saucer, which had been constructed along the tops of these sand hills.

Just before dawn the soldiers 'stood-to' — the first of many times during the war.[58] It was hot and the strengthening light revealed the total desolation of the battalion's new 'home'. Not a tree, bush or blade of grass was visible: there was no sign of any animal or bird life. On the other hand, flies, snakes, a 'robust and virile breed' of scorpion, and sand insects were present in plague proportions.[59]

Water was at a premium and the soldiers were told that it was an offence to use it for anything but drinking. There was a restriction of one water bottle per man per day; unknown to the men the water pipeline from Ferry Post had been turned off the previous day and water reserves were critically low. After 'stand-down' the battalion instigated a system of sentries and patrols, leaving the remaining men to maintain the defences, a recurring chore as the drifting sand would quickly fill a trench.[60] These activities came at a terrible physical cost to the thirsty soldiers as Private Harpley recorded: 'The second day in the trenches was so hot that quite a large number of men were carried away exhausted.'

Gradually the troops became used to the routine of trench life: working by day and fruitlessly waiting for the Turks by night. The awful heat and trying conditions made this a period of great hardship. Private Campbell wrote from a support area just to the rear of the front line:

> We are out of the trenches at present and for the last couple of days the temperature has been 130°F [54°C] in the shade and 170°F [76°C] in the sun … The inside of a sandbag dugout … is about all the shade there is as there is not a tree within miles. Water is very scarce. We fall in and have our bottles filled every 24 hours … a lot of the [men] when, off duty, having drunk their water walk back about 7 miles to a camp called Railhead and fill their water bottles and return to camp. Fancy walking 14 miles through drift sand for a drop of water. We are not allowed to wash or shave – any man doing so is crimed.[61]

The battalion was relieved by the British 53rd Division during the morning of 30 May and trudged back to Ferry Post Staging Camp. This march was completed in good order despite the heat and the arduous going. On reaching Ferry Post the unit was greeted with the news that the 5th Division would shortly embark for France, although the exact timing remained the subject of ongoing speculation. Everyone was energised by the anticipation of 'real' soldiering on the Western Front and of an imminent escape from sand, heat and insects.

The 55th spent the next couple of weeks engaged in rigorous training and being issued with the equipment necessary for the Western Front. But life was not all work and no fun. When they were not tramping through the desert or practising marksmanship and bayonet assaults, the men spent the daylight hours swimming in the Suez Canal. Evening concerts, with the band as the star attraction, became a feature of battalion life and, according to the War Diarist, were 'A great assistance in maintaining good spirits and filling the leisure hours after the day's training was over.' For those men with money, canteens supplied tinned fruit, chocolate, lemon squash and beer. After their tribulations in the front line, the greatest comfort of Ferry Post was the ready availability of fresh water.

On 15 June, the level of anticipation reached fever pitch. Private Gill wrote in his diary: 'Heat terrific. Believe that we are leaving tomorrow morning for Moascar. Hope so for then we shall soon be out of this place. Too hot for a very white man.' On the same day Private Harris scribbled: 'Leaving tomorrow morning for sure … the Colonel shouted a case of beer … All bustle and excitement.'

The battalion crossed the Suez Canal for the final time the next morning and occupied the camp at Moascar; the long-awaited move to France had begun. The troops entrained at Moascar on 19 June and, after an overnight journey on open rail trucks, arrived at Alexandria. By 8.00 am the entire unit was aboard the transport ship SS *Caledonia*. Private Harris was underwhelmed by the prospect of the voyage: 'Crammed in like sheep ... Fearful heat below. Everyone looked like they had a Turkish bath at tea time.' Once all were aboard, the *Caledonia* cast off and made its way to the outer harbour. Here they waited for two hot and frustrating days for the remainder of the convoy to assemble, the tedium broken only by occasional swims in the ocean.

On 22 June, the *Caledonia* finally made its way out of Alexandria and linked up with its destroyer escort. The delay had done little to improve Private Harris' attitude: 'Much growling by the men. This pigsty of a ship. Hot trip, poor food.' Harris was not alone in his disgust at the conditions, Private Winter observing: 'Having a good trip but the boat was wrongly named as it should have been called "starvation ship". The food was awful and very scarce.' Owing to the presence of enemy submarines, lifebelts were worn continuously during the day and used as a pillow at night, adding to the discomfort of the voyage.

Seven days later the *Caledonia* docked in Marseilles; the 55th had finally arrived in the major theatre of war. After months in the desert wastelands of Egypt the soldiers were awestruck by the beauty of the green hills and soft mists enveloping the French city. Even the usually dour Private Harris was moved to write: 'It is the most beautiful country that anyone could see. No wonder they call it la belle France.' In little more than a fortnight over 80 of the men gazing in wonder at the French countryside would be dead, while hundreds more would be prisoners or wounded.

Chapter 2: Fleurbaix, July 1916 – 'a short apprenticeship'

The battalion disembarked in Marseilles on the morning of 30 June to a muted reception from the populace. Private Robert Harpley was relieved that his Mediterranean cruise was over: 'We were pleased too [sic] leave the cussed boat. I never had such starving in my life – the last three days we had only hard-ship biscuits and tea and a very small portion of "burgoo".'[1] After disembarking from the *Caledonia*, the troops made their way to a nearby railway station where they boarded a train for the front; there was no time for sightseeing. Private Herbert Allen chronicled the sentiments of the French and Australians: 'Amid much playing of the Marseillaise and the cheers and emotions of a French crowd we entrained. The Battalion was one big smile.' In the afternoon the train began making its slow way north.

The 55th was destined for the British zone of the Western Front. Like all newly arrived AIF units, the battalion's initial tour in the front line was to take place east of Fleurbaix, a village in northern France. The marshy ground around Fleurbaix made the area unsuitable for large-scale military operations and the sector had developed a reputation for quietness, at least by the standards of the Great War. Fleurbaix was the perfect nursery in which to acquaint inexperienced troops with trench warfare and prepare them for more vigorous operations elsewhere.

For three days the train wound its leisurely way through the French countryside, passing through numerous cities and towns. Occasional stops near villages allowed the men to stretch their legs and provided the opportunity to procure a bottle or two of the local vintage. The Australians, after months in the desert, were captivated by the French countryside. Private Allen was moved to comment: 'The Battalion ... wondered if this was the Garden of Europe. Not an inch of ground was wasted. Tilled land, rich crops, lush grass, tall trees, broad streams, prosperous villages and towns everywhere.' Private Frank Brown recorded similar sentiments: 'Never did anyone enjoy 63 hours travelling more than I. The glorious green country, rich with orchards of cherry, pear and other fruits seemed like home. Everyone became enraptured with the miles of vegetation … and a more happy crowd of troops would be hard to find.'[2]

The train reached its final destination, the little French village of Thiennes, in the early hours of 3 July. A grim reality check greeted the alighting soldiers as Private Harpley observed: 'As soon as we got out of the train we could hear the guns roaring – they seem to be firing almost incessantly.' In darkness interrupted by the occasional flicker of distant artillery, the battalion marched along narrow lanes until the men arrived at their allotted billets in farms and barns. They were 30 kilometres behind the front line and Private Herbert Harris 'could see the flashes of the bursting shells and hear the growl of the guns.'

The 55th recommenced its training the following day. Emphasis was placed on honing skills in bayonet fighting, musketry, and gas protection. Route marches of progressively longer distances were undertaken. The soldiers found the program exhausting, as they had lost much of their fitness during the journey from Egypt.

While the battalion was putting the finishing touches to its preparation, the 14th Brigade was ordered to occupy the front line near Fleurbaix on the night of 11 July. The first stage of the journey to the firing line required the unit to march to Estaires on 9 July. On receiving the news of the impending move, Private Harris wrote: 'So we are off at last to the bullets and shells.'

The troops began moving out of Thiennes at 8.00 am, Private Allen recording the unit's departure: 'The village waved handkerchiefs to the Battalion as it strode out on a long march to the trenches. Headed by the C.O. and inspirited by the band, we set off in gallant fashion.'

There were 22 kilometres to cover before they reached Estaires. The day was oppressively hot and many of the soldiers were not feeling especially gallant by the time they finished the journey. To their surprise, they found Estaires relatively untouched by the hostilities. Private George Gill described the market town:

> It is all narrow streets, few nice shops ... We are now very near the firing line but outside the town everything looks peaceful enough ... Have had a look around this place but are not impressed at all ... Estaires is

> not altogether free of gas attacks ... most of the inhabitants appear to be possessed of gas helmets but do not appear to have much fear of gas.'

The battalion departed for the village of Sailly sur la Lys during the evening of 10 July. Enemy aircraft prowling the night sky kept progress slow along the ten-kilometre route but, by midnight, the troops were in their billets. While in Sailly sur la Lys the men were ordered to keep close to their accommodation; wandering around the wrecked town was forbidden. Sailly sur la Lys was within range of the German guns and the need to remain concealed from German observation balloons and spotter aircraft was apparent to even the most rookie troops.

The next day was spent cleaning weapons and completing the administrative and logistical preparations necessary before any tour in the trenches. An edgy Private Harris found an opportunity to update his diary: '11th July. Tonight into the firing line. God this is a rotten game ... so if this is my last entry good bye.' Sergeant Francis Armstrong was feeling less anxious, writing: 'Don't feel least bit excited or afraid.'[3]

The battalion formed up on the cobblestone roads outside the billets at 8.30 pm and commenced marching frontward shortly after. Private Allen described the move to the firing line:

> On the night of 11th July we set out on the last stage of the journey to the trenches. The Battalion went slowly and cautiously.
>
> The position of the firing line was easily seen miles away on account of the numerous flares or Very lights which both sides send up to see what's doing.[4] Gradually we approached the 'fireworks'.
>
> The artillery was strangely silent. The Battalion had expected to come into a welter of dreadfulness – guns booming and biffing, blood flowing and searchlights glaring but the crackle of a small gun or the cough of a big one was very occasional. We saw one searchlight – far away.
>
> The Battalion filed down a trench and was warned to keep its head low. In small [groups] for greater safety it arrived at the dead-end called the firing line.
>
> Some men were detailed for duty. The rest were stowed away in dugouts and fell soundly asleep. The Battalion was up against it.

Private Harpley also described this first trip to the trenches:

> We moved out along the side of the street in file and kept close in to the trees on the side of the road the whole way. It was about 5 miles, and the 'swankers' [officers] did it in one stage, not even a blow did they give us. When we got about 2 miles from the front line the artillery began to fire and the bombardment was terrific for about an hour, but luckily we took a bye [sic] track and missed the shrapnel. We entered a communications trench and made towards the main firing line. We had about a mile to do. One could hardly hear the other for the roar of shells. On all sides of us had evidently been a town but it was levelled to the ground. We walked like blazes and the cussed pack dragging our shoulders out, but we eventually got to our destination named 'Dead Dogs Avenue'. It's a jolly good possy [position] ... I took charge of a [firing] bay and we were potting away [shooting] all night.'

After travelling thousands of kilometres and undergoing months of training, the troops were finally just 300 metres from their major adversary: the Germans.

The men stood-to as first light approached and Private Allen was 'astonished at the peacefulness of war. Beyond a hail of bullets from machine guns, which was the usual good-morning greeting, there was no noise to report.' In this part of the line the German machine-gunners fired interminably, but the unit soon adapted to the constant presence of machine-gun bullets. Dawn also revealed that the 'trenches' were passages between high sandbag breastworks rising above the surrounding countryside; the high water table around Fleurbaix made it difficult to construct earthworks below ground level.

Private Jim Maynard, who had previously served at Gallipoli with the 3rd Battalion, wrote of his experiences during this first tour:

> This part of the line was reckoned a real 'rest home' ... Our cookers [field kitchens] were only just behind the actual front line, and across the way Fritz's could be seen stoking up after stand down in the morning.[5] Rations and rum were obtainable without difficulty and were plentiful.
>
> [We] had a few fatigues to do of a night. Sometimes pushing the ration truck up a little light railway, or perhaps a bit of mending of a broken parapet – but generally speaking our nights were peaceful.[6] 'Parapet Joe', the German machine gunner, played a few tunes at intervals, perhaps at Jim Magee coming in from one of his nightly prowls – he was the company's sniper.[7] Small patrols went out occasionally, ready for trouble, but seldom finding any – Fritz didn't believe in spoiling a good thing.

> But a very few miles away the sky was lit up every night and the roar of the guns was unceasing – that was Ypres they told us. Our days were spent mostly in sleeping – with an occasional sentry posted here and there with a periscope. The heavy fatigues that were such a feature of the days at Gallipoli were absent here. Now and then our 18 pounders gave Fritz a bit of peppering, and he replied in kind – generally knocking a few more bricks off the old monastery, which was just to the rear of the line. We tried to reach Fritz's trenches with rifle grenades ... and on occasion the Stokes mortar specialists, as was their custom, fired a dozen or so of their 'little beauties' [mortar bombs] and made a hasty 'get-away' leaving the PBI [poor bloody infantry] to take the consequences.[8]

The Germans periodically pounded the battalion with artillery. For the majority of men, this was their first time under sustained shellfire. One afternoon Private Harris had 'Just finished dinner [lunch] when Germans began sending sausages [shells] over. Got so hot were ordered into dugouts. The shells bursting not 20 yards away and covering us with mud – a lot of men hit.' Describing another bombardment, Private Gill wrote:

> Early in the evening we were entertained by a big German gun or guns which I think was trying to fix up our battery. It fired about sixty shells which must have been about 9 [inches] at least. We could trace the course of the shells which went over our place and landed some distance away and would burst with an awful explosion. Although it was some distance away, we could feel the rush of the wind after each explosion.

During one artillery barrage a shrapnel ball from one of the shells smashed into the eye of Private Edward McKenna, a cook, killing him outright.[9] As the 55th's first fatality, McKenna's death sent shock waves through the battalion. The incident is mentioned in a number of the diaries even though few of the diarists would actually have known the dead man.

After the initial trepidation and nerves, the men settled into the routine of trench-life eloquently portrayed by Private Harris:

> Have been working all night on fatigues ... since coming into the trenches have had six hours off in 34 and haven't fired a shot yet, nothing to fire at. Where I am sitting one fellow had his hand smashed. ... Have got a little dugout of my own. At any rate it is drier than the last one where the walls were dripping wet and slugs climbed all over [my] face.

On a night when he was not rostered for work parties, Harris crossed another threshold in the life of a soldier, making his 'First effort to kill a German who was sending up flares. If I did not get him at any rate made it too hot for him for no more came from there. It's flukey shooting at night.'

In late afternoon of 15 July the Canadians, several kilometres away from the Australians, launched a gas attack. The wind direction changed unexpectedly, blowing some of the poisonous vapours over the section of line held by the 55th. In another 'first', the troops were forced to don their gas masks.

Several hours later the Germans launched a raid against the 58th Battalion located in the line to the right of the 55th. The Germans supported their operation with massed artillery and *minenwerfer* fire.[10] Private Harris updated his diary the following morning:

> I never expected to write this after last night. [The Germans] started bombarding the 58th. It was hell. My position adjoined theirs and of course came in for some of it. I never thought anyone could live through such lead and explosives ... The 55th carried themselves like veterans ... the shells were so close that once my hat rose up. The metal was flying around in tons.

Despite the heavy bombardment, the 6th Battalion Shropshire Light Infantry took over the front line from the 55th in the early hours of 16 July. The Australians made their way to billets in Bac Saint Maur, having suffered no casualties from the German bombardment or during the relief operation.

The 55th's first tour of duty in the trenches was over. For Private McKenna and the dozens who were wounded, Fleurbaix had proved anything but a 'nursery'. The few days in the front line had been sufficient to introduce the men to the reality of the Western Front: sniping, patrolling, bombardments, blood, death. They had entered the domain of war and emerged as better soldiers. In the words of Private Jim Maynard: 'Looking back, it seems Fleurbaix served the purpose of a kind of short apprenticeship for the more deadly parts of the front we were presently to visit.' They had little time to wait before they would be visiting these 'more deadly parts'. Private Allen recalled: 'On a hard floor of a village factory the Battalion as it turned in at night [16 July] heard whispers of a stunt.' Rumours circulating within the unit indicated that the 'stunt' was to be undertaken in a nearby sector, close to a village called Fromelles.

Chapter 3: Fromelles – 'the whole turnout was absolute hell'

The Allied war strategy for 1916 involved near-simultaneous offensives against the Central Powers by the Russians in the east, the Italians to the south and the Anglo-French forces in the Somme Valley.[1] However, the Allied planners had not consulted the enemy. Far from being willing to remain on the strategic defensive, the Germans launched their own massive assault against Verdun in February 1916. The French defenders of the fortress-city were stretched to their limits, but did not break. The Verdun offensive soon settled into a grinding affair of attack, counter-attack and endless artillery bombardments that feasted on men and materiel.

Originally intended to sunder the German Army on the Western Front and permit the hoped-for strategic breakthrough, the proposed Anglo-French offensive now took on the additional objective of drawing enemy resources away from Verdun. The Anglo-French push commenced on 1 July 1916 with substantial casualties and mixed results. The anticipated breakthrough remained elusive, but the British High Command persisted in its belief that the German Army was close to cracking under the relentless onslaught.

To maintain pressure on the enemy, the Commander-in-Chief of the British Expeditionary Force, General Douglas Haig, sought proposals to prevent the Germans augmenting their forces on the Somme. General Herbert Plumer believed the Germans were susceptible to an attack south of Armentières near where his Second Army shared a boundary with the First Army.[2] Accordingly, Plumer communicated with General Charles Monro, his counterpart in the First Army, proposing a joint operation using elements of both armies in the Fromelles/Aubers area.[3] On 8 July, Monro directed Lieutenant General Richard Haking, commanding XI Corps of the First Army, to draw up proposals for his corps to launch an assault.[4]

Lieutenant General Haking drafted a plan to capture the strategically significant Aubers Ridge. Situated to the south of Armentières, this gently rising and barely discernible feature was a little over a kilometre behind the German forward lines. Occupation of the ridge gave the enemy a significant point from which to observe the British to the west. Haking's proposal was a virtual replica of an idea previously executed with no success during the Battle of Aubers on 9 May 1915.[5]

General Monro initially vetoed Lieutenant General Haking's plan but the seed of eventual disaster had been sown. The British High Command agreed to undertake an 'artillery demonstration' in the vicinity of Aubers Ridge that would be a useful diversion and assist in operations to the south. This demonstration was to take the form of bombardments of enemy positions to persuade the Germans that a massive infantry assault was imminent and thus convince them to keep their troops in the area.

However, on 14 July, the British High Command reversed its earlier decision and approved a modified plan put forward by Lieutenant General Haking that included an infantry assault as well as the artillery demonstration. Haking's new plan restricted the infantry's objectives to capturing the German support trenches; there was never any intention to launch a subsequent attack on Aubers Ridge. Precisely how Haking thought the addition of a limited infantry assault would add to the confusion of his enemies remains unclear.

A fortified enemy salient, known as the 'Sugar Loaf', dominated the area selected by Lieutenant General Haking for the operation. This strongpoint lay approximately 1500 metres in front of Aubers Ridge and three kilometres north-west of Fromelles.

Lieutenant General Haking was allocated three infantry divisions for the enterprise: two British (the 31st and weakened 61st) from General Monro's First Army and the novice 5th Australian Division from General Plumer's Second Army. In terms of artillery, 258 field guns and howitzers, and 134 heavy and medium trench mortars were placed at Haking's disposal.[6]

The battle plan called for the preparatory artillery barrage to begin on 14 July and continue for three days. However, poor staff work saw Haking advised as late as the morning of 14 July that the quantity of artillery ammunition allocated to his attack was not as copious as he had anticipated. To make up for the shortfall in shells, Haking reduced the frontage of the infantry attack to 4000 metres and concentrated his artillery until he had one field gun or howitzer for every 15 metres of front. The net result of this compression was that the 31st Division was removed from the operation.

The build-up for the offensive was disjointed. The weather closed in, hindering the movement of artillery and supply convoys. The rain was not alone in wielding a significant influence on preparations; doubts were raised at various headquarters concerning the wisdom of the attack. Delay followed delay, orders were issued then quickly countermanded. The day chosen for the infantry attack was 17 July, but arrangements were not completed and the advance was postponed. There was talk of cancelling the entire undertaking.

Finally, on 18 July, Lieutenant General Haking unequivocally ordered the attack to occur the following day. The plan of operations was by now substantially revised from what had originally been approved. The preliminary artillery bombardment was reduced to just seven hours' duration and the start of the infantry assault was moved from 11.00 am to 6.00 pm, three hours before nightfall.

The 5th Division was allotted two objectives: the capture of the German front and support lines from where the Laies River intersected the Sugar Loaf to a point near Delangre Farm, followed by consolidation of the German support line.[7] Critically, Major General James McCay, commander of the 5th Division, directed his troops not to consolidate the German front line. The division was to attack with all three brigades abreast; on the right the 15th was to secure part of the Sugar Loaf, the 14th was in the centre with the 8th on the left, a vulnerable position as the left-hand flank of this brigade was 'in the air'. The 61st Division was to assault the Sugar Loaf from the right of the 15th Brigade.

Based on aerial imagery and common tactical practice, Lieutenant General Haking believed that the German support line would be found 100 to 150 metres beyond the front line. Given the limited objectives of the operation, he directed that only two battalions from each brigade were to undertake the actual attack. The commander of the 14th Brigade, Colonel Harold Pope, selected the 53rd and 54th battalions. To describe the majority of men in these two battalions as 'green' is an understatement; the 54th had spent just part of a day in the front line, the 53rd two days. The 54th was the left battalion, with the 53rd on its right.

The 55th was designated the support battalion for the 14th Brigade. The role of the battalion evolved in the days preceding the operation. Early instructions assigned two of the 55th's companies to build two communications trenches from the Australian front line across no man's land to the German front line once this was secured by the assaulting battalions. The remainder of the battalion was to provide carrying parties for ammunition, water, engineering stores and so forth. Final orders confirmed that C and D companies, under overall command of Major Selwyn Holland, were to be employed as carrying parties.[8] A and B companies, however, were no longer to dig communications trenches but were to garrison the Australian support line, known as the 300-yard line, once the attacking battalions had vacated it. The carrying parties were to move behind the assault waves and, on delivering their loads, were to return to the Australian front line for further tasking.

Similar to Fleurbaix, the Australian and German defences in the Fromelles area were not entrenchments but above-ground breastworks. A narrow communications trench, named 'Brompton Road', serviced the front line and 300-yard line in the 14th Brigade sector. In turn, Brompton Road linked up with two good quality roads, Rue du Bois and Rue Petillon, running roughly parallel to the forward lines and which were the major thoroughfares to and from the rear areas.

No man's land in the 14th Brigade sector was flat, generally featureless and approximately 150 metres wide. It was devoid of trees but was covered by a knee-high mixture of grass, self-sown crops and rank weeds. Shell holes abounded, many of them difficult to distinguish from the surrounding ground by virtue of the tall grass growing in them. Shallow drainage ditches criss-crossed the area. Rue Delvas, the main road from Sailly sur la Lys to Fromelles, ran straight across no man's land and formed the boundary between the 53rd and 54th battalions.

No attempt was made to conceal the British and Australian preparations for an offensive. From their observation points in Fromelles and along Aubers Ridge, the Germans scrutinised the movement of men and materiel in minute detail. The rushed planning, flat terrain and tight timetable removed any opportunity for the Australians to achieve surprise.

The *6th Bavarian Reserve Infantry Division*, on whom the blow would fall, made arrangements to meet the coming storm. The Germans thinned out the numbers in their front line (to reduce casualties from artillery), strengthened their strongpoints, reinforced their pool of reserves and rehearsed their counter-attack plans. Most importantly, they pre-positioned huge quantities of hand grenades and other munitions in locations accessible to the counter-attacking troops. Lieutenant General Haking's objective of persuading the Germans that an infantry attack was imminent proved a resounding success. The problem for the Australians was that they were about to undertake the very operation their enemy had been led to expect.

It was in their billets at Bac Saint Maur that the troops first became aware that something major was afoot. Officers briefed the men on what they knew. In his diary entry for 16 July, Sergeant Francis Armstrong cryptically recorded: 'Arrived at Bac St Maur at 3.30 a.m. this morning. Billeted in school grnds [grounds]. Went out to view route tonight. Big attack by Australians in preparation.'

There was little rest after their spell at Fleurbaix as the diggers were soon tasked with carrying stores to the forward area. Private Archie Winter was not enthusiastic about the incessant work parties: 'Out of the line doing fatigues, carrying ammunition to the front-line sometimes with gas helmets on; this fatigue is far worse than holding the line.' Private Herbert Harris, on the other hand, simply accepted his lot: 'Was out this morning carrying ammunition. 5 miles out and 5 back and almost 1 mile to the trenches. Did two trips. Feel tired and hardly fit for what is in front of us, but it's no use not being fit, you just have to do it.'

Late in the afternoon of 18 July word was passed that the attack was confirmed for the following day. The men began to receive detailed briefings. Corporal Merton Austin wrote that:

> All NCO of A Coy were instructed by the OC [Officer Commanding], Lt H.L. Palmer, as to how to dig a comms trench connecting enemy front-line to our own ... McConaghy [the CO] informed the company that the position might easily be taken but the holding of it would be desperately disputed by the enemy in a probable counter-attack later in the night.[9]

Lieutenant Colonel McConaghy's comments proved eerily prescient.

Despite the impending operation, there was little sleep. Throughout the night of 18 July, Private Harris was 'carrying munitions ... 60lb bombs etc.' By morning, however, the entire battalion was back in its billets.

The recently promoted Lieutenant 'Bob' Chapman recorded the feeling within the battalion in the days leading up to the assault: 'News came that we were to take part in a certain operation of storming a portion of the enemy's trench. The Battn [Battalion] was anxious for this as we all wanted to do our share.'

Private Charles Hardy shared Chapman's view: 'There was the glory of being the first Australians to go into pitched battle in France.'[10] Private Bert Bishop was less upbeat, noting: 'There was a feeling of suspense, of something unpleasant about to happen, over everything.'[11]

At the stroke of 11.00 am on 19 July 1916 the British and Australian artillery fire, which for the preceding days had been limited to desultory shelling and target registration, began building up to a fully fledged barrage, the intensity of which even the Gallipoli veterans had never experienced. Lieutenant Chapman was awestruck:

> For about 6 hours our artillery stormed the enemy trench, boom, boom, boom it thundered. All afternoon, the windows in the houses where we

> were situated rattled in their frames, while great clouds of smoke rose from the bursting shells.

Private Bishop observed that not all his mates were enthusiastic about the artillery storm: 'All day a barrage of our massed artillery had pounded the German front and second lines. "That's to let them know we're coming," Nugget had growled, "and give them time to get ready."'[12]

The battalion was not to commence moving from Bac Saint Maur to its assembly positions until mid-afternoon. It was a fine, warm day and, while the roaring artillery precluded sleep, the enforced rest period gave men time to ponder what lay ahead. Private Harris reflected: 'The shells are flying around like ants. It's awful. This is the big day and God knows how many of us will come out of it alive.' Sergeant Armstrong was more relaxed: 'Will probably be in action tonight. Everything ready. Don't feel least bit excited. Boys all anxious to get in.'

Just after midday the German retaliatory shelling began in earnest, concentrating its fire on the Australian front line, the 300-yard line and Brompton Road. Private Hardy recalled that the Germans also:

> Replied by counter battery fire which was very successful. In some cases our gun crews had to be removed and one gun was struck and the whole pit was wrecked. We had an initial bout of bad luck for a German [shell] found one of our bomb reserves which went up in yellow smoke.

At 4.00 pm final preparations began. Private Robert Harpley of B Company recorded: 'We were all fallen in with each 200 rounds of ammo and about 15 sandbags and every man had to carry either a pick or a shovel and also 48 hours rations. So we were well loaded up.' Officers used the opportunity to remind the troops of their tasks. Private Robert Micklewright wrote: 'Captain Gibbins [Officer Commanding B Company] lined us up and tried to impress on us what we really had to do and he finished up by wishing us every luck and a safe return.'[13]

Private John Bain recalled that Captain Gibbins also found time to have a private conversation with him. 'His last words he said to me before going into action were "Well! Bain, you are going into action that will be pretty hot. But if you get out of it alright you will have something to talk about all your life."'[14]

Sixty minutes later the battalion was in its assembly positions. D Company carrying parties had led the way from the billets and many troops had already reached the 300-yard line, although those from C Company making their way to the same destination were held up in Brompton Road. Soldiers from A and B companies, along with Battalion Headquarters, were dispersed along Rue Du

Bois in readiness to advance and garrison the 300-yard line after the attack had been launched.

The journey frontward was horrific. The Germans were flaying the Australian trenches with high explosive and shrapnel shells. Private Harpley described some traumatic experiences in Brompton Road: 'These were shelling continuously [sic], we were climbing in many places over mangled bodies ... it was an awful sight.' Lieutenant Chapman wrote:

> We marched along the road single file under cover of the hedgerows as much as possible at about 5 minutes interval between platoons till we got to the communications sap [Brompton Road] leading to the main trench. In there we were slightly congested owing to supplies going forward and wounded coming back. The first wounded man I saw was one lying on the road with a bullet wound through his stomach. The sight seemed to being [sic] to me the first indication that we were actually going into battle, a slight feeling of sickness crept over me and I felt annoyed with myself but it soon passed. In the sap a shell landed among our front party but we could not stop. One poor chap was blown to pulp, bits of arms and legs scattered about. I trod on his head by mistake as I hurried by and it gave under my foot like a sponge. Others were lying about moaning and groaning but all feeling had left me now. I passed dead men without pitty [sic] or remorse.

During the move to the front Private Sydney Bell[15] was killed by a shell-burst, the shrapnel from the same explosion ripping Sergeant Jack Doyle's right foot so badly he never again walked properly.[16] Sergeant George Blunt had his legs shattered by a shell; he was carried back to an aid station where his lower limbs were amputated but the wounds proved fatal.[17] Lieutenant Roy Goldrick had his face and neck lacerated by shell splinters.[18] For most of the battalion this was a true baptism of fire. The previous stint in the front line at Fleurbaix had been nothing more than a curtain raiser to this, the real show.

The troops who reached the 300-yard line found it cramped, with three men squeezed into every metre of trench frontage. Soldiers of the third and fourth waves of the assaulting battalions were still there, suffering from the shellfire scything through their ranks. The congestion in the 300-yard trench, and the constant stream of wounded men moving rearwards down the confines of Brompton Road continued to impede progress for many of the 55th's carrying parties trying to push forward. In the 300-yard line the overcrowding was only eased when the assault waves moved to the front line at 5.25 pm.

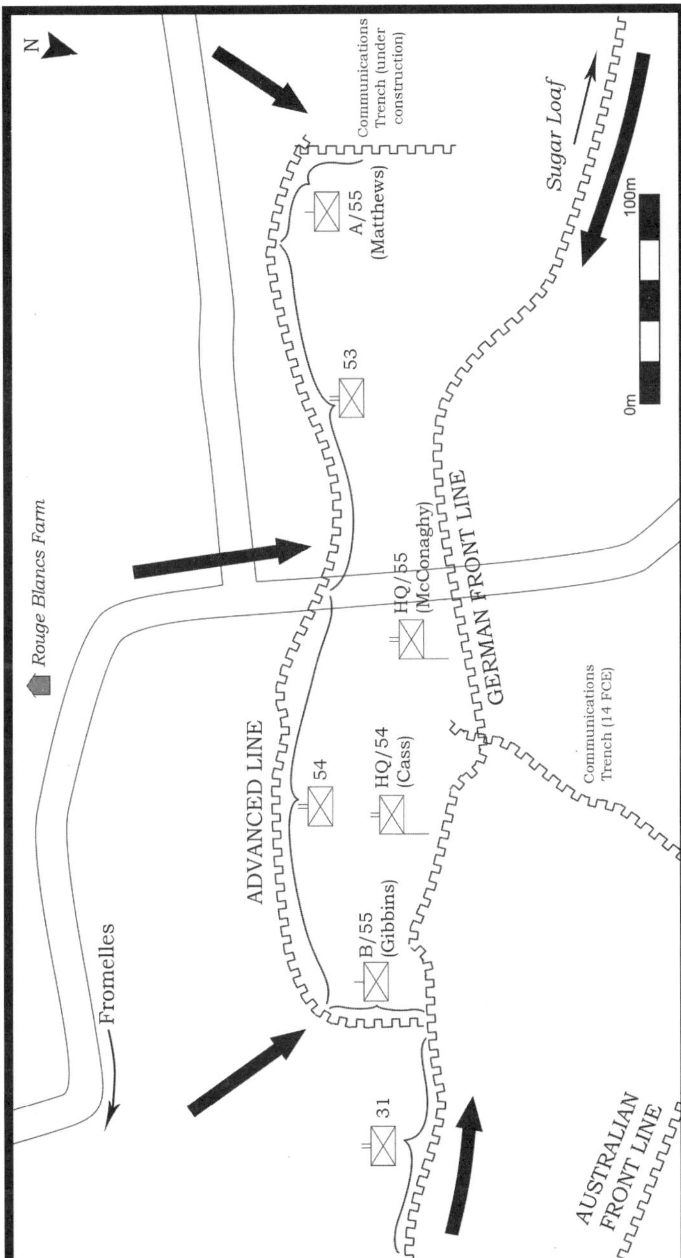

Map 3: Fromelles (dark arrows indicate direction of German counter-attacks).

The covering bombardment lifted at exactly 6.00 pm and the 53rd and 54th battalions surged towards the German trenches. It was broad daylight. The men in the 300-yard line shouldered their burdens and began making slow progress over the broken ground towards the Australian front line. Shells fell among them but most moved through unscathed. A planning oversight meant that many of the barbed-wire entanglements situated between the 300-yard line and the front line remained in place, disorganising the carrying parties as individuals tried to pick the best path through the obstacles.

Private Cyril Gray wrote of his experiences in one of the carrying parties: 'About half an hour after the attacking party had gone over we, being in supports, were given our ammunition, and sent over. We reached our firing line safely after going up to our waists in water, and dodging the bullets and shells.'[19]

On reaching the Australian front line, the heavily laden men clambered down the parados then climbed over the parapet into no man's land.[20] Almost immediately, casualties began to occur. Second Lieutenant Stuart Munro was leading an ammunition party when he was hit by several machine-gun bullets and killed; his platoon sergeant, Albert Noldart, was killed in the same burst of fire.[21] Private Horace 'Horry' Blackburn's section was carrying sandbags and had moved some 30 metres when machine-gun fired ripped into them, bullets striking eight of the 13 men in the party and mortally wounding Blackburn.[22] Private Gray was also wounded at this time:

> After a rest [in the front line] we started for the other side. I threw my ammunition on to the parapet, and climbing over had barely gone ten yards when I stopped a bullet in the left leg, just above the knee. I tried to walk back but fell over each time, and then decided to crawl, and by rolling over the parapet, was helped in by an officer, who is attached to our Battalion. He bound up my leg.

The majority of fire was coming from the Sugar Loaf. The billiard-table flatness of no man's land offered the German machine-gunners positioned there the opportunity to bring enfilade fire on the thousands of khaki-clad figures making their way across no man's land in the bright summer sunshine.[23] It was a gunners' picnic.

The departure of the assaulting battalions and work parties allowed the men stuck in Brompton Road to begin moving forward again, albeit still at a reduced rate. Further back, the soldiers of A and B companies made their way to the 300-yard line, picking their way past dead and dying. Private Bishop was one of the C Company men struggling along Brompton Road, unaware of the true awfulness just ahead:

> Shells were landing all about us. We met the first wounded coming out, often two men helping each other to be mobile. Then a batch of German prisoners squeezed past us. A hold-up stopped them and we interestedly looked at each other. One big German was bared to his waist. In his top back was a hole large enough to put a fist in. Germans unable to walk were crawling along the duckboards, blood dripping from them. The duckboards were covered with blended mud and blood.
>
> Some of the Germans could speak a little English ... A man nearby said to a German: 'Well, Fritz, what do you think of things now?'
>
> He nearly fell over when the German replied: 'I'm quite happy, but I'm sorry for you poor bastards.'
>
> Slipping, pushing, we slowly progressed.
>
> Then the communications trench suddenly debouched in what had been our front-line.
>
> I stood, I saw, a queer numb feeling took possession of me. At first it was shock, then it was terror, then it was unbelief. This just could not possibly be true. We were supposed to be civilised. The most beastly and cruel animals on earth wouldn't do this to each other. I was dazed and numb, my brain wouldn't work. A thought somehow forced its way into my thinking. It was not true, it was utterly unbelievable. I was asleep, having the worst nightmare of my life. Nightmare, that was it, I clung to the thought.
>
> 'Get along there,' someone shouted. 'What the hell do you think we're here for?'
>
> I was pushed forward. I stepped over dead bodies, I stepped over live and smashed bodies, I stepped over pieces of what had been bodies. Wounded were crying out in agony. Shell-shocked men crawled and clung to each other, some blubbering like babies. And every few seconds another shell would burst into my nightmare.[24]

Bert Bishop had no idea that his nightmare was just beginning.

At 7.24 pm Lieutenant Colonel McConaghy received an order from the 14th Brigade: 'Please take your headquarters and your two companies in 300-yard line quietly to our old front-line and await orders. Do not leave our old front-line.' Shortly thereafter A and B companies proceeded up Brompton Road to the firing line; they were all in the front line by 8.00 pm. Private Percy Larbalestier,

Captain Gibbin's batman, described an incident on the way forward: 'When we were going to the front-line we met another battalion coming back. "No good" said their officers "we can't get up there". "The 55th can" said our skipper [Gibbins] and we did! [Larbalestier's emphasis].'[25]

Despite the 14th Brigade's explicit directive that no-one from A and B companies was to leave the front line, two platoons of B Company did just that. The continually changing orders in the lead-up to the battle created uncertainty. Reverting to their original instructions, Lieutenant Chapman and 2nd Lieutenant Ken Wyllie led their platoons into no man's land and began scratching out a communications trench to the German lines, commencing where Brompton Road terminated in the Australian front line.[26] A burning British ammunition dump and bursting shells sent smoke and dust swirling across no man's land, partially cloaking the labouring men. However, concealment did not offer protection from the German machine-guns and these troops began to suffer casualties.

After catching their breath, the moment came for Private Bishop and the rest of his carrying party to make their way to the German front line:

> Men were trying to dig a trench across No-Man's-Land. We got into it, but in a few yards it fizzled out. We got out of it, and bullets hissed about us. 'Keep going on top, we've got to get our stuff over,' our officer was yelling. No-Man's-Land was littered with dead and wounded. We'd race a few yards, lie down, race again, down again. Our platoon was scattered, we had to do our own thinking. I had sheltered in a crater at the edge of the German wire. I studied the wire closely, looking for a passageway. My course planned, I tore my way through that wire. Bullets hissed everywhere. Not looking into the German trench, I heaved myself over their parapet, dead Germans and sandbags all jumbled together. Tumbling over the top I landed full-length on a German body. The body grunted. Standing up I looked at him. His waxen colour meant he had bled to death. He was only a boy, his face was the most handsome I'd ever looked at.[27]

Not all C Company soldiers were as fortunate as Private Bishop, with several fatally wounded in the crossing of no man's land. Private Sam Lennard,[28] one of three brothers in C Company,[29] was bringing up sandbags with Sergeant Claude Fuller[30] and Sergeant Charlie Matterson when a machine-gun bullet hit him above the left knee.[31] His companions crawled back and bandaged his wound. Later in the evening Fuller sent Corporal Augustus Lennard, one of Sam's brothers, out to retrieve his wounded sibling. It was too late — Sam had succumbed to his injuries.

However, most men in the carrying parties made it safely across no man's land. Shouldering their loads they followed the assault waves past the German front-line trench. By now the residual elements of the attacking battalions, failing to find the anticipated German support lines, had fallen back to a waterlogged ditch some 150 metres to the rear of the German front line. At one time this ditch may have been a trench, but the Germans now used it as a sump in which to pump water from their underground dugouts. In the absence of anything resembling proper support lines, the 53rd and 54th battalions set about consolidating this ditch and turning it into an advanced line.[32] It was here that the 55th's carrying parties caught up with the assault troops.

By now, C and D companies had become disorganised. The experience of D Company's Private John Coates is representative of what happened to many of the men in the carrying parties once they made it to the German lines:

> We were ordered forward from our own lines with reserve ammunition between 7 and 7.30 p.m ... On reaching the [advanced] line we handed over the ammunition to the officer in charge who ordered me to remain where I was and take part in the defence of the drain we were holding.[33]

The 53rd and 54th had suffered grievous losses and the remaining officers of these two battalions were only too keen to plug the gaps in their ranks with whatever troops were available. In truth, numerous officers and men of the carrying parties were happy to remain in the German lines; returning across no man's land for further supplies was a dangerous proposition and they were keen to participate in the fight. All along the advanced line, small groups of men from C and D companies joined with their sister battalions, assisting in building defences and fighting off the enemy incursions.

While many men in C and D companies chose to remain in the German lines, where a strong officer or NCO remained in charge, some carrying parties returned for additional loads. Private Bishop was in one such party:

> I scrambled into a trench that ran from the first German line back to their support line. It was a shambles of duckboards, smashed men, mud, water, blood. Quite a few of our platoon were getting together again ... We reached the [advanced line] finding our men spread along to left and right.
>
> Our sergeant divided us. 'You lot go along to the left as far as you can, you lot get along to the right' ... Dumping our loads we made our way back. It was safe going till we reached what had been the German front-line.

> Our officer appeared as we grouped together. 'There's nothing for it but to get up top. Just do the best you can'. [Crossing back across no man's land] There was no safety now.
>
> German machine-gunners got onto us from either flank, letting us know that our men held very little of their support line and even of the front-line. We crawled, ran to shell-holes, got our breath to run again, bullets zipping around us. A shrapnel shell burst above me. I turned a somersault, for some minutes I could not move.
>
> Our own wire was only a few yards away, and having picked a track through it, I ran for my life ...
>
> Our platoon officer and sergeant were waiting together to gather up the stragglers. When no more came we were ordered to go back to V.C. Corner[34] for our next load.[35]

Lieutenant Colonel Walter Cass, CO of the 54th Battalion, was the senior surviving officer of the 14th Brigade in the enemy lines and had assumed overall command of the brigade's elements *in situ*.[36] The situation was fluid; the Germans were mounting localised counter-attacks and, given the severity of Australian losses, Cass was uncertain how many effective troops he had with which to repulse them. Appreciating the dire nature of his situation, Cass entreated Colonel Pope for reinforcements.

Colonel Pope acquiesced to Cass' requests and Lieutenant Colonel McConaghy received a message from Brigade at 8.55 pm: 'Please send one company to report to OC 53rd Battalion and one company to report to OC 54th Battalion.' McConaghy ordered A Company, led by Lieutenant John Matthews, to augment the 53rd and B Company to join the 54th. Private Harpley described B Company's passage across no man's land:

> We were each issued with three hand grenades ... then came the supreme moment. The order was then given 'leap the parapet and charge'. My God, it required all the nerve available. I set my teeth and over I went, we had only gone about 20 yards when we encountered our barbed wire having to delay here in getting through we lost a few for the schrapnel [sic] and machine gun and rifle fire was terrific. However on we went, there were dead and wounded everywhere. About 300 yards and we were into the German front-line, some were left to clear this up and on we went through more barbed wire to the [advanced] trench. It was in this last stretch ... that a lot were lost. Poor Penfold was shot there through the head.[37]

Both A and B companies traversed no man's land relatively unscathed; it was dusk and the poor visibility reduced the accuracy of the German machine-gunners and artillerymen. Lieutenant Chapman, a Gallipoli veteran, clinically observed the fallen as he strolled across the grass of no man's land: 'Our road was strewn with dead, men lay as they had fallen mostly face downwards with the heads towards the enemy. Their yellowy white complexion, blue fingernails and clear staring eyes gazing into vacancy told that death had for some time taken his toll.'

As ordered, A Company made its way to the 53rd Battalion area. Here everything was in upheaval; the officer ranks of the 53rd had been culled and the battalion was struggling to regain its equilibrium. Machine-gun fire from the Sugar Loaf swept over them, confounding the Australians, for this was where the 15th Brigade was supposed to be. Lieutenant Matthews placed his troops in the locations specified by Captain Charles Arblaster, the acting CO of the 53rd. Some groups of A Company men were positioned in gaps along the advanced line. Matthews and the bulk of the company, including all the trained grenadiers, were given responsibility for the extreme right of the advanced line. Arblaster put several of the 55th's Lewis guns to the right of Rue Delvas, a weak point in the defences and a logical enemy counter-attack route.

Corporal George Stringfellow[38] was in charge of one of these Lewis gun teams and related his actions:

> When we reached the [advanced] line we received orders to consolidate. And we did so. This work was extremely difficult as we were working in water and had to fill sandbags with a very wet and sticky clay. Nevertheless, the work was completed. The men worked splendidly and in a very short time everything was in readiness. Lieutenant Agassiz[39] was in the trench during the whole of the operation.

On the brigade's other flank, Captain Gibbins took B Company to the extreme left of the 54th. To his surprise, this flank was 'in the air'; the 31st Battalion (8th Brigade), unable to find the German support line, had retired all the way to the enemy front line. This left a gap 250 metres wide in the Australian line between the 54th's left flank and the right flank of the 32nd Battalion (8th Brigade), which was consolidating some 150 metres beyond the German front line to the left of the 31st. Gibbins strung out most of his troops between the advanced line and the German front line. Unlike Lieutenant Matthews, who was expecting the 15th Brigade troops to be in position, Gibbins could see that the 31st Battalion had retired and realised this opening in the Australian front would soon be swarming with enemy.

As night fell sappers from the 14th Field Company Engineers recommenced working on the communications sap across no man's land started by B Company. To assist in this undertaking, the engineers employed troops of the 56th Battalion, although some reports indicate that stragglers from the 55th also took their turn with a pick and shovel.

The decision to commit A and B companies meant that most of the battalion was now in action. Lieutenant Colonel McConaghy was chafing to enter the fight and, at 9.30 pm, he sent a runner to inform Colonel Pope that the move of A and B companies was complete. The message concluded: 'I am now in our front-line organising parties but I think I might be better over in Hun Lines. Unless you order otherwise I will proceed to the Hun line at about 10 p.m.'[40]

There was no possibility of Pope being able to 'order otherwise' — McConaghy knew that the chaotic state of communications made it impossible for a refusal to reach him in the 30 minutes available. By 10.10 pm McConaghy and his key staff had crossed no man's land and he was where he wanted to be — making a closer acquaintanceship with the Germans.

Lieutenant Colonel McConaghy entered the German front line where it was intersected by the half-finished communications sap. He came across two young officers of the 53rd Battalion, but they were overwhelmed and unable to tell him what was happening. A veteran officer such as McConaghy took little time to assess the situation and, as Bean commented, '[McConaghy] at once received the impression that matters were far from well.'[41] Exactly how 'unwell' the situation was he could not yet imagine. He established his headquarters in a small dugout in the German front line just to the right of the sap head and set about bringing order to mayhem.

McConaghy's first task was to meet with Lieutenant Colonel Cass who was occupying a well-appointed dugout around 100 metres from McConaghy's own. The update he received from Cass did not assuage McConaghy's initial alarm. Cass had not been to the advanced line and was relying on the oft contradictory messages from the forward troops to inform his decision-making. McConaghy decided to make an independent assessment of the situation and, returning to his own dugout, he began sending out reconnaissance parties. Major Cowey was despatched to the right flank and Captain Percy Woods, the Adjutant, to the left. McConaghy also sent an 'officer's patrol', comprising one officer and three soldiers, to the right along the German front line to establish contact with the 15th Brigade.

Major Cowey and his batman, Private Ernie Smith, cut across open ground until they reached the advanced line.[42] Here they split up; Smith moved north to find B Company and Cowey turned south towards A Company. In 1926, Cowey penned a description of his ramble along the advanced line:

> This was a perilous journey made partly through a ditch nearly full of water and partly over the open with bullets and shells adding a good spice of danger to it. The Germans had many places marked by snipers. I eventually reached the right of the line and there found our men under Lieut Matthews of 55th Bn [Battalion] being hard pressed and partly surrounded by German bombers.
>
> The Germans had excellent bombers and a profusion of bombs whilst we did not have sufficient bombs or sandbags at the time. The men were too crowded. Lt Cumming[43] [was] there in charge of a part of the line ... I returned to where Lt Col Cass was, after thinning the line somewhat and setting organised parties digging a sap back to HQ [i.e. from the advanced line back to the German front line].'

The officer's patrol sent by Lieutenant Colonel McConaghy did not return and, still uncertain of what was happening along the German front line to his right, he tasked Corporal Henry Anson to investigate.[44] Anson moved along the trench but was driven back by enemy grenades. Speculating that they might be from a cut-off German garrison, an undeterred Anson went out again, this time using the concealment offered by darkness to move across the open ground. He discovered that the Germans were holding the front line for some distance to the right of the 53rd before he was again forced to retire by enemy grenadiers.

McConaghy formed an opinion, based on Anson's report and his own intuition, that either the 15th Brigade was not 'in' on the right flank or that a gap existed between the 14th and 15th brigades that the Germans were exploiting. McConaghy's hunch was correct; no member of the 15th Brigade had reached beyond the German wire and most never made it that far.

The folly of Major General McCay's directive that the German front line was not to be consolidated now became apparent. The 15th Brigade's failure to capture its objectives, and the placement of virtually every man of the 53rd, 54th and 55th battalions in the advanced line, left the almost undefended German front-line trench as the perfect avenue along which enemy counter-attacks could surge to the rear of the Australians. Realising the danger, Lieutenant Colonel McConaghy ordered the construction of two 'bomb blocks' to prevent further German progress up their front line from the right.

Sandbags from both sides of the breastworks were pulled down and piled up to construct these 'blocks'. Bombers and marksmen were positioned to prevent the enemy bypassing the obstacles.

As the full extent of the calamitous circumstances on the right became apparent, McConaghy received a note from Captain Gibbins on the left flank. This note reveals as much about the character of Gibbins as it does about the tactical situation:

> We hold front-line with the 54th. CO 54th (Cass) in next trench in rear. 53rd on our right. 31st and 32nd on our left. Consolidating positions as fast as possible. Sending back to rear trench by parties of ten under NCOs for ammunition and sandbags. We want Very pistols, flares and sandbags (plenty). Have 54th Lewis guns and five of our own under Sergeant Colless for counter-attack.[45] Each of my men has three bombs, but require more. Expect a counter-attack shortly. Anyway, we can hold them easily.
>
> [Signed] N. Gibbins, Capt. OC B Coy.
>
> Sergeant Colless doing good work – my officers also of course. Would you say something to this man (i.e. the bearer of the message). He is doing splendid work.

Private Archie Winter was in one of the work parties sent back by Captain Gibbins to collect stores and wrote later of this enterprise:

> My mates were scattered in all directions ... Vin Baker,[46] Fred Carpenter[47] and self managed to keep together ... We were not there long before we volunteered to go back for sandbags. Being between the German lines we had to move when getting across the parapets. There were dead men everywhere. It being our first actual fight and seeing some ghastly sights, made us feel peculiar for a while but soon wore off ... Fritzy was making it interesting with his artillery. We made several trips back for sandbags.

Around 11.00 pm Major Cowey returned to Lieutenant Colonel Cass' dugout after his adventures on the right of the Australian position. Here he found Captain Woods, Captain Gibbins and probably Lieutenant Colonel McConaghy. The officers had gathered to exchange information and plan the next step. Gibbins was upbeat: he reported that B Company was in touch with the 8th Brigade on the left and that they were strengthening the advanced line, as well as constructing a low sandbag wall to link the advanced line and the German front line in case the situation deteriorated in the 8th Brigade area.

The tone of Major Cowey's report of conditions on the right was less heartening. He described how small groups of men, widely spaced, were digging in along the length of the advanced line; the water in the trench was up to half a metre deep and the constant traffic of men had turned it to a muddy bog. Cowey also told of his conversations with Lieutenant Matthews and Lieutenant George Folkard[48] on the extreme right: 'They [Folkard and Matthews] had not too many bombs and the mud was so thick the [sand]bags and bombs had to be thrown along the trench [i.e. thrown from man to man to place them where they were needed]'. The conference broke up with a commitment to consolidate inside the brigade's existing perimeter.

The localised enemy counter-attacks faded around 11.00 pm. The enemy had contained the Australian encroachment with minimal losses and now needed time to reorganise before launching an attack that would destroy the incursion altogether. The Germans planned to launch counter-attacks from the early hours of the morning with simultaneous assaults on various points of the 14th Brigade's perimeter.

Along the advanced line the Australians grabbed the opportunity offered by the respite in German activity to work on their defences. The troops, often waist-deep in water, used their entrenching tools and whatever shovels and picks they could acquire to fill sandbags with the sticky soil and construct a parapet. Private Harpley recalled: 'We worked all night on this and were well dug-in by 1 a.m. on the morning of the 20th.' Lieutenant Chapman was equally satisfied: 'By morning we felt a bit safer than when we got into this muddy little trench.'

Despite progress in some areas, many parts of the advanced line remained little more than a filthy ditch with no defensive preparations. There were simply not enough men, stores and time for the position to be properly consolidated. A start had been made digging two trenches to link the advanced line with the German front line, but these remained little more than scrapes. The most critical oversight was on the right flank, where the ad hoc bomb blocks in the German front line had not been strengthened.

There were two significant achievements during this pause in the fighting. By midnight, the 14th Field Company Engineers had completed a communications sap across no man's land. Given the circumstances, this was a remarkable feat. The trench was up to 1.5 metres deep in places and offered a safer and easier way of moving stores and exchanging messages than running the gauntlet across no man's land. By mid-morning hundreds of men would owe their lives to the sappers and infantrymen who had constructed this link.

The other achievement was the connection, for a brief period of 15 minutes, of a telephone line between Lieutenant Colonel Cass and the 14th Brigade Headquarters. This short window permitted requests for specific stores and ammunition to be sent to the supply parties in the Australian front line. It also provided Brigade Headquarters the means to inform Cass that it was by no means certain that the 15th Brigade had occupied the German trenches. If indeed it had, there was no indication of exactly what parts of the trench-line were held. The penny finally dropped in Cass' mind that his right flank might be wholly open. Lieutenant Colonel McConaghy had arrived at this conclusion at least 90 minutes beforehand but had not managed to convince Cass of the danger. In fairness to the latter, he had been fed a stream of contradictory messages concerning the exact status of the 15th Brigade from the moment he arrived in the German lines. However, he had failed to take active measures, such as despatching patrols, to clarify the situation.

While most members of the unit dug in, work parties from C and D companies continued to transit no man's land with their precious stores. Private Bishop described the hazards of this enterprise: 'I didn't receive a scratch all through, though I had many narrow escapes. Once a shell burst over my head and blew me up into the air, and I couldn't move for a few minutes afterwards; while I was often smothered with dirt and rubbish they kicked up.'[49] Private Luther Chadwick[50] and Private Jim Perkins[51] both won Military Medals for their bravery in crossing no man's land under heavy fire to bring ammunition forward. Lieutenant Norman Pinkstone and his carrying party crossed no man's land at least a dozen times.

In the early morning the anticipated German counter-attacks began in earnest. A major enemy push came from Rouge Blancs, a ruined farm on the left front of the 54th Battalion. Other German attacks came from several points on the right flank, including along the enemy front line, the shallow drainage ditches running parallel, and from the direction of the Sugar Loaf and Delaporte Farm, another German strongpoint. For the sake of clarity, this narrative will first focus on the events on the right flank.

Just after 2.00 am, Lieutenant Matthews sent word to Lieutenant Colonel McConaghy requesting more bombs and bombers. Major Cowey later wrote of the events precipitated by Matthew's request:

> Lieutenant Colonel McConaghy ordered me to go forward to the assistance of this officer [Matthews]. My previous journey had been so difficult, slow and dangerous that I decided this time, for the sake of

> quickness, to get forward from the right after traversing the German original front-line. I started along this line with a couple of men, and found 19 Germans hiding in a dugout. These I sent to the APM[52] under an escort after thoroughly frightening them with some revolver and rifle shooting. I then proceeded on my way preceded by two men, all the time passing men who were hurrying in the opposite direction and warned us not to go further. We continued on our way, and when the two men who were with me turned a certain traverse I heard a cry of 'Hands Oop'! Hands Oop!' immediately followed by the explosion of a bomb, I should judge six feet away. I waited for some sign of what had happened, did not observe any, so concluded that the Germans had accounted for my two men, and would continue along the trench. I went back to Lieutenant Colonel McConaghy, between whom and myself was not a single fighting man, and told him that, unless he left his dugout, he would be captured. He would not believe me, and thought this was a ruse on my part to avoid going to the assistance of Lieutenant Matthews. He apologised later.

After briefing an incredulous and impolite Lieutenant Colonel McConaghy, Major Cowey stormed out and went to Lieutenant Colonel Cass with the same information. Cass was willing to believe Cowey, as he had also been receiving reports of German advances along their front line on the right. The Germans were now just 80 metres from the sap head, where the communications trench across no man's land entered their front line; the Australians were facing imminent encirclement. In desperation, Cass ordered Cowey to collect bombs and bombers to resist the German infiltration. This directive was easier uttered than obeyed, for Australian reserves were non-existent. Appreciating this, Cowey obtained permission to return to the Australian front line to seek the assistance so desperately needed.

In his dugout, Lieutenant Colonel McConaghy's anger cooled and, as reason re-asserted itself, he came to accept the truth of Cowey's report. Realising the danger posed by the German bombers, McConaghy summoned help from Captain Gibbins, who despatched a platoon under command of 2nd Lieutenant Ken Wyllie to the right flank. Lieutenant Chapman described the next few moments: 'As he [Wyllie] got up to go thud came something against his side and he rolled over grasping his side. 'They've got me Chappy, they've got me' he said as I held up his head. They carried him to the main German trench and from there to our own trench.'[53] The remainder of Wyllie's platoon moved to the right and flung themselves into the fray.

It was not just along the enemy front-line trench that fighting began to intensify as the German counter-attacks closed with the Australians. Second Lieutenant Eric Farmer[54] and 30 men from the 55th had been tasked to construct a trench from the extreme right-hand point of the advanced line back to the German front line. Farmer's party suddenly found themselves under sustained German bombing attack from the front, flank and rear. One of Farmer's soldiers, Corporal Austin, recorded:

> The first intimation we had that the enemy was in our rear was when he began firing flare pistols and orders were passed along to find out what was meant by them. We were soon convinced it was the enemy as we could easily discern his spiked helmets and later on he 'rained' grenades on our trench.

Second Lieutenant Farmer's group, augmented by grenadiers sent by Captain Arblaster, traded grenades with the Germans in an escalating exchange. At this stage of the war, grenades were regarded as a specialist weapon and only selected men had been trained in their use. This specialisation had its consequences as Farmer discovered that the grenades supplied to his group were without detonators.[55] As a result, he was forced to sit in the mud, at night and in the middle of a firefight, screwing detonators into grenades until the arrival of fresh supplies with detonators attached.

Over the period of an hour the German bombers gained the ascendancy, forcing the remains of 2nd Lieutenant Farmer's group back to the advanced line. Here they linked up with Lieutenant Matthews and the troops defending the right of the Australian position. Corporal Austin recollected that:

> The sector of our trench connecting the German [front line] was ordered to be evacuated as the casualties were so heavy and in the remaining portion of the trench [the advanced trench] for an hour or so later there was desperate fighting with hand grenades and rifle fire on both sides.

With 2nd Lieutenant Farmer's withdrawal the Australian right flank had been turned. The enemy were now behind the advanced line where soldiers of the 53rd and 55th battalions fought on, many unaware of the menace to their rear.

The Germans kept pressing their advantage and commenced rolling up the advanced line, forcing the defenders to creep to the left. In spite of the stout resistance of Lieutenant Matthews and A Company, the security of the right flank became problematic; a grenade killed Lieutenant Berrol Mendelsohn during this fighting.[56] The desperate Australians made numerous attempts to dislodge the Germans, but their uncoordinated efforts were easily driven back. Historian Peter Pedersen captures some of the drama of the battle:

> Arcing flares and the flashes of exploding grenades, whose dull crumps often drown out the harsh cacophony of machine-gun and rifle, light up the area. Mud-covered figures splash along the ditches in response to the unceasing call for more grenades and men to throw them. The desperation in the voices leaves no doubt that things are critical.[57]

The Germans made profligate use of flares, often firing them so as to silhouette the Australians. Conversely, the Australians suffered from a shortage of flares; officers who were now lying dead or gravely wounded in no man's land carried most of the flare pistols. The men resorted to using captured German flares where these were available.

In the confusion of the battle raging around the Australian perimeter, most men on the right did not realise the gravity of their situation. Corporal Stringfellow wrote:

> The first indication I had of Germans behind us was [when] Lt Agassiz drew my attention to them occupying from the right at about 3.30 a.m ... When the Germans retook their front-line (about 3.30 – 4.00 a.m.) we were entirely exposed as there was no cover at all from the rear of the trench and consequently ... it was no time at all before our fighting strength was considerably reduced. During this time our supply of ammunition was becoming alarmingly short and to make matters worse the majority of our rifles and [Lewis] guns were covered with mud etc.

A strong enemy party infiltrated along Rue Delvas, set up a machine-gun in a ditch between the front line and the advanced line, and commenced firing into the backs of the Australians. One of the 55th's Lewis gunners, Private Henry Mayer,[58] swung his gun around and, resting the gun's barrel on Corporal Stringfellow's shoulder, began engaging the Germans, whose spiked helmets were clearly visible. When Mayer was shot through the head, 2nd Lieutenant Agassiz took up the gun and, repositioning it on Stringfellow's shoulders, continued to fire on the German party, suppressing them for more than an hour.

The sustained German bombing of the advanced line from the front, flank and rear caused the already scattered defence of the right flank to fragment further. German bravery, coupled with their familiarity with their own trench systems, saw them constantly outmanoeuvre the Australians. Slowly the reality dawned in the minds of the Australians that they were attempting to defend the indefensible. Corporal James Skelly[59] recalled that at this point of the battle,

'Lieutenant Albert Bowman [53rd] went along the trench ... He told us he was going to find the Colonel. When he returned he said "We're in a hell of a mess and I don't know how we are going to get out of it."'

At 3.45 am an anxious Lieutenant Colonel McConaghy informed 14th Brigade Headquarters that 'We have been strongly counter-attacked on right (where 15th Bde should be). I have given 20 yards of trench. Please tell me if 15th Bde are in position. Reinforcements urgently wanted. Could you send 2 or 3 organised bombing parties?' From this message it is apparent that Colonel Pope had yet to tell his battalion commanders that the 15th Brigade attack had failed. Given the tactical importance of this information, the reason for Pope's omission is unclear.

At this time Lieutenant Colonel Cass also sent word to Colonel Pope that the 53rd had started 'dribbling away' to the Australian lines, perhaps joined by some members of the 55th. The defenders on the right flank, experiencing ferocious bombing from three sides and with the enemy occupying the trenches behind them, began looking to save themselves. In an attempt to stabilise the situation, Captain Arblaster led a forlorn charge designed to eject the Germans from their front line, thereby opening a withdrawal route. The attack was a failure, with Arblaster mortally wounded in the process.

Lieutenant Colonel McConaghy was feeding whatever troops he could find to the bombing battle along the right of the German front line. The fighting was frantic with neither side able to prevail. For the second time the Germans were able to close to within 80 metres of the sap head.

Just before dawn, McConaghy met with Lieutenant Colonel Cass to discuss the situation. Cass ordered McConaghy 'to drive the Germans back with rifle fire and then with the bayonet.' This directive led to a verbal stoush between the stressed and exhausted officers. McConaghy argued that bombs, not rifles and bayonets, presented the only means of halting the Germans. He had noticed that the enemy bombing parties were always covered by rifle or machine-gun fire to prevent Australians moving to outflank the enemy bombers in the trenches. Cass remained committed to a bayonet charge and, as the senior officer, his will prevailed.

However McConaghy wrested one concession from Cass — time to prepare the counter-attack. Cass, annoyed by McConaghy's belligerence and delays, ordered a nearby sergeant from the 54th Battalion, Frank Stringer, to collect a dozen men and launch an immediate bayonet attack.[60] Stringer obeyed and, with a mixed party of men from the 54th and 55th battalions, managed to force the Germans back 20 metres. Eventually casualties sapped the momentum of Stringer's attack and it petered out.

McConaghy again turned to Captain Gibbins to provide the troops needed for the counter-attack. Gibbins despatched a force comprising Lieutenant William Denoon, another officer and about 50 soldiers.[61] On arriving at McConaghy's headquarters, each man in Denoon's party grabbed an additional three grenades.

After making their way a short distance along the German front line, Denoon's troops left the protection of the trench and crept into the open along both sides of the German breastworks. They were unobserved by the enemy; it was 4.30 am and dawn was still 30 minutes away.

Moving past the ongoing bombing battle, the group made its way alongside the trench for another 60 metres before the Germans spotted them. Suddenly aware of the danger, the enemy began to pour fire into the Australians.

Realising that he was losing the element of surprise, Lieutenant Denoon lined his men up along the German breastworks. At his signal, each man threw a grenade into the German front line. The grenades had scarcely detonated when the party clambered into the trench and cleared along it with the bayonet. The astonished Germans on the right of the Australian incursion fell back 20 metres before stabilising; those on the left found themselves pressed from the front and rear and made the choice to submit or die.

Lieutenant Denoon's attack forced the Germans back 80 metres — not as far as had been hoped, but sufficient to temporarily relieve the threat to the communications trench. These gains came at a cost: Denoon lay on the German parapet, alive but with a bullet wound through his throat and scapula.[62] Corporal Eric Hancock was shot in the chest and died in the German trench.[63] The other officer accompanying Denoon was also gravely wounded.

Far from subduing the enemy, Lieutenant Denoon's attack provoked even more vigorous bombing. In the face of this onslaught, it seemed that once again the Australians would be forced to retire.

Salvation came from a most unexpected direction. Major Cowey, after a difficult journey down the communications sap across no man's land, reached the Australian lines. He sought out Captain Fred Fanning, the officer in charge of the two companies of the 56th Battalion garrisoning the front line. Cowey asked Fanning to provide men to return with him to the battle. Fanning, quite rightly, refused Cowey's appeal; Major General McCay had issued specific orders that the 56th Battalion was not to become involved in the fighting. Cowey eventually convinced Fanning to give him a section of grenadiers under the command of Sergeant Bill Hurley.[64] Realising that they were heading straight into the cauldron, Hurley's party grabbed every spare grenade they could and accompanied Cowey to the German trenches.

On their arrival, Major Cowey and Hurley's men entered the melee, joining the remnants of Denoon's party and the other Australians nearby in defending the gains made by Denoon's attack. The tempo of the fighting immediately escalated. Charles Bean described the ensuing grenade duel as 'herculean'.[65] So intense was the bombing that there were, reportedly, almost a dozen grenades in the air at any one time. After the battle, a German grenadier told the German Crown Prince he had thrown over 500 bombs.

Such concentrated hostility could not be sustained. Casualties and sheer exhaustion eventually brought the warring parties to a standstill. The German push faded and petered out. All was not yet lost on the right; with great sacrifice the Australians had secured more of that most precious commodity — time.

Thus far the narrative of the night's events has focussed on the action on the right flank. This focus will now shift to the activities on the left of the 14th Brigade.

Captain Gibbins kept B Company on the extreme left of the 14th Brigade line. Prudently, he positioned several Lewis guns where they could provide enfilade fire along the front of the 31st Battalion now resident in the German front line. As the nearest officer, small parties of 31st Battalion men had placed themselves under his command and these Gibbins either positioned in the German front line or linking his right-hand elements with those parties of the 54th Battalion in the advanced line positioned farthest to the left. Sergeant Frank Law[66] was one member of the 31st Battalion who found himself reporting to Gibbins:

> Just at dusk, an outwork, covering about 70 yards of a very broken portion of the trench, was commenced under the supervision of Capt Gibbons [sic] 55th Btn [Battalion], and we rendered some help by expediting supplies of picks and shovels and sandbags across No Man's Land for this officer. This outwork, or sandbag bank, ran out from the captured [front-line] trench near the right flank of the 31st Btn position, curving to parallel the trench at about 40 yards distance, the further end not being connected to the trench.

B Company spent the rest of 19 July fighting off German probes and constructing defences.Lieutenant Chapman described a harrowing sight that confronted him during a pause in the fighting:

> Gib [Gibbins] and I were sitting on the parapet of the front trench we had captured while the men were busy filling sandbags with earth and mud building the parapet, when in our rear staggering through the gloom we saw a man – he came about 10 yards towards us, and then staggered and started to crawl. I thought it was one of our own men so went out to him. Poor beggar I have seen worse looking mess ups but he was bad enough – his left eye was gone – as for the rest of him I could not tell what else was wrong except that he was a mass of blood and looked as if he had been through a sausage machine. He pleaded something in German – I don't know what, it was hardly a plead – it was a moan, or a prayer – so I gave him my hand to hold and said nicely as I could 'All right old chap'. He kept pushing towards the trench all the time and as it was rather awkward getting along on one hand and two knees while I held the other hand, I let it go. Whereupon the poor mangled brute got up on his knee – put his hands together and started to pray! 'Oh cruel – cruel' Gib said when he saw the poor beggar – Gib was with him all the time also. But as I looked at him the thought struck me: 'How can men be so cruel?' I got on one side of him and Gib the other and together we helped him along. [...] He was covered all over with wet cold blood.

Private Winter described the German attacks that commenced soon after midnight:

> Early next morning (20th) Fritz started his counter. God knows which way he came – we don't. He appeared to come from every direction. We were unsupported. Consequently Fritz could come in on our flanks. They had snipers everywhere and our men were falling fast. Then we got to close quarters with bombs but we were only a handful and Fritz was there in his thousands – it was here in the bombing fight that Fred Carpenter was wounded.

The German attacks were repelled but the unceasing enemy pressure began to affect the morale of the Australians.

Just after 3.00 am rumours swept along the Australian lines that a general retirement had been ordered. Lieutenant Colonel McConaghy had been despatched by Lieutenant Colonel Cass to ensure that the 14th Brigade had linked up with the 8th Brigade on the left. Still conducting his review,

McConaghy, with the support of Captain Gibbins and other officers, was able to steady the men and keep them in their trenches. However, not all Australian troops were as blessed in the abundance and quality of their officers. The 31st Battalion fell to pieces; troops left their trenches and scrambled back to the Australian front line. With their right flank now in the air, the remainder of the 8th Brigade retired. In the pre-dawn gloom the men of the 14th Brigade were forced to watch as their mates abandoned them to an uncertain fate.

The Germans were quick to exploit the opportunities created by the whispers. They re-entered their front line and began pushing bombers in both directions along this trench. With a mounting sense of dread, Lieutenant Chapman witnessed the German reoccupation:

> And then coming through the dusk [sic – he means dawn] on our left we saw Germans. Our machine guns opened fire – but word came from the right that they were our men [falsely]. However – although we accounted for a good many, the enemy got in on our left. Then came the sound of bombing.

Recognising the danger the collapse of the 31st Battalion posed to his own position, Captain Gibbins moved to shore up his defences. He pushed Corporal Anson and some men along the German front line with instructions to bomb the enemy back. Anson, assisted by an artillery barrage, was temporarily able to hold the German advance. This was only ever going to be a brief respite; the enemy numbers and momentum were too great for a few brave men to withstand.

Captain Gibbins began assembling a counter-attack force with what members of his command he could spare. There were few available; half of B Company (Wyllie's and Denoon's platoons) had been sent to the right flank and the remaining two platoons had been whittled down by casualties. Of the officers in B Company, only Gibbins and Lieutenant Chapman were uninjured, although 2nd Lieutenant Norman Robinson from C Company had joined them.[67] Chapman managed to secure additional grenades and the counter-attack party grabbed as many as they could carry. All up, Gibbins could muster no more than 20 men, a mixed force from the 31st and 55th battalions.

Captain Gibbins placed his group in a single line extending into no man's land, positioning an officer on each flank with himself in the middle. At his word, the small force charged, bombing the Germans from outside the parapet and forcing them back over 150 metres. Private Harpley was a member of Gibbins' party: 'Our captain drew us out of our trench to charge them, but we could do very little good with the rifle so we sent for grenades and started to bomb them ... We drove them nearly out.'

Leaving Lieutenant Chapman to consolidate their gains, Captain Gibbins moved back along the front line to find the reinforcements necessary to hold the recaptured length of trench. He came across Sergeant Law and despatched him to Chapman. Law later wrote:

> Shortly after daybreak, I was informed that the 31st Btn had retired to our own lines. While making an investigation, I was met at the entrance to the outwork, by Captain Gibbons [sic], who ordered me to get my few men [Law also grabbed several 55th Battalion troops] and hold the Btn sector on his left as he expected enemy attacks from that quarter.
>
> Having established and manned a temporary post about 150 yards to the left of Capt. Gibbons' position, we attempted to seize and hold ... a further 100 yards to our left ... but discovered a strong party of the enemy in possession. A furious grenade fight now took place, the enemy making five successive attempts to storm along the trench, but each, in turn, were met and defeated by the gallantry of the eight men with me.

The actions of Lieutenant Chapman, Sergeant Law and their small band brought only a fleeting respite; the enemy attacked repeatedly and the party was forced to cede ground. During a lull in the fighting, Captain Gibbins moved back along the front line to see how Chapman's group was coping. Appreciating the pressure they were under, Gibbins prepared another counter-attack. In a mirror image of his previous endeavour, Captain Gibbins led the same officers and almost the same men in the assault. Lieutenant Chapman wrote:

> Bombs and bombers were called for and still more bombs ... but still the [German] bombers came. 'Get as many bombs as you can and come with me' said Gib – so I got all the bombs I could – called some men to follow and Gib led the way outside the parapet. We shifted those [German] bombers – but poor Gib got a wound on the head and had to retire.

Again the Germans were thrown back over 100 metres, but with such limited numbers, Captain Gibbins' counter-attack soon lost momentum. With blood streaming down his face, Gibbins instructed Lieutenant Chapman to hold until his return. Lieutenant Chapman recalled: 'I took charge of the bombing party and as the Bosches had dropped bombs for the present and taken to rifle fire, we had to take shelter in the trench. We waited there for perhaps a quarter of an hour ready to bomb Fritz should he come again.' Sergeant Len Davis (31st Battalion) stayed with Chapman: 'When ourselves running short of bombs we got behind the trench and for about an hour engaged them [the Germans] in hand to hand duels and bomb fights.'[68]

The exact cause of Gibbins' injury is uncertain — one unconfirmed report asserts that he was hit on the forehead by a German grenade but avoided the subsequent blast. Private Peter Ferguson was not so fortunate; an enemy grenade exploded behind him, lacing his back with metal fragments.[69] Ferguson's mates dragged him to the front line and made an unsuccessful attempt to find stretcher-bearers. With able-bodied men at a premium, they left the mortally wounded man to his fate and returned to the fighting.

Before leaving to get his injuries dressed in a nearby dugout, Gibbins despatched 2nd Lieutenant Robinson to the Australian front line to garner reinforcements from the 8th Brigade. Robinson made two trips across no man's land. The results were disappointing: he found four willing men on his first trip and only three on his second. The 8th Brigade troops, safe in their own lines, were reluctant to return to the fight, particularly one the Australians were so evidently losing.

That the entire operation was a failure was also beginning to occur to Australian and British senior officers. At 5.00 am Major General McCay travelled to Lieutenant General Haking's headquarters to discuss the coming day's operations. General Monro was also present. The meeting had just started when McCay received several calls from his headquarters. These contained news of the desperate plight of the 14th Brigade and confirmation of the 8th Brigade's retreat to the Australian lines. McCay passed this news to Monro and Haking; Monro decided to abandon the attack and ordered the immediate withdrawal of the 14th Brigade. On receiving these instructions, Colonel Pope despatched runners to Lieutenant Colonel Cass with directions to retire.

While the generals talked, dawn ushered in a sunny summer's morning. Improved visibility altered the nature of the engagement; artillery on both sides improved in accuracy and machine-gunners and riflemen could now fire on discernible targets. It was no longer possible for the infantry of either side to move across the open without the expectation of stopping a bullet or shell fragment. Australians in the German front line were comparatively safe, but those in the advanced line were now exposed to enemy fire. Corporal Austin recalled: 'After dawn, the enemy commenced and kept up heavy machine gun fire from both front and rear, and many became casualties when firing in one direction they exposed half their bodies to the enemy in the other direction.'

Daylight reduced the tempo of the high intensity battles that had been raging most of the night. The bruising defence of both flanks and robust resistance along the advanced line had briefly blunted the German attacks. Lieutenant Colonel Cass was able to inform Colonel Pope at 6.15 am: 'Position much easier and improved. Have driven enemy back by counter-attack and grenades well out of bombing distance. Capt Dunoon [sic] 55th did good work leading his men forward to attack until wounded.'

Despite an easing in the crisis, Lieutenant Colonel Cass was under no illusion that his position was tenable and realised that his reprieve was momentary. It was inevitable that the German flanking attacks would overwhelm the Australians and link up, excising the 14th Brigade from its supply route and allowing the enemy to pick off the Australians at their leisure. Cass knew that it was time to get out before the discipline of the brigade collapsed.

Anticipating an order to withdraw, Cass had been making preparations, briefing his officers and sending back Vickers and Lewis guns individually so that their absence would not be noticed by the Germans.[70] He also ordered Lieutenant Colonel McConaghy to organise a rearguard to cover the brigade's retirement. Withdrawing in contact with the enemy is the most difficult and risky of all military undertakings, and McConaghy appointed his most trusted and courageous officer to lead the rearguard: Captain Gibbins.

Lieutenant Colonel Cass intended Captain Gibbins to remain in his position on the left flank until the formal order to retire was given, at which time he would be responsible for the rearguard. McConaghy had misunderstood Cass' orders and, believing that the rearguard was to be immediately formed, sent Gibbins the following message: 'Capt Gibbins and 55th Officers. You must prepare for an orderly retirement. We are unprotected on our flanks. Hold first Hun line [the German front line] until further orders.'

On receiving McConaghy's direction, Gibbins, his head swathed in bandages, grabbed the nearby Lewis gun teams that constituted the rearguard and went with them to the head of the communications sap. Here he placed the gunners in positions covering both flanks and the front. Gibbin's redeployment of the Lewis gunners resulted in the weakening of the left flank, a situation the Germans exploited by infiltrating men to the rear of the advanced line.

While the rearguard was being organised the Germans had forced their way into the advanced line where it intersected Rue Delvas, virtually encircling the 53rd. Captain John Murray, the acting commander of the 53rd Battalion following Arblaster's demise, had witnessed the German incursion and seen a group of Australians being taken prisoner.

Shaken, he went to Lieutenant Colonel Cass and, in Charles Bean's delicate phrasing, 'represented that the troops in his sector were being gradually faced with a situation in which they could only die or surrender.'[71] Cass berated Murray for advocating capitulation. Major Cowey was in Cass' dugout during this confrontation and took an even dimmer view of Murray's behaviour:

> I flourished my revolver at him and endeavoured to make him realise that I was more ferocious than any German. I did intend to have him arrested at the time in case he caused disaffection in the firing line, but thought that perhaps he'd had his nerves badly used and that my bullying would save court martial business. I was astonished later to see him wearing the MC [Military Cross] for this engagement.

A chastened Murray returned to the advanced line.

Sensing imminent victory, the Germans again intensified their attacks. Isolated pockets of diggers, often leaderless, were forced to choose whether they fought on or surrendered. A decision to throw in the towel by one group was often not embraced by other Australians in the vicinity as Private George Gribbon recalled: 'Somebody hoisted a white flag in the trench on our left rear. Lieutenant Agassiz promptly fired on the flag with the machine gun and ordered the men to open fire on it with their rifles.'[72]

Other small groups elected to risk a break towards the Australian lines, realising that this meant having to fight through the Germans. One of the parties that chose to try their luck was led by Corporal 'Nutsy' Bolt.[73] After the battle, one of his mates, Private Frank Johnston, recalled Nutsy's fate: 'He and I were close together, when we were attacked by the Germans. He got more than six of them with his bayonet and the butt of his rifle, when he was killed by a bullet through the head.'[74] Shortly after this incident Johnston himself was shot in the chest, but was assisted back to the Australian lines.

The critical message from Colonel Pope directing Lieutenant Colonel Cass to retire did not reach Cass until 7.50 am; the seventh runner despatched by Pope was the only one to make it to the beleaguered brigade. On receiving the directive, Cass sent his own runners to inform the men that it was time to begin thinning out.

The Australians in the advanced line made their way to the sap head as best they could. For most, this involved a scramble across the open ground between the advanced line and the German front line. Others were able to make their way along shallow depressions which offered some prospect of protection from the German fire.

In 2nd Lieutenant Agassiz's party, Private John Reay[75] and Private Gribbon volunteered ('on their officer's suggestion' according to Charles Bean[76]) to make their way to the same communications trench Agassiz and his Lewis guns had targeted earlier in the morning to see if it was still occupied by the Germans.

It was.

Gribbon was shot through the shoulder and Reay mortally wounded; both men were abandoned and left to fall into German hands. Where Gribbon and Reay had failed, another man, Private Joe Grimes, succeeded in finding a promising ditch further to the left.[77] Agassiz ordered his troops to use this and fight their way out as best they could. In a post-war interview with Charles Bean, Corporal Stringfellow recollected: 'It looked stiff but we decided to try.' After running and crawling 50 metres, Stringfellow and Grimes were surrounded and captured.[78] Agassiz and a few others escaped and made their way back to the Australian lines. Agassiz avoided physical injury, but he never recovered psychologically from the battle; he was returned to Australia in 1917 suffering severe shell-shock.

To cover the withdrawal of his comrades, Sergeant Stan Colless kept the crews of two Lewis guns working in the advanced line to protect the retreat of B Company. The crews remained in action until enemy bombing parties drove them out.[79] Private Ray Bishop also offered to stay behind.[80] His cousin, Private Bert Bishop, wrote of Ray's final moments:

> About daylight the word was given to retire to our own trenches again, and it was then that Ray got hit. He and another chap[81] volunteered to try and shift a party of German bombers near their [Lewis] gun, who were giving a lot of trouble, and so give the gun a better chance of getting away. Each took a load of bombs and made for the Germans. Ray's mate was killed before he got three yards, and Ray himself was hit before he got halfway. He fell, and then crawled on his hands and knees closer to the Germans and threw all his bombs into them. He then commenced to crawl back, and before he got far the Germans, who were coming on in scores cut him off. It was impossible for his mates to do anything for him ... There are any amount of our chaps in the same position, wounded and having to be left behind in the retirement.[82]

Haemorrhaging from a bullet wound to the thigh, Ray bled to death before he could be captured.

Private Hardy observed that some men refused to believe that a withdrawal was underway, assuming that this was just another rumour: 'It was with difficulty that our men could be induced to retire from some places as they did not understand the general position.' Other men did not receive the order to extract themselves and for many the order arrived too late — they were surrounded, short of ammunition and their rifles clogged with mud.

Those from the advanced line who reached the German front line were confronted with another life-and-death decision: whether to attempt a passage down the overcrowded communications sap or make their way across no man's land, risking the machine-guns and artillery. There were few leaders left — it was now a matter of every man for himself.

Private 'Vin' Baker chose to run the gauntlet; the moment he clambered above the parapet a bullet smashed into his left temple, the projectile going through his steel helmet and out behind his right ear. Private Percy Geason was also shot as he broached the parapet.[83] One of his colleagues, Private Norm Ford,[84] described Geason's fate to the Red Cross:

> I did not see [Geason get hit] but I saw him lying wounded in the first German trench. He was wounded in the head, lying on his face. We were holding the trench at the time ... I was right next to him, and we turned him over, and a man in the 53rd or 54th, who was with me, dressed his wound with the field dressing. I held him up – he was unconscious and would not, in my opinion, live long, unless properly attended to at once.

Private Harpley also chanced no man's land:

> The order came for us to retreat to our line, this we had to do across about 200 yards of open country under heavy machine gun and rifle fire and shrapnel. Oh! The whole turnout was absolute hell. The bullets were cracking all around us yet most of us got over ... There I got into a bay with some of the 56th Batt. There were eight of us, four were on the fire step and the remainder were sitting down when a shell burst right on the parapet and killed too [sic] and wounded all the rest except myself. I was knocked silly for a time but managed to crawl out.

Gradually the remnants of the 14th Brigade withdrew, a large number passing through the communications sap. Private Winter was one of these:

> Got the signal to retire. This must have been about 8 a.m. That's where a good many men were killed. In the retirement I found myself with the Colonel [McConaghy] and RSM.[85] They went along a sap and we who were there followed. The sap was half full of mud and nothing but dead men. The boys suffered even when they attempted to get across the parapet.

Shell fragments were striking men as they made their way along the trench. Some of the dead were placed on top of the trench to provide protection for the living. Frustrated by their lack of progress, a number of soldiers climbed out of the trench and made a dash for the Australian front line; most of these were mown down by German machine-gun fire.

Lieutenant Chapman, in command of those elements of the 55th and 31st battalions still on the left flank, described his retirement:

> I borrowed a rifle from one of the troops passing and sniped at Fritz 'til he got up to me with his bombs. It was then time to go ... My return is a bit blurred. I remember picking my way through barbed wire with rifles cracking around me – at one place the grass in front was shaking and quivering. I looked at it for a second and realized that a machine gun was playing through there – so I jumped and hurried on. I got in all right – as the trench was becoming too crowded I sent what 55th men I could back to support. As soon as the enemy saw our men making use of the sap they opened fire with high explosives. 'Crack – Crack' came whistling over our heads – but we leaned against the parapet and were comparatively safe.

At the beginning of the battle Captain Sidney Pinkstone, the Officer Commanding D Company, found himself 'officer without portfolio' as his company had been broken up into carrying parties. He had spent a large part of the night near the sap head, directing the movement of the stores parties once they arrived in German lines. However, as the battle neared its conclusion Pinkstone had: 'Joined a party of 56th Battalion bombers [Sergeant Hurley's group], who were holding a barricaded trench. We threw Mills and German bombs and kept the trench clear till instructed by Colonel McConaghy to go back to our original front-line.'

After Pinkstone's party withdrew, the defence of the front line on the right passed to Major Cowey who, along with Captain Woods and a few men, had been ordered by Lieutenant Colonel McConaghy to prevent enemy interference with the retirement by whatever means possible. Cowey wrote of the action that followed:

> We had gradually been withdrawing our men in accordance with orders received, but there were still many men to come. I decided to try getting on the parapet again, but with bombs this time and accordingly led a party of men ... parallel to and a little distance away from the parapet and out of sight of the Germans, and then closed in on the trench and threw our bombs in. I was slightly wounded upon the chin in doing this,

> evidently by a sniper. We then returned to the sap head where we had started out from and awaited the return of the Germans, hoping to repeat the trick on them, but we had evidently inflicted casualties and they must have been puzzled as they ceased bombing and did not approach again before we left the position ... I kept about six men there with me as the remainder of our men evacuated the position via the sap or clambered out of the trench alongside Lt Col Cass' dugout. The Germans followed them up with bombs. When there appeared to be no-one else to back and it was useless to remain, I told our men to get back to our trenches, and started out myself ... I attempted to return along the sap, but it was congested with wounded and escaping men, and for some distance I went alongside the sap in the open, jumping down into it from time to time to avoid the murderous machine gun fire with which the Germans were now sweeping No Man's Land and the vicinity of the sap. In this way I regained our trenches.'

The decrease in Australian fire and the sight of troops scurrying from the advanced line and streaming across no man's land alerted the Germans that a retirement was underway. The shortage of defenders led to a fracturing of the left flank; the triumphant Germans crowded along the front line, encircling the troops of the 54th and 55th battalions still in the advanced line. Inside their perimeter, the Australian discipline was steady, with men continuing to engage the enemy until ordered to attempt a withdrawal.

The gradual compression of the Australian perimeter placed increasing demands on the rearguard. Captain Gibbins knew his time was running out; the Germans were visible bombing along the line and fewer and fewer men filed past him on their way to the Australian lines. As the grenades came closer it was apparent that nothing more could be achieved and Gibbins decided it was time for the rearguard to depart. Sergeant Bert White,[86] in charge of one of the Lewis gun parties comprising the rearguard, captured this moment:

> [Gibbins] was quite cheery and moving freely amongst his men, although he was wearing a bandage around his head. At the time Capt Gibbins was but a few yards from me and most of the infantrymen had retired. I was suddenly brought back to my senses by hearing Captain Gibbins call to 'Come on all you gunners'. I immediately picked up my spare parts and followed him.

The two Lewis gun crews under the command of Sergeant Frank Hocking were the last of the rearguard to leave the German front line.[87]

The final stragglers made the relative safety of the Australian lines by 9.00 am. Captain Gibbins was not among them. Like others before him, he had become irritated by the slow progress along the communications sap and had chosen to make a dash for the Australian trenches. Just as he reached the parapet he was shot in the head, dying instantly. Captain Gibbins' valour and leadership during the battle received no official recognition. Perhaps Lieutenant Colonel Cass blamed Gibbins for the untimely removal of the rearguard contingent from the left flank, a shift which Cass believed led to many of his battalion being cut off and either captured or killed.

Fearful of a German counter-attack on the disorganised Australian trenches, the exhausted soldiers were kept in the forward positions under heavy artillery fire, the Germans pounding the point where the communications sap cut through the Australian parapet with high explosive and shrapnel. Conditions were appalling. In his biography of Brigadier General Pompey Elliot, commander of the 15th Brigade, Ross McMullin writes: 'There was no more distressing sight for Australians in the whole war than the scene in the AIF front-line after the Battle of Fromelles. Devastated trenches were packed with dead and dying men.'[88]

In a post-war article, Sapper Smith of the 14th Field Engineering Company described an incident he had witnessed in the Australian front line:

> Colonel H Pope (14th Brigade) and Lieutenant Colonel Cass (54th Battalion) came along the new trench [the communications sap] and through our cutting in the breastwork. Both of these officers had been across to the enemy lines. Col. Cass was obviously over-wrought and distressed. He and Pope were having a heated argument about the attack and Col. Cass unburdened his mind. 'I tell you that it was wholesale murder, they have murdered my boys'. 'Oh pull yourself together man, this is war!' 'This is not war they have murdered my boys.'[89]

While this conversation has a ring of authenticity, there is no record of Pope's presence in the front lines at this or any other time during the battle. It is more likely that the 'Colonel Pope' referred to by Smith was in fact Lieutenant Colonel McConaghy. Certainly Cass reports meeting McConaghy at around this time: 'When I reached our own front-line I found McConaghy going for my men for leaving me behind.' The tensions and occasional flare-ups between both commanders showed no sign of abating.

Major Cowey strove to impose order on the diggers in the forward trench: 'I spent some time clearing our end of the sap in order that slightly wounded men might get in quickly from No Man's Land, and in getting wounded men from dugouts as there was great congestion there, and the Germans were shelling the spot regularly.' Private Gray was one of those who benefitted from Cowey's efforts. Gray had been shot in the leg early in the attack and had crawled back to the Australian front line. He had remained here throughout the battle, later recording:

> I had a close shave while sitting in the trench. One of the Ashfield boys was standing talking to me, when a piece of shrapnel flew past, and missing my head by inches, hit his water bottle puncturing it. It was a close shave for him also. I was put in a dugout which was none too safe, and during the night a piece of shell came through, and hit the chap who was sitting beside me in the chest. We expected a shell to lob on the roof at any minute, and send us up in the air. We remained here until afternoon, and were taken down by our stretcher-bearers, who were doing great work. I was then put in an ambulance and taken to a Dressing Station, where we were given cocoa and had our wounds dressed. We were next taken to another hospital, where we again had our wounds dressed, and from there went by train to Boulogne.

Not all Australians made it back to their own lines by the time the rearguard retired. Scattered in pockets along the advanced line, tired men continued to be pressed by the Germans. On realising the game was over, most of them surrendered; the sight of their mates laying down their weapons hastened the demoralisation of the remainder. Lieutenant George Folkard, nursing a shrapnel wound in the left hand, described the circumstances of his capture:

> Towards morning the Germans made several counter-attacks which were repulsed ... Owing to our men being up to their waists in mud and slush, to shortage of ammunition and bombs and to most of the rifles being clogged with mud, we were unable to make any counter-attack or to return to our lines ... About 10.00 a.m. I was made prisoner ... The enemy came over in swarms. They were all around us and we were completely cut off and unable to put up a fight.

Lieutenant Matthews, who had stoutly defended the right flank, had been badly wounded in both legs by a grenade and was also captured by the Germans, along with many of his men. His wounds were so severe that a partial amputation of his right foot was necessary.

In places, men chose to fight to the death rather than be taken prisoner, and sporadic flare-ups occurred as the Germans destroyed the last vestiges of resistance. By 10.30 am the only Australians on the German side of the front were either dead or captured. By midday the artillery fire had also ceased, the worn-out gunners on both sides incapable of further effort. The Battle of Fromelles was over.

This realisation dawned on the Australian commanders as well. Early in the afternoon, the men of the 55th received the order to move back to their billets. Major Cowey, somewhat ruefully, described an incident on the way out of the forward area: 'In passing the advanced dressing station in the Rue du Bois I took my final part in the operation by being rude to the AMC personnel there. This had the effect of getting stretchers up to the front for the removal of the large number of wounded who were lying there.'

The mass casualties suffered during the attack overwhelmed the medical system. Private Hardy had been shot in the left forearm and, like many of the walking wounded, made his own way rearwards:

> Two parallel roads led to the front. They were visible and were swept by gunfire. There were advanced dressing stations on each of the roads in the ruined farmhouses we called Rifle Villa and Beacon House. [Brompton Road] emerged near both and soon a ceaseless stream of wounded began to pour into these dressing stations, me being one. All around the dressing stations they [the wounded] were lying in the open where I saw many of my pals whom I had served months in the line with go west with terrible wounds.

Stretcher-bearers worked throughout the entire day to bring the wounded from the front line to the dressing stations. Many of the battalion's wounded lay in no man's land and a number of them used the concealment offered by the long grass to make their way to safety. The following days and nights witnessed sustained efforts at casualty recovery. Stretcher-bearers and others, risking their own lives, prowled no man's land collecting the wounded. Many of the suffering were gathered during these forays, but others could not be located and succumbed to their injuries. Of those brought in alive, gangrene or other infections had often taken hold, resulting in amputations and, in some cases, death.

The uninjured survivors made their way to Bac St Maur and the billets they had left just over 18 hours previously. It was a distressing time. Private Harris found an opportunity to update his diary:

> Thank God I am still alive and not wounded except for a slight bang on the finger from a splinter of shell. My steel helmet saved me five times and how many escapes I had could not be counted ... Nearly all our officers are dead or wounded and the Battalion is about half a battalion now. The sights I saw will never be forgotten. It was like a butcher's shop ... nearly all our lieutenants are gone, also our sergeants and corporals ... we look a sorry crowd. Covered in mud from head to foot ... it was a ghastly sight stepping over dead men in the trenches, some of them only half there.

Private Bishop described the traumatised battalion:

> I couldn't help noticing the look on the faces of the men as they came out. They all had an ashen grey pallor, and looked years older than when they went in. You'd see mates shaking hands with each other in silence, unable to speak, some of them crying, but no words were needed, each understood the feelings in the hearts of the others, feelings which could not be expressed in words.[90]

Private Hardy observed the exhaustion of his companions:

> When the fight was over men lay about in a battle stupor caused by the weariness. They were quite oblivious to the danger of passing gun teams or motor ambulances and one could roll or move them to places of safety without waking them.

For many men, such as Private Harpley, the roll call that took place late that afternoon was one of their most poignant memories of the battle: 'When we got back [to billets] we had a muster parade and could muster half our company and one officer. The exact number is 110 out of 204 ... Nearly half the battalion is missing. Oh! The sights were awful – mangled bodies everywhere.'

Around half the battalion had become casualties: 87 killed (including those who died in captivity), hundreds wounded and 103 captured. Of those who returned to Bac St Maur, unscathed did not necessarily mean unaffected. Many troops suffered from some form of anxiety disorder. Most recovered, but for others the assault on their senses was too great — these men were quietly removed from the unit.[91]

In his diary on the day following the battle, Private Harris wrote of the strain with which he and his fellow soldiers were coping:

> All around me are sleeping exhausted men, some moaning and others talking, the events of the last two days seem like a bad nightmare. I am the

> only one in 9 Section and there are nine [left] in the platoon. We muster about half a company all told in 'A' [Company]. All my chums are dead or wounded and the guns are still blazing about 1½ miles away ... Men are scattered about in little groups discussing the [battle] and telling one another about this one gone and that one wounded. It's almost unbelievable to think of fellows with us a day ago are now in the cemetery.

Private James Barker lamented: 'Terrible slaughter on both sides … lost nearly all my mates.'[92]

General Hacking, in his post-action report to First Army Headquarters, asserted that 'the blame for failure lay with the attacking troops' and the 5th Division troops who, although attacking 'in the most gallant manner ... were not sufficiently trained to consolidate the ground gained.' Haking went on to write that, although the attack had failed, it had done the 5th Division 'a great deal of good'.

The division had been in France for just over three weeks and had been rendered ineffective in a little over half a day. It was hard to see any 'good' coming from the 5th Division's 5000 casualties.

The unit's War Diarist noted: 'losses were heavy but the battalion, four-fifths or more of whom were strangers to battle, acquitted itself honourably in its first engagement.' It is hard to argue with this self-assessment. The 55th had done everything asked of it and, as Colonel Pope informed Charles Bean, the 14th Brigade was 'the only brigade who didn't come back until we were told to'.[93] Not that his brigade's achievements spared Pope; the day following the engagement he was relieved of his command by Major General McCay, ostensibly for drunkenness but more likely due to personality clashes between the two men.

Lieutenant Colonel McConaghy was awarded a Distinguished Service Order for his actions in defending the right flank. Despite his key role at many critical parts of the battle, Major Cowey received no official recognition. In early 1917, Cowey was sent on a Senior Officers' Course in England to prepare him for command of a battalion. His course report is instructive, describing Cowey as 'energetic and keen but inclined to be over-confident ... very capable to a point but lacks imagination.' He was deemed 'not fit to command a battalion' without additional experience. After his involvement at Gallipoli and Fromelles, Cowey would have appreciated the absurdity of being critiqued by British officers for a failure of imagination. To add to the irony, Cowey returned to France in time to command the 55th at Bullecourt.

The battalion was broken by the disaster at Fromelles. It would take six months, significant reinforcements and determined leadership to restore the combat efficiency of the unit. The psychological impact of Fromelles on the individuals who were there lasted far longer than the time it took to rebuild the battalion. In 1927, John McCallum,[94] a lance corporal in B Company on 19 July 1916 but destined to become a long-serving Federal senator, composed some lines to describe Fromelles:

Many years have glided by
Since I saw Fromelles
Still I see it each July
Hear the guns! Fromelles!
See duckboard doused with blood!
Hear the five nine's crunching thud!
See that body in the mud!
Cordite stinks! Fromelles.

Lightly out of Bac St Maur
We went to Fromelles
Kissed the girl behind the bar
Gaily for Fromelles
Wended down a winding sap
Halt. A murderous thunderclap
The front platoon is off the map
Hell opened at Fromelles.

Gasping 'neath the parapet
Over boys! Fromelles
Three hundred yards and then we're set
Broken ground. Fromelles
Three hundred more. Consolidate
Dig for love and dig for hate
God! The night is growing late
This gutter's wet. Fromelles.

Passing words along the line
'We're anchored here at Fromelles
The Fifteenth is fast, the Eighth is fine'
We lied at old Fromelles
Steadily digging through the night
We're filling bags, we're making it right
Tomorrow sees us sitting tight
What! Cornered at Fromelles.

Bombing at us from both ends
Fritz knew his Fromelles
55th on you depends
The keeping of Fromelles
We're holding what? Two hundred yards
Both flanks in the air? Pooh, pooh canard
What's that? Retreat! We're running hard
Sauve qui peut![95] *Fromelles.*

We're back behind the parapet
Gasping from Fromelles
Many a fellow's sun has set
Bloody mess Fromelles!

Chapter 4:

Fleurbaix, July–October 1916 – 'I am starting to feel the strain'

The British offensive on the Somme was maintained throughout the summer of 1916, placing enormous demands on the available manpower. For General Haig, who still harboured the dream of a strategic breakthrough, it was imperative to keep the Germans from reinforcing their defences in this contested sector. This mission was allocated to divisions too weakened for major combat operations. As a result, only hours after the Fromelles engagement, the depleted and traumatised 5th Division was ordered back to the trenches with the task of tying the Germans down around Fleurbaix. With the limited resources available to him, the 5th Division's commander, Major General James McCay, chose to pursue a policy of aggressive harassment: subjecting the enemy to intermittent artillery barrages, ad hoc mortaring and the occasional infantry raid.[1]

In the 14th Brigade, now commanded by Brigadier General Clarence Hobkirk, it fell to the 55th and 56th battalions to return to the firing line; the 53rd and 54th had been so badly mauled that they were, to all intents and purposes, ineffective.[2] Robert Harpley, promoted to corporal after the Fromelles engagement, recorded:

> This evening [22 July] we got orders to go out and take over another portion of the trench. I can tell you the boys seemed very reluctant about going, but the thing had to be done ... On taking over the line we had to combine A and B Companies together to make anything like company strength. We only have one officer left.[3]

C and D companies were also amalgamated.

The first few days in the firing line were uneventful; the Australians were exhausted and the Germans appeared amenable to the maintenance of an uneasy calm. Corporal Harpley observed that the shortage of men and the demands of front-line routine left little time for recuperation: 'Things fairly quiet although we are nearly worked to death. We get no sleep of a night and of a day we are three parts of the time on fatigues.'

Still, the long, warm summer days offered opportunities for the soldiers to relax. Writing in his diary, Lieutenant 'Bob' Chapman captured the surreal atmosphere of one of these moments:

> It's a beautiful Australian day – Aeroplanes are droning overhead far up in the sky like huge dragonflies – the hum of their engines has a peaceful sound. Occasionally a rifle cracks out its message and vicously scatters some clods of earth from off the parapet. Far down on the right towards the Somme the sound of big guns rumbles over the ground booming out their message of death. The English are advancing there but here our men are sunning themselves.

That evening Chapman made another diary entry which told that the realities of trench life had again risen to the fore:

> Fritz has been returning a little present of iron rations that our artillery presented him before sunset. Twenty seven shells came over into our sector, but no damage was done. Well I suppose it would not do to let Sunday pass without paying a few compliments ... Behind the trench are a number of stagnant pools. They hum [smell] a bit – also breed mosquitoes. The reason they smell slightly is that dead men are buried all over the place – fatigue parties filling sandbags come across the remains. Not far from my dugout is a peculiar small mound that springs up and down as one walks on it. This mound excited my curiosity 'til someone explained to me that a dead man rested underneath – poor chap! His rest is a bit disturbed.

As July progressed, Major General McCay's policy of aggressive harassment was implemented with increasing vigour. Artillery exchanges intensified, trench mortars made their way more frequently to the front lines, snipers became active and patrol clashes punctuated the night. In his diary entry for 30 July, Private James Barker wrote of the effects of the change in battlefield tempo: 'Had a heavy bombardment for three hours. I am starting to feel the strain. Looking for a day or two spell. Tucker worse.'

Private Herbert Harris' diary entries for the period capture the nature of this sojourn in the trenches:

> 31st July. The Huns thought there was an attack on owing to the bombardment. Blazed at us all night with rifles and machine guns ...
>
> 1st August. Had a lively night again then our officers got the jumps. They expected the Huns over ... but barring firing tons of lead at us the night was quiet enough.
>
> 2nd August. Paterson[4] wounded in the thigh while out on outpost duty ... No gas last night. Pretty lively bombardment this evening but not much

> damage done so far ... We smash Fritz's trenches and he turns 'round and does the same to us.

Most days brought casualties. Privates Robert Buckenham[5] and Harry Smith were both killed by German artillery.[6] Lieutenant Chapman wrote of the personal impact of the wounding of his batman, Private Percy Larbalestier[7]:

> They [a trench mortar team] came here yesterday … fired two shots, and then went away laughing. Fritz answered – first with bombs then with shells; things got interesting for a few minutes, the consequence being that my batman got wounded in the back and arm. He walked down to the dressing station, although you could almost put your fist in the hole in his back, and sent word to me that he had been plugged. I went down to the dressing station as soon as I heard about it. He was lying on his face on a stretcher looking very white, but as soon as he saw me he started to laugh and joke. Of all the officers who came over [to France] in B Company, four that is ... and their four batmen I am the only officer left – and only one batman is left.

On the night of 5 August the battalion was relieved in the firing line. For the next fortnight the unit was billeted in Fleurbaix and the hamlet of Croix-Maréchal. When not sleeping, training or on work parties, the troops had time to sample some of the pleasures of life in wartime France. Private Bert Bishop made the most of these opportunities:

> We had spells back in billets in Fleurbaix. We became civilised creatures. The shops sold us all sorts of souvenirs, chocolates, writing pads, pencils, ink. The estaminets were well-stocked with wine and all kinds of drinks I'd never heard of.[8] Other places sold eggs and chips, and there was always the lovely countryside to walk through and admire.[9]

A steady stream of reinforcements trickled into the battalion and a number of officers were commissioned from the ranks. At least one of the lessons from Fromelles had been learned as, according to Private Herbert Allen, the battalion 'Spent its spare time in bomb throwing practices and in learning how to use the many kinds of bombs.'

Day by day the 55th was being reconstructed. It was a far different battalion to the one that had arrived in Marseilles just seven weeks before. One night at Fromelles had transformed the men into veterans; they retained their cockiness and confidence but were honing the ruthless military edge that would reach its sharpest at Polygon Wood 12 months later.

On 20 August, the unit was notified that it was to move into the front line that evening. The handover occurred without incident and, before midnight, the men found themselves back in their old haunts. According to Bert Bishop, it took little time for the routine of trench life to reassert itself:

> A few days in the front line, manning fire steps at night, sleeping or doing fatigue work by day. Always there were dugouts to be mended and drainage to be improved by day, and at night, as well as duty on the fire steps, there were rations to be brought up, patrols to go into no man's land and shell-torn wire to be mended.
>
> One of the worst parts of our duty was the lone vigil on the fire step at night. It was two hours on, four off. Gazing across at the German line, complete silence around one, it was so easy for the rubbish to assume different shapes, and as star shells turned darkness into vivid light, and the flares slowly floated to earth, everything turned into mixed shadow and light, full of movement.[10]

During the 55th's previous stint in the front line the weather had been dry and warm; this time they contended with frequent rain and thunderstorms. Mud became a constant companion. Private Harris described the lot of the infantryman:

> Put in a miserable night last night. Stand to at 9 p.m. Raining. On sentry 10-12 p.m. Raining hard and black night lit up now and then by flares which made it darker when they went out. The duckboards are like ice and we could hardly walk on them.[11] Then stand to again at 4 a.m. On sentry 5-7 a.m. On sentry again 11 a.m.-1 p.m. and again 5-7 p.m. Stand to 8.30-9.30 p.m. Sentry 9-11 p.m. and again 3-5 a.m. 'Oh this is the life'.

Bert Bishop wrote of the ubiquitous rats which tormented and sometimes provided amusement for his mates:

> We got some fun occasionally from the rats. Now and then a nest would be found amid the work we did. Bayonets were drawn and we chased the rats everywhere. The loathsome creatures slept with us, they crawled over us, they gnawed our haversacks hunting for scraps of food.[12]

Days turned into weeks and there was still no word of the battalion being relieved. Lance Corporal George Gill wrote in his diary:

> We are having a long spell in the trenches this time. Well we are giving the Bosches a lovely time lately, and at the same time they are giving us one. It is nothing but bombardments and bang banging and the worst of it. Their shells are coming around our place so much ... what makes these little affairs so trying is that we cannot get any or much sleep. They come over at any time of the day or night.

The mental strain was unrelenting and Gill noted that he was beginning to falter: 'Another bombardment. Wish we could get a spell out of this. My nerves, [shaken] with poor food, hard work, no sleep, are going to pieces.'

Several possible instances of self-injury are suggestive of the damaged emotional and mental state of the unit in the post-Fromelles period. The social stigma and harsh penalties associated with self-harm deterred most soldiers from taking such a drastic step to escape the trenches, but some were willing to pay the price. While men wounded in ambiguous situations were always given the benefit of the doubt, several events during this tour are tainted with suspicion. Private Bishop wrote that one night 'Reg', a member of his platoon, placed his hand on top of the parapet, where it was hit by a German bullet. By way of explanation, Reg offered the excuse that he was adjusting sandbags at the time. His platoon sergeant (Harlock[13]) was sceptical of Reg's explanation for the wound:

> 'What's happened here?' he demanded of Reg.
>
> He was told ...
>
> Harlock looked at the hand.
>
> 'You'll have to go out,' he said, 'but I'd like to give your backside the hardest kick it ever had. And you know why.'
>
> Our platoon never saw Reg again.[14]

Private Arthur Luby was fatally shot as he was cleaning his rifle: his service record documents his death as unintentional.[15] However, Lieutenant Colonel James Dick, Officer Commanding the 1st Australian Casualty Clearing Station, was unconvinced of the accidental nature attributed to Luby's wounding: '[Luby] was admitted to this Station at 10.30 p.m. on the 31st August 1916 in a very low condition, suffering from an 'accidental' GSW [gunshot wound].'

The circumstances of Luby's wounding are interesting. He was sitting in a dugout with Private Bertie Hilton and both men were cleaning their rifles.[16] In the cramped conditions, (at the time they were sitting 50 centimetres apart) it would be difficult for an individual to position a rifle and pull the trigger to wound himself in the thigh. Possibly Hilton, who like Luby was an inexperienced soldier, accidentally shot Luby. Another explanation is that Luby deliberately injured himself and Hilton concocted the story to protect his mate from official retribution.

Eventually, the exhausted men were relieved. On 21 September, after more than a month in the front line, the 53rd Battalion replaced the 55th. Even leaving the trenches was an ordeal as Private Frank Brown attests: 'It was a very black and moonless night and leaving the trenches by long saps, greasy and tricky ... 'twas a hard job to get through without breaking your legs.'

This tour had been very similar to its predecessor: endless fatigues and shelling interspersed with minor patrol clashes. For all its banality, the front line in even this 'quiet' sector had proved to be a dangerous place: in addition to Private Luby, Private Harry Banwell,[17] Private Francis Beaven[18] and Private Patrick Walsh had lost their lives during the stint and over 20 men had been wounded.[19]

Private Harris was thankful to be back in the billets at Fleurbaix: 'What a treat to wake up and find that the shells are a long way away. Have had about 17 days almost continuous bombardment and a little more of it would have wrecked my nerves. [I] am standing it better than a lot of other fellows.'

Lieutenant Chapman provides an insight into the psychological changes he noticed during this period: 'After a fellow has been here some time and has seen men picked off under all sorts of conditions a belief in fatalism seems to be the natural outcome.'

While most troops spent the time out of the line resting, for others the period was one of intense training in preparation for a raid on a group of enemy trenches north of Bas Maisnil, colloquially known as Clapham Junction. The operation was to occur on 30 September.

The attack plan called for the artillery to deliver a 'box' barrage along all four sides of the area to be raided to prevent enemy troops entering or leaving; trench mortars were to smash the barbed wire protecting the German trenches.[20] The covering bombardment was to commence at 10.00 pm. While the barrage was under way, the raiding party was to proceed across no man's land to a position close to the enemy lines. At 10.10 pm the shelling of the enemy front line was to cease and the raiders were to enter the German trenches while the bombardment continued on the other three sides of the 'box'. The raiding party was to spend a maximum of ten minutes in the enemy lines, killing Germans and hopefully capturing a prisoner for intelligence-gathering purposes, before retiring across no man's land.

Captain Sidney Pinkstone was the designated raid commander, giving him overall responsibility for the planning and execution of the operation. His duties would see him remain in the Australian lines rather than taking part in the raid. The leader of the actual assault was Captain Eric Stutchbury assisted by Lieutenant Tom Gitsham.[21] The raiding party comprised the two officers and 60 specially selected soldiers, the majority drawn from D Company. Private Arthur Buckingham related: 'All the boys, excepting one, volunteered for a raid to be made to the Huns lines, which is a very risky job.'[22] Buckingham was one of those chosen to take part in the endeavour.

The raiders spent the fortnight leading up to the operation at Sailly sur la Lys practising all aspects of the forthcoming mission. On the two nights prior to the raid, patrols were sent out to examine the condition of the German wire where the incursion was to be made; these returned with news that the entanglements were well cut.

On the evening of the operation the raiding party travelled by trucks to Fleurbaix, where the 55th's acting CO, Major Robert Cowey,[23] organised for a superb dinner to be served — part of the meal consisted of a wedding cake sourced from a nearby village.[24] Non-alcoholic toasts were made to the success of the mission.

After the meal the raiders made their final preparations, checked their weapons and completed a final run-through of the plan. Faces and hands were blackened. To assist in identifying friend from foe, the men wore a white armband which was covered with a piece of sandbag as they made their way across no man's land. To disguise their nationality in the event they fell into German hands, the raiders were dressed in old British uniforms and were without any form of personal identification.

When all was ready, the raiders remounted the trucks and travelled the short distance to the communication saps. Here they alighted and filed down these trenches to the front line. Raider Private Archie Winter described how events initially unfolded:

> After the usual tot of rum we crawled through the sally ports.[25] We worked our way across No Man's Land. When about half way a machine gun opened up and made it pretty warm. One of our lot (Raymon[26]) was shot dead and another (Young[27]) died later. It was then that our artillery barrage started … when the barrage lifted we hopped into Fritzy's trenches.'

Private Robert Harpley witnessed the supporting barrage from the Australian trenches: 'The bombardment commenced at exactly 10 o'clock last night and for about 400 yards along the German trench shells of every calibre were hurled at it. The whole trench seemed to be on fire … then the guns lifted and over our boys went.'

The raiders discovered that the German front line had been badly damaged by the bombardment; three semi-naked German corpses were seen on the parapet, their tunics torn off by the blast from exploding shells. Entering the enemy trenches, the raiders fanned out looking for trouble.

There were only a few living Germans above ground. Private Louis Wood encountered one of these and overpowered him in the ensuing struggle.[28] An enemy soldier wielding a rifle and bayonet attacked Lance Corporal Bill Hitchenor, who was armed with a revolver. Hitchenor fired, wounding the German, and then proceeded to club the injured man to death with the butt-end of his weapon.[29] Lance Corporal Claver Sheedy[30] killed a German in a communications trench, having the presence of mind to remove the dead man's epaulettes which were later useful in identifying his unit. Private Edwin Brookes,[31] assisted by Private Edward Butler, captured a prisoner and returned with their trophy to the Australian front line.[32]

The raiders began hunting those Germans still in their underground shelters. Private Len Miles,[33] Private Al Browne[34] and Private James Cox bombed five shelters between them.[35] In one of the underground shelters, Miles reported hearing someone 'squeak' before the blast from his grenade forever silenced the occupants.

Their mission complete, Captain Stutchbury gave the order to withdraw.[36] The raiders had been in the enemy trenches for just eight minutes. In the after-action report, Lieutenant Gitsham wrote of the retirement:

> No difficulty was experienced in returning to our lines. Owing to enemy fire being heavy I collected stragglers of our own party and detained them in a ditch in front of our parapet. When enemy fire ceased the men entered our lines. Private Ross reported to me that one of our men was wounded near enemy wire.[37] I led a party consisting of [Lance] Corporal Sheedy, Private Ross and Private J. Dawes and searched the right sector of ground and found the body of Private Raymon near enemy wire and brought it in.[38] We went out again and searched left sector and brought Private Young in wounded. A further search of 'No Man's Land' was made but we could not find anyone.

The German retaliatory bombardment had commenced soon after the Australian artillery opened fire. At 10.30 pm Captain Pinkstone ordered the supporting barrage to cease; the Germans halted their own artillery fire shortly after.[39] By 11.00 pm the raiding party was back in the Australian lines.

Senior officers judged the raid a success: 20 Germans had been killed by either raiders or artillery fire, and a prisoner had been taken. Major General McCay, accompanied by the commander of II Anzac Corps, Lieutenant General Alexander Godley, met with the raiding party on the day following the operation to offer their congratulations.[40] Success, however, had come at a price. Private Robert Raymon and Lance Corporal John Davies[41] had been killed; the next day Private Jim Young and Private Roy Cameron were also to die of their wounds.[42] Five men had been wounded. Private Buckingham emerged unscathed: 'I am reported killed but am a long way from it.'

On 8 October, the battalion re-entered the front line, returning to the same sector as the previous two tours. The weather was worsening as the slide into winter commenced in earnest.

Private Farrall had rejoined the battalion as a reinforcement during the period out of the trenches.[43] Long after the war, he related the experiences of his first trip to the firing line:

> Every step now was a step nearer to the frontline of the trenches ... I can still hear the first shell from the German artillery that passed over us.
>
> All movement in an area within reach from German guns had to be carried out in the dark, and that's how I was when I first experienced trench warfare at Fleurbaix. The weather was good, the trenches clean and the dugouts comfortable ...
>
> The ever present threat of death flying around didn't make for any certainty about the future, and therefore gave rise to a feeling of anxiety, a state of living which laid the basis for the disability of neurosis.
>
> In the dugout where I lived ... the roof ... was made of sheets of corrugated iron and sand bags, which was fairly good protection against the shrapnel shells, but, like any part of the frontline, was vulnerable when high explosive shells landed.
>
> It was in the bombardment from the German artillery one night, the 12th October 1916, that we in the dugout experienced some anxious moments, though in fact they seemed like an eternity. We were being subjected to a merciless bombardment, the heaviest we had experienced at Fleurbaix.[44]

The German shelling was more persistent than during the battalion's previous two stints in the trenches. Six men were killed in the five days the 55th was in the front line.[45] On 13 October the men made their way again to Fleurbaix, having been replaced by the 3rd New Zealand Brigade.

Private Farrall recalled: 'The night Gus[46] died, we were relieved by the New Zealanders, and taken out of the front line there, and unbeknownst at any rate to me ... we were started on the way to the Somme.'

Corporal Harpley's diary relates his discussions with members of the relieving force: 'They [the New Zealanders] are just back from the Somme and had a way of describing it in as few words as possible. They term it Hell, which by all accounts is the most appropriate word for it.'

The next day the battalion made its way by foot and motor transport to Oultersteene. Here, in the final fine and warm days of 1916, the troops participated in physical training and attended lectures on the peculiarities of the Somme battlefield. Private Harris reflected: 'Right away from the front now. We spell here for a few days then on to the Somme where a great number of us will stay for good.' After three months of comparative tranquillity in a 'quiet' sector, the unit was returning to the real war.

Chapter 5:
Somme winter, 1916–1917 – 'like hell let loose every day'

By now the Somme campaign had been in progress for almost four months. It had begun on 1 July 1916 with an attack on a front of around 80 kilometres. While the British enjoyed mixed success, the British Commander-in-Chief, General Haig, was sufficiently encouraged by the results to launch a series of bitter stand-alone battles over the summer and into autumn.

With the capture of Flers and Lesboeufs at the end of September, the British gained control of the third line of German defences in the Somme sector. However, while there had only been three defensive lines at the start of the offensive, the Germans had not been idle during the slow Allied advance. General Haig's forces now confronted a fourth line of defences stretching along the ridgeline running from Le Transloy to Bapaume. Behind these trenches, other defensive works were in various degrees of construction.

General Haig, believing that German morale was cracking and undaunted by his previous reversals, pursued his aim of a breakthrough. He ordered the Fourth Army, commanded by Lieutenant General Henry Rawlinson, to capture the fourth defensive line in what became known as the Battle of Le Transloy.[1]

The offensive, which opened on 1 October, began well for the British. But, with the arrival of the autumn rains, the battlefield, pounded by artillery fire into a fine dust over the preceding months, turned into a quagmire. Lieutenant General Rawlinson, unaware of conditions at the front, kept his army pressing forward, mounting further attacks on 12 and 18 October. He planned his next thrust for 23 October.

At this time no Australian infantry were engaged on the Somme and the 5th Division remained in the Fleurbaix sector.[2] However, the Fourth Army's interminable operations required the ongoing replacement of battle-weary divisions by 'fit' divisions. As a result, on 9 October, all four Australian divisions then on the Western Front were ordered to concentrate on the Somme. The 5th Division, deemed by General Herbert Plumer, commander of the Second Army to which the division was then attached, to be 'the one most fitted for offensive operations'was chosen as the first division to head south, with the others following.[3]

On 17 October the battalion left Oultersteene and entrained for 'Hell'. The next morning found the unit at Pont-Rémy. Corporal Bob Harpley described the train journey in disparaging terms: 'Our trip down to here [Pont-Rémy] was not a very pleasant one as we all packed into cattle trucks 35 in each and it was bitterly cold, rain all the time but the one thing that goes to make the journey through France pleasant ... is the beautiful country.' Leaving the train, the stiff and weary troops marched across a bridge to their temporary billets.

Private Bishop considered his new environment a pleasant change:

> The Flanders flat country had given way to hilly country, beautiful country. In between drilling and marches I found time to explore our surroundings. From any hilltop a glorious view of little villages, rolling uplands, rich valleys under cultivation, small woods was obtainable.
>
> We stayed at Pont-Remy three days, a lovely place, but full enjoyment was marred by that never-ceasing rumble, like continuous thunder, from the east.'[4]

While at Pont-Rémy members of the unit made their final preparations for the move to the war zone. Several of the officers went forward to reconnoitre the area in which the battalion was to operate, returning with stories of mud and deprivation. These tales added to the disquiet felt by many about what lay ahead.

The journey forward began on 20 October. The men were woken at 4.30 am and, according to the recently promoted Corporal Sinclair Hunt, '[The men were] busier with the size of the bacon for breakfast than with the thoughts of the march ahead or that the evening of the day would find them on the Somme battle-field.'

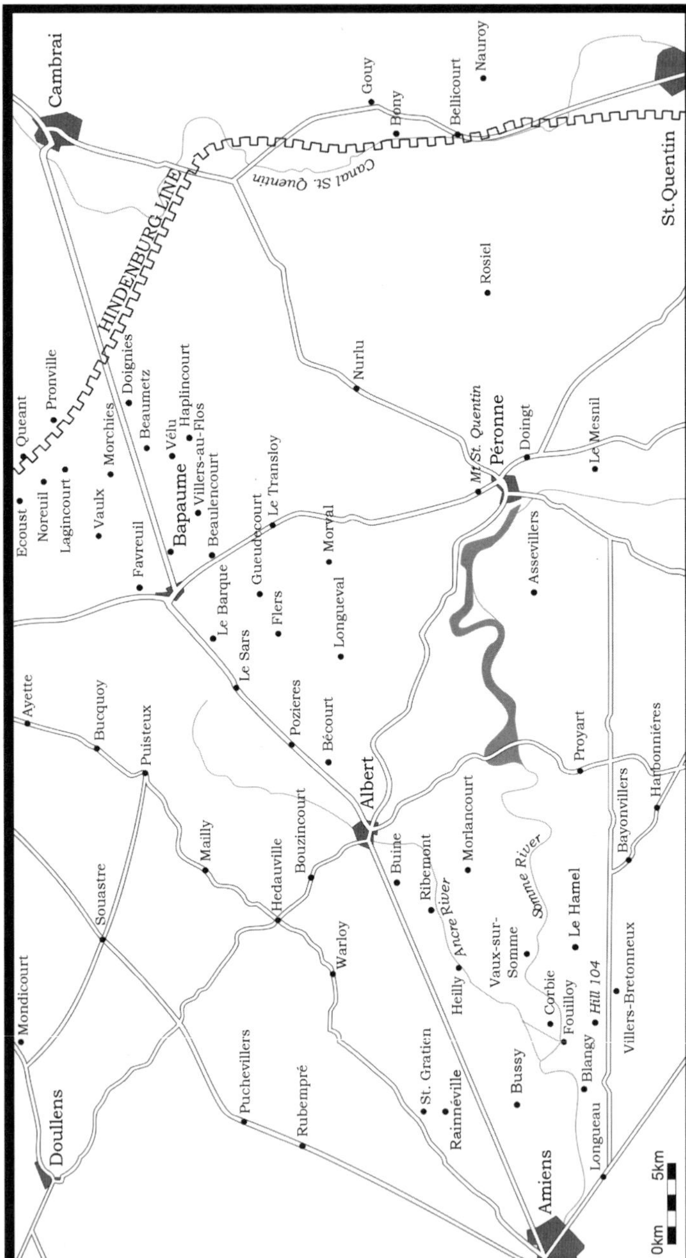

Map 4: Somme region.

After breakfast, and still in darkness, the 55th marched the ten kilometres to Ailly-le-Haut-Clocher. Here the battalion linked up with the rest of the 14th Brigade before tramping a further six kilometres to Mouflers. Private Bishop noted that, by the time they made it to this commune, they were exhausted:

> We carried loads of about eighty pounds and the usual ten minutes' spell each hour did not accompany us. For the first time we were experiencing a forced march. We tramped for four hours. Exhausted, lame, starving, we came to a main road. Almost as one man we just fell by the roadside. Never had we seen such traffic as that main road carried. Columns and columns of troops, tens of thousands of them. Hundreds of motor lorries, all waiting to take us closer to that now horrible, dreadful noise. Horse-drawn wagons and limbers, horse and motor-drawn artillery, army service wagons going to and fro. Now and then an ambulance fighting its way back with a load of wounded, others fighting their way up for another load ... Weary and hungry, we just watched.'[5]

However, the men were in for a treat. French Army buses arrived to transport the troops the 50 or so kilometres between Mouflers and Buire, the debussing point.[6] Ellis wrote of the bus ride: 'The drivers ... drove at a rapid rate, and the convoys, many miles in length, made an impressive sight as they dashed through village after village and swung out with open throttles along the broad high-roads of the Somme Valley.'[7]

The convoy arrived near Buire in the afternoon, the diggers tired and hungry. Bert Bishop caught the mood:

> It was now almost twelve hours since we'd gormandised that almost invisible breakfast ... We were expecting a meal at last, we knew there was a limit to what we could do. Quite a few men, forced to fall out because of blisters or inability to keep up the pace [during the morning march], caught up and sprawled with us on the ditch bank off the roadway, most of them having been given a lift by a driver who had some humanity.
>
> Our platoon officer, who had gone along the road to find out something of what was going on, returned.
>
> 'I'm sorry, chaps, but there's no food for us. We resume marching in a few minutes.'[8]

The buses departed and the battalion started walking. The officers insisted on long stretches of marching punctuated by a few short rests. According to Private Fred Farrall, 'the roads were shocking, torn up with traffic and difficult to walk on.' As the battalion made its way through Dernancourt and Fricourt, the day's accumulated fatigue began to tell, the pace of the column slowing. Bert Bishop witnessed an exchange on the side of the road:

> Men were dropping out along the line. Our colonel [McConaghy] on his horse came by, making an inspection of his wonderful battalion. He spoke to a man who had just fallen out.
>
> 'I can't go on, Sir, I've passed my limit.'
>
> 'Pass the word along to all men dropping out to get along as best they can. We are to bivouac at Montauban for the night.'
>
> Looking over us he stood up in his stirrups.
>
> 'And please, men,' he said in a louder voice, 'remember I'm not the man running this bloody war.'[9]

As they approached the battlefield, the nature of the country changed. Gone was the rustic autumn landscape. Private Bishop wrote of the transformation:

> We were then not far from the line and the country all round was in a frightful state. Battered trenches, shell-holes, dugouts, barbed wire everywhere, and everything from six inches to two or three feet of mud. One could easily see what terrific fighting had taken place by the way the country was torn and blown about, and rough graves, both Germans and ours, everywhere. And in front the continual bombardment was in full swing right along the line, now and then gathering in intensity till it was like thousands upon thousands of terrific thunderstorms rolled into one, and the heavens over the lines seemed one mass of darting, shooting electricity. It was a grand but awful sight.[10]

Around 7.00 pm, in the pitch black, the lead elements of the battalion reached their final destination, Pommiers Redoubt, a bivouac camp two and a half kilometres west of Montauban.[11] Corporal Hunt recollected: 'And when ... the camp was reached, there were as many lying on the road as formed up before the Colonel.' The unit's laggards drifted in over the next few hours.

It had been a hard day for the battalion and many compared it for sheer exhaustion with the desert march from Tel el Kebir. The men had trudged 24 kilometres carrying heavy packs along, or adjacent to, muddy tracks. Their mood became even more bitter as rumours, probably correct, circulated that the French officers in charge of the bus convoy had made it known that the buses could have picked the battalion up from Pont-Rémy had they been asked to do so.

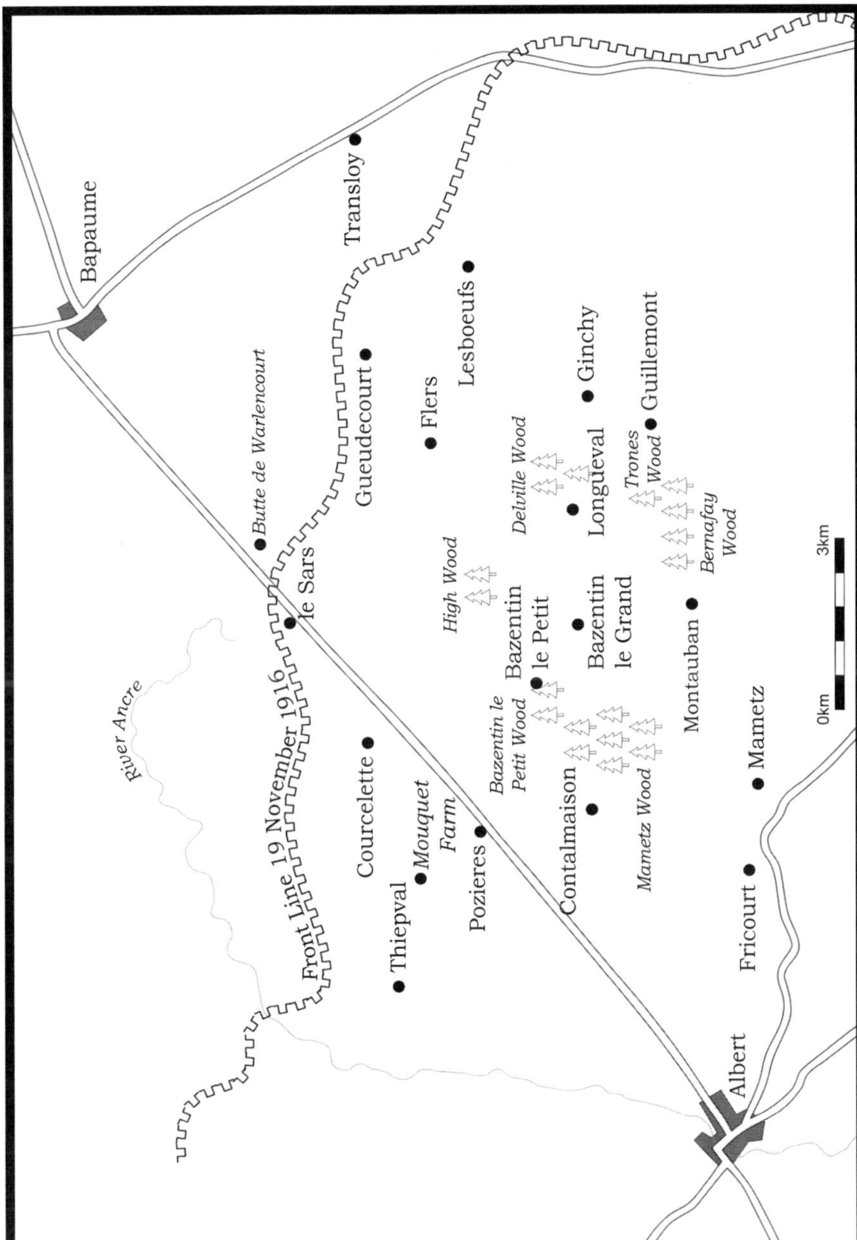

Map 5: Somme area of operations, late 1916 to early 1917.

Yet, the privations of the day were not over. At Pommiers Redoubt there was covered accommodation for 200 men: the battalion numbered almost 1000. Every house, shed or barn in the surrounding area had been obliterated by months of bombardment. There was insufficient engineering material to construct shelters and the 'camp' was little more than a muddy bog with no kitchens and few latrines. The men had no option but to sleep rough.

Corporal Hunt spent the night without shelter: 'There were no tents, it was inky dark so we felt with our hands for a place where the water was not lying and in the mud spread out waterproofs to spend our first night on the Somme.' Private Farrall likewise 'roughed' it: 'There was no camp as such, and that night I slept in the open with Bing Hall.[12] When we slept out, we would have one waterproof sheet and one blanket under us, and the same over us, and we slept together with all our clothes on, boots and all. All that would be taken off would be our tin hats.'

It was a freezing night and, by morning (21 October), the first frost of the coming winter covered their blankets and groundsheets. Cold and cramps meant that many soldiers had little sleep despite the exertions of the previous day.

At 8.00 am the 55th's CO, Lieutenant Colonel David McConaghy, attended a briefing session at 14th Brigade Headquarters. There he received orders for his battalion to move immediately into the front line north of Flers.[13] The unit was unprepared for such a move and scrambled to get ready. Corporal Hunt recollected:

> [We] woke next morning shivering and stiff to be greeted by orders to pack up at once and proceed to the front line. After a hurried breakfast we threw on our packs, passed over[14] our wet blankets and at 9 o'clock left the muddy road to drag our weary way over that awful quagmire, the "Somme Battle-field".'[15]

Fifty years later, Private Farrall still remained distressed by one of the new 'security' measures deemed necessary before going into the trenches:

> [We were] told that all personal belongings, all letters and all photos had to be destroyed. This came as a shock to everybody, it was totally unexpected, and caused us great anxiety as to what was in front of us. Nobody knew why we had to burn our treasures ... The cruelty of it still appals me. Men carry such things as photos and letters into battle as comfort, and they were stripped from them at the last moment.[16]

Leaving A Company at Pommiers Redoubt as work parties, the remainder of the battalion made its way to the firing line by mid-morning. Passing through the ruins of Montauban, the men followed a muddy road skirting Bernafay Wood before proceeding to Longueval and then along the north-west edge of Delville Wood. Dense lines of slow-moving vehicles were using the road, forcing the infantrymen 'to pick a precarious way among the wheels and hoofs by which they were surrounded.'[17] From Delville Wood the condition of the road deteriorated until, around Flers, it became indistinguishable from the morass of mud surrounding it. From here they followed communications trenches to the front line.

Everything about this first journey to the Somme trenches made an indelible impression. Corporal Harpley wrote of the ordeal:

> Poor frozen devils we were too, we had to advance across all open country with scarcely a yard of sound ground ... Please excuse the writing for I am too cold to hold the pen. I am in the trench trying to write this on my knee. It took us all day to do about 8 miles, some places where the shell fire was too heavy we had to go through a sort of trench up to our knees in mud and in some places over our knees ... However we got to our destination ... at dark, just a deep drain. We could not lay down and had to sit everywhere, wet, cold and uneasy.

Corporal Hunt was similarly affected:

> Over this roadless, trackless tangle of trenches, barbed wire and shell-holes, terrible enough in dry weather, we already tired, gradually worked our way line-ward after the whole had been drenched by the rains of early autumn ... One thing I shall never forget – the sight of the men we relieved. They had made three unsuccessful attacks in a fortnight ... They crawled out and we struggled in.

The battalion had suffered a casualty even before reaching the communications trenches: Private Stan Bingley was mortally wounded by an overhead shell-burst.[18] The German gunners knew the location of the roads, communications trenches and tracks and subjected these to regular shelling.

The relief was completed by 6.00 pm. There were no engineering stores at the front line due to the difficulty of transporting them across the bog. With the exception of battalion and company headquarters, this meant there were no dugouts or other forms of shelter.

For reasons never explained, the troops had handed in their blankets that morning. As the temperature plunged below freezing point, the men found themselves in sodden clothing, their lower legs immersed in icy, muddy slush. Very quickly a combination of exhaustion and hypothermia reduced many of the officers and men to a torpor-like state rendering them almost incapable of thought, let alone taking the smallest steps to ameliorate their own suffering. Sleep was out of the question: fatigue permitted dozing until this was interrupted by cold and shellfire. Private Fred Farrall described his first few hours in the front line at Flers:

> The British Regiment who'd been holding the line ... left behind one chap [a guide] that we met up with, and we hadn't hardly got properly into the front-line [when the Germans] landed a shell almost in the bay where we were with disastrous results. The British soldier who'd been left behind had his leg badly shattered. Sam McKernan ... was badly wounded.[19] Dave Vass ... was badly knocked about in the face ...[20]
>
> That was the first few hours on the Somme, and the first experience I had had of such casualties: the British soldier with his leg almost off, had a tourniquet put on to stop the bleeding, or a lot of it; Sam ... was covered in blood, as was poor David Vass ... I got into this pocket hole in the side of the trench and wedged myself in there for the time being. Sergeant Neville soon made it clear that you couldn't do that sort of thing, you had to be up and about doing something.[21] His cool, calm and collected way of dealing with a situation like this still amazes me. Neville ... set an example that took a bit of following. He carried out his duties in the front line and under heavy fire very much the same as he did on the parade ground.

Patrols were sent out. Sergeant Denis O'Dea, Sergeant Henry Anson and several other men established a bombing post in no man's land that engaged a German party, killing two men.[22] This patrol remained *in situ* for 48 hours in the knee-deep mud.

After a long night, the bright but bitterly cold dawn revealed that the front line lay on a slight forward slope which, a little over a kilometre away, dropped to a shallow valley where the ruined settlements of Le Barque, Ligny-Thilloy and Thilloy could be seen. Beyond the valley the ground rose and Bapaume, some five kilometres distant, was discernible. The Butte de Warlencourt, an ancient burial mound alongside the Albert—Bapaume road, was visible around two kilometres behind the enemy lines. They were surrounded by hundreds of German and British corpses in various stages of decomposition. Corporal Harpley wrote of the conditions: 'I am that cold and shivering that I can scarcely think, let alone write. My God: this is asking too much of human beings, but we are all cheerful and cracking jokes and shells are bursting everywhere.'

The shelling rarely ceased and casualties mounted. Private Alex Connelly[23] was blown to pieces as he sat in his tiny dugout, while another shell buried and killed Private Arthur Passmore[24] and Private Patrick Barry.[25] Private Farrall had to perform some unpleasant tasks:

> During the morning the Germans landed a shell in the bay next door to us and killed three and wounded five. Of the three that were killed, two of them were brothers ... the Mcllwains.[26] ... So, that was a pretty frightening experience, as far as I was concerned. But it became worse, really, towards evening when Sergeant Neville, who seemed to be able to cope with this situation with great ease ... said, 'Well, we've got to have a burial party tonight to bury the chaps next door, and you and you and you and you will be in it'. Which included me. I was scared stiff already and, of course, didn't want to be on anything like that. However, there you are, that's what you're supposed to do ... They [the dead] had to be picked up, to some extent in pieces, and put on a ground sheet and in the darkness, because nothing was done in daylight ... this is how the burial had to be carried out. All it amounted to really was to put them in a shell-hole and cover them over.[27]

During the day, some men unearthed a British soldier who was still alive despite being buried for four days. While physically intact, one can but speculate on the mental condition of the entombed man. Evacuating the wounded from the forward areas could only occur during daylight and this was extraordinarily difficult even then due to the poor condition of the trenches, the vast seas of mud, and the constant sniping and shelling. Many of the wounded men who died may not have succumbed to their injuries in more benign conditions.

Supplies were scarce. Due to the atrocious battlefield conditions, every form of store was brought forward by mules — until the mud proved impassible to even these hardy animals. Then the stores were man-packed forward. The intense physical effort involved meant few supplies actually reached the front line. Hunger became such an issue that one of the 55th's sergeants ventured out into no man's land to rummage through the haversacks of rotting cadavers.

Despite the sunny start to the day, by afternoon Private Farrall was lamenting: 'The weather broke and the rain came, in bucketfuls. The ground that had been torn up for the past two years was just like a quagmire.'[28] According to the War Diarist, the 14th Brigade's senior officers were well aware of the deteriorating state of the unit and a decision was made in the evening that 'Owing to the conditions in the front line it was advisable to relieve the Battalion.' Before they were replaced at 5.00 am on 23 October, the troops endured another night, this time standing in the cold rain with their feet still encased in freezing slush. Private Farrall recalled of the relief:

> We had to go down the narrow communications trench, and the men who were taking over from us were in the 53rd Battalion … It is raining and everybody's nerves are in a state of irritation. There was the 53rd trying to get in, there was the 55th (that was us) trying to get out, in as big a hurry as we could. And so, in the darkness … all this collection jammed up fellows coming in and those going out, and there was a lot of swearing and cursing done on both sides.[29]

The filthy, unshaven troops made their way back to Pommiers Redoubt in small parties. Many of them had cut off their greatcoats at the waist to make it easier to move through the mud. On arrival, the soldiers were served a hot meal and tea; this was the first warm sustenance they had received since breakfast on 21 October.

In the two days they had been absent, sufficient tents and makeshift shelters had been erected to shield the majority of men from the rain. Daylight permitted those men without 'formal' shelter to scrounge sufficient materials to build a semi-dry humpy. The War Diarist described the pitiful condition of the returning troops: 'Majority of men being wet and muddy from the waist down. Day spent trying to dry themselves and clothes – everywhere is mud and water now owing to the adverse weather conditions.'

While they may have been able to clean up and to some extent rest, the men were in no position to relax. Soon after they arrived at Pommiers Redoubt Camp, Corporal Harpley recalled: 'We were fallen in ... and told we are to go into the trenches this evening to attack about 5 a.m. on the morning of the 25th. It will be Hell for it has been raining all night. It is still raining.' At the last minute, however, they gained a reprieve, as Private Bishop described:

> The Battalion paraded to hear battle orders again. We were lined up in pouring rain. About to move off, a dispatch rider ... slid to a halt among our battalion heads. He handed a message to the adjutant, who handed it on to our colonel, who read it ... 'Attack postponed indefinitely. The battalion will dismiss'.[30]

Corporal Harpley soon realised that, far from bringing relief, the delay added to his stress. He confided in his diary: 'Word has just come through that our attack has been postponed ... I wish we would go in at once and get it over with.'

Lieutenant Colonel McConaghy had been careful with his words for, while the attack had been postponed, it was definitely not cancelled. That evening the 14th Brigade's commander, Brigadier General Clarence Hobkirk, gathered the COs of the 55th and 56th and briefed them on their battalions' roles in a forthcoming advance.

The plan called for the 55th to re-enter the line the next morning (24 October) and occupy its former positions. The following day, 25 October, the 55th Battalion, as part of a multi-division operation, was to capture the German trenches to its immediate front and then exploit forward, seizing as much of the high ground south of Le Barque as they could. Here they were to consolidate against a likely German counter-attack. Fortunately, the rain continued to tumble down, resulting in a further postponement of this ill-considered undertaking.

Private Frank Brown wrote of his despair at the conditions in Pommiers Redoubt Camp: 'We are up to our necks in mud and former conditions were child's play to this. Words are not expressive enough for the place.' Lance Corporal George Gill was 'Lucky enough to get a place in one of the tents. We sleep at night packed like sardines. It is a difficult matter to turn over. Last night Fritz sent over a few good shells, the most of them turned out to be duds.'

It was not just the mud plaguing the Australians; the artillery batteries next to the camp maintained an almost unceasing barrage. As the rain continued to fall, the officers became increasingly concerned about the health of their troops. Day by day the toll of sick mounted, and trench foot soon made its first appearance.[31] Corporal Harpley recorded the effect of the appalling conditions:

> Still living in mud and slush we are up to our ankles the whole time. There are 18 of us in a small tent and the ground is slopping wet. How it is that we haven't all got pneumonia I can't work out for the conditions are most awful ... Every day of our lives makes us prouder to be an Australian ... but I'm very much afraid the cold wet weather will break them up ... for quite a number are suffering from frost-bitten feet already.

As their health deteriorated, so too did their morale.

Late in the morning of 30 October orders were received for A and B companies to return to the front line that afternoon; C and D companies were to join them the following day. The launch of the oft-postponed attack had been confirmed for 1 November and the unit was required to concentrate in its jumping-off position.

After lunch the men floundered their way forward over the slimy mud, taking over the same segment of trenches they had held during their first tour.

If possible, this trip was worse than the previous one. Rather than immerse themselves in the freezing slush they preferred to walk in the open under enemy artillery and sniper fire. Corporal Hunt wrote of his move forward:

> We started out ready for a stunt but it poured on the way up. We were drenched to the skin, one officer was wounded, another was lost, and the third lost us, and in the end we found ourselves anywhere but where we should have been. The day was spent in wading through the trenches with sticky mud to the waist. We go into position: it rained all night, and the stunt was abandoned. The weaker became bogged, just like our starving cattle in the creeks. It was nothing to see half a dozen men industriously digging themselves out with spades. Three of us spent two hours on one unfortunate. He was up to his armpits and the suction had him caught as if in a vice.

Corporal Harpley was unwell and, when told he was going back to the line, he wrote despondently: 'I have developed a bonser [sic] cold and really it would be a blessing if one could get something and go to hospital. I don't expect to battle through this myself and I am as hard as nails, but not hard enough for this.' But Harpley proved hard enough and the following day (31 October) he was sufficiently recovered to write:

> Coming into the line yesterday it poured raining [sic] the whole time. We were sopping wet and coming down the saps on the slopes we were plodding through water and mud (<u>combined slush</u>) [Harpley's emphasis] up to our waists at times. The front line is mud up to your knees and we have no place to lie down or even sit – unless in the mud. Putting in a night is really hell. One can manage during the day but we dread the night coming on.

There was no relief for the men from the constant exposure to the elements and enemy artillery fire. Despite continual labouring to improve their living conditions, the rain washed the removed soil back into the trenches and saturated walls and dugouts simply collapsed. Troops were rotated between the front and support trenches but the same living conditions were to be found everywhere in this dismal swamp. When asked after the war how he had coped with such awful surroundings, Private Farrall answered:

> Well, I didn't cope with it at all well. The weather conditions for a start were actually more than I could cope with, and, of course, the heavy shell fire never ended, and the thought that went through one's mind at least a couple of times a day, for instance, when the sun, if it got up in

> the morning, you would wonder whether you'd see it set at night, and when it did, you'd wonder whether you'd see it get up in the morning. That was the attitude I suppose, of a good many and although we had to put up with that sort of thing and carry on, because that's what we were there for and that was the job and that was it ... We were scared ... rather frightened. Although there were those rather rare ones like Sergeant Neville and Sergeant Peters,[32] who seemed to be able to handle the situation remarkably well.[33]

While the conditions were not conducive to large-scale military operations, there was scope for patrolling and sniping. Captain Sidney Pinkstone was unsuccessfully recommended for the Military Cross for organising these activities. One of the advanced bombing parties he positioned close to the enemy trenches accounted for at least five Germans.

And still the rain continued, further delaying the launch of the attack. By now the diggers were approaching breaking point and were no longer in any condition to assault the Germans. The unit was rotated out of the line on the evening of 1 November and moved to reserve trenches near Bernafay Wood. Private Farrall wrote of an incident during the journey rearwards:

> The communications trench ... was not all that wide, and ... to get out, we had to pass through where D Company had occupied a part of the front line, and there lying in the trench was Corporal [Frank] Hunt, who'd been shot through the head, when apparently he took a risk, and looked over the top, and there he was lying in about anything up to a foot of mud and so how do we get past him?[34] In the dark ... the only way that I could negotiate was, I could get one leg down between the side of the trench and him, as quickly as possible and get over ... In some cases, he might have been walked on. The question that always bothered me about it was why he'd been left there like that.[35]

It took the profoundly exhausted men over six hours to travel four kilometres to the reserve trenches. They found conditions there on par with those of Pommiers Redoubt, accommodation consisting of holes in the soggy ground. There was little respite from the rain, cold and mud. Lance Corporal Gill used his time out of the line to bring his diary up to date: 'Had a hellish time ... No one could possibly imagine how dreadful it is.' Even the War Diarist was drawn to observe on 2 November: 'The conditions have been most trying over the last few days and 60 men have been sent to hospital.' Corporal Sinclair Hunt described the disintegration of his battalion:

> The number of sick from the line grew apace, till the procession of sick vied in numbers with that of sound men. Trench feet threatened to sap the vitality of the whole force. The damage had been done before the 'heads' were aware, and even for those who escaped in the first trial, the seeds have been sown which bore fruits under the future exposure [once afflicted by trench foot, a recurrence of the condition becomes more likely].

Hour by hour the battalion's combat effectiveness ebbed under the combined assault of illness and trench foot.

Welcome news arrived on 3 November: the attack had been cancelled and the troops were to be sent out for a rest. By mid-morning the men began making their way to Fricourt Camp. The trek was slow due to the omnipresent mud, but the rain held off, making the journey seem less of an ordeal. Arriving at Fricourt, they were placed in huts for their first dry night's sleep since arriving on the Somme. Private Herbert Harris, now a Regimental Policeman, observed the condition of the unit as it made its way to Fricourt: 'Half the men nearly dead ... If they keep us in France there will soon be no Australian Army left.'

The following day the unit moved again, this time to a camp at Ribemont.[36] The battalion spent a few days here before being transported to the Corps Training Area at Rainneville on 8 November. Private Winter was unimpressed at the standard of the accommodation: 'The billets were simply rotten. Not fit enough for a pig little enough for a man to sleep in.'

The next ten days were spent training, but time was made available for relaxation and many members of the unit walked the eight kilometres to Amiens. General Haig inspected the battalion on the afternoon of 11 November, although he left no record of his impressions.

The rest period ended on 19 November. The battalion was transported to Ribemont and then marched the 15 or so kilometres to Mametz Camp through a snowy landscape. Captain John Kennedy, the 53rd Battalion's padre, found Mametz: 'Most decidedly uninviting. The camp consisted of elephant huts erected on piles on a sea of mud.[37] The approaches were awful and the huts themselves were ... in a most filthy condition.'[38]

For the next week, the troops were engaged in work parties, unloading stores and constructing a tramway to facilitate the movement of stores and the evacuation of casualties. Mud, slush and rain remained constant companions. The prevalence of chest infections, pneumonia and other cold-related illnesses soared, placing enormous stress on the medical system, often forcing doctors to make decisions at odds with their professional judgement. Private Farrall was one of those afflicted:

> [I] got some sort of infection ... something like the 'flu, plus some sort of nasal trouble. I was sent back a few miles to a dressing station. I was there for a day or two, but that was all, and then back up to the Battalion, because the Battalion at this time, instead of having a full complement of 1,000 men, had only about 400, or, at the most, 500. It was only at about half strength, and that meant that everybody who could stand up was expected to be used.[39]

After their short sojourn at Mametz, the depleted battalion returned to the forward areas on 30 November. Since the unit had last been in the front line, the 5th Division had been moved to a new sector about half a kilometre in advance of the ruined villages of Gueudecourt and Lesboeufs. The Australian line was on a forward slope facing the German-occupied villages of Beaulencourt and Le Transloy. A series of reserve and support trenches had been dug to the rear of the Australian front line. There were several communications trenches leading to the front; these were little more than muddy drains and in all but the most advanced areas the troops used the duckboard tracks criss-crossing the desolate landscape to move about.

The battalion moved to the front line after four days in the support trenches. During the relief operation, Private Philip De Jongh, a runner, was making his way rearwards when a shell exploded nearby. [40] The blast tore De Jongh's legs apart and threw him, senseless, into a crater. Some passing men pulled him out of the shell-hole, but seeing the extent of his injuries, left him for dead. When he recovered consciousness, De Jongh began to crawl towards Battalion Headquarters.

At some stage along this nightmarish journey, he took out his knife, cut off one of his legs and applied a tourniquet; the offending limb had been hanging on by a few sinews and shreds of muscle. Captain Eric Stutchbury was heading to the front line and, on finding De Jongh, organised some Lewis gunners to act as a stretcher party. It took them five hours to carry the wounded man the one and a half kilometres to the Regimental Aid Post (RAP). On arrival, De Jongh asked Captain Hugh Wyllie, the RMO, 'Now tell me, Doc, have I got a chance, have I a sporting chance? It's all I want to know.' He then asked for a cigarette. The microbes didn't give De Jongh a sporting chance; he died four days later of gangrene.

The battalion settled back into the routine of front-line life. Private Winter described the trenches in the new sector as 'bad', although 'in better condition than last trip'. Private Farrall wrote of the incessant bombardments that caused a steady stream of casualties: 'The Germans were shelling us and our artillery was shelling them ... Our shells would go over the top of our trenches and into the German lines, and the Germans' shells would go over us and behind us, so there was a ceaseless roar of bloody shells overhead. It never stopped!'[41] Sergeant Eric Biden won a Military Medal for laying and maintaining a telephone cable under artillery fire.[42]

One night, Private Winter found himself and a few of his mates 'on advanced bombing outpost about 40 yards from Fritzy's. Oh it's lonely there watching out over the little strip of No Man's Land. There was dead men to no end, some of them making part of our parapet.'

The men spent much of their time on various types of work parties. Disliked in the best of conditions, fatigues during the Somme winter were noteworthy for their difficulty as Corporal Hunt recollected:

> [I] disliked carrying rations worst of all. It was hard at the best of times but especially when the ever-appearing reinforcement officer got lost ... The duckboards ended a mile or more from the line and it was from there that the real trouble began. The path was marked by tape which soon became covered with mud and invariably became invisible until at last it took an experienced hand to find the track. This was often done by a series of objects remembered in succession – spades, rifles and other less pleasant things. The long winter nights were pitch black so that the work simply became a string of slips, slides, falls, each bringing the precious food and water a little closer ... As the line of men neared the trenches, talking ceased – one even forbore to swear – and as each flare went up and lit the night, everyone dropped flat, hastening on as soon as the brightness gave way to still darker night. At times, however, I have seen parties too exhausted even to take this precaution. They just walked up to the front line, handed in their rations, and walked away.[43]

Work parties were also dangerous, as Private Arthur Buckingham noted: 'One of our mates killed. Frank O.T. Detmers.[44] While carrying rations to the front-line was wounded in the hip and while being carried to the Aid Station was killed outright with one of the bearers. The other had his leg blown off.' Not surprisingly, the following day Buckingham confided in his diary: 'Not feeling well. Both of us, Roy and myself, a bit shaky.'[45]

The battalion was replaced on 8 December and made its way to Delville Wood Camp. This stint in the front lines set the pattern for all the unit's subsequent tours through the winter months; a few days in the support or reserve lines and then three or four days in the front line before moving to a rest area. Four days was considered the maximum period troops could tolerate in the front line given the bitter cold, ever-present mud, relentless rain, hard work, incessant shelling, deficient food and the near absence of sleep.

Private Harris had been born and raised in country NSW, but was unimpressed with the behaviour of his countrymen when they were out of the forward areas:

> The more I see of the Australians the more I am convinced that they are the most foul mouthed, bragging, malingering crowds in this war and all they think of is food, women, drink and to see who can curse the worst. My honest opinion is that they are mostly a damn lot of wastrels ... always talking about how to get to Blighty or hospital and they [the British leaders] expect to beat the Huns with this material (I don't think).'

The unremitting strain and dismal conditions resulted in a resurgence in the blight of self-inflicted wounding. On 10 December a newly joined member of the battalion, Private Les Coleman, in a moment of despair, acted on an impulse.[46] At his subsequent court martial, he was found guilty of 'An act to the prejudice of good order and military discipline in that by carelessly discharging his rifle wounded himself in the foot.' Coleman was sentenced to 18 months' imprisonment with hard labour, commuted to 90 days' field punishment.

After a few days spent resting and on fatigues, the unit moved forward to the reserve trenches on 12 December and to the front line two days later. The War Diarist recorded in his understated manner: 'Weather wet and cold and holding of line rather difficult in consequence.' Private Farrall was far less restrained, describing in detail his dreadful experiences on this tour:

> He [the RMO] saw my feet and it was pretty obvious that at this stage I definitely was on the way ... to becoming a casualty with trench feet. But as the Battalion at this time was down to half its number, everyone who could walk about and was standing up was considered to be fit enough to go on with the job ... The doctor was just doing what he had been asked to do [by the military authorities], and he simply said to me, 'Well, when you come out of the line we will send you away to a rest camp, but, for the time being, yes, you go up tonight.'
>
> So that was it, up I went to the front line.

> I had got into rather bad graces with Sergeant O'Dea, who was now the Acting Sergeant Major ... So, when we got up to the front line that night, three men had to be delegated [to establish] an outpost. This was some distance from the front line, in No Man's Land, and I was one of the men detailed by O'Dea to do this.
>
> It would have been bad enough in the front-line trench which had a duckboard bottom to it, and was kept reasonably clean, but when we went out on to this outpost, there was nothing there at all, nothing at all. It was out in the open, and in the mud. This was disastrous for anybody, let alone for one who was already fit to be a casualty with trench feet.
>
> We were there for two or three days ... when we were finally relieved and ... we got to the front line ... I found that my feet, once they were out of the mud, were useless.
>
> Well, it seemed now that nothing could be done, because it had been known to the soldiers, long, long before, weeks and weeks before, that, if you got so badly wounded in the front line that you couldn't get yourself out, the possibility of being carried out was remote ...
>
> It now looked as if that was the situation I was in. So I said to my mates, 'Well, I can't walk. I don't seem to be able to do anything, and I don't know what to do about it, other than that I stay here.' This was in the bottom of the trench. 'Well,' they said, 'we're going to take you out. We are going to take you out, because if you stay here for a few hours, you will be dead. So, we will do what we can, even if it is question of dragging you along.'
>
> This they did, my mates, Wally Osborne,[47] Roy Wild,[48] and Corporal Wally Smith.[49] Somehow or other, during that night – because this was being done in the dark ... they had to more or less drag me along.[50]

Private Vince Irwin[51] also reflected on the trials of time in the trenches:

> I consider that I am the luckiest man in the world to be able to tell the tale, because where we were was like hell let loose every day. My mate had a lucky escape one night, we were together when shrapnel burst over our heads, a lump of shell hit him on the steel helmet and flattened the crown right on top of his head. He was not hurt ... All the boys are living in hope that the war will end this spring. My opinion is that it will not see spring out. It is nearly impossible to go on much longer. The loss of life will be too great.'

Private Brown wrote of the effect of the ongoing physical deprivation on him and his mates:

> I am down to 8 [stone] and cannot even throw a shadow, at least that's what my pals tell me ... The cold however makes us miserable and five of us are sharing two blankets under a star boarding house [the open sky] so we don't sweat much ... The crawl back [from the front] was on a wet night and I was and still am mud from head to foot. Yesterday I had my first wash and shave for 20 days.

The 55th came out of the line on 16 December and was billeted at Montauban Camp, the War Diarist noting: 'Dry floors and braziers in each hut make matters more comfortable for the men and enabled them to dry their clothes and boots and prepare for the march back to the rest area [conditions] wet and extremely muddy.' Private Harris adds his own description of life inside the newly erected accommodation: 'The hut is full with men ... some standing others writing on the floor in the mud. It's wonderful how dirty one gets in the huts ... we eat [mud] now as well as sleep in it.'

The battalion spent two days at Montauban Camp before making its way to Buire, arriving on 22 December. On the day the 55th left Montauban, a change was announced that would have a significant impact on the fortunes of the unit in the coming years. Major General Talbot Hobbs replaced the 5th Division's unpopular commander, Major General James McCay.[52]

Publically, McCay was said to have left due to ill health: the reality was that he had paid the price for having his division fall apart. Plummeting morale, combined with a high incidence of trench foot (regarded by senior commanders as a sign of ill-discipline and poor leadership by junior officers) had literally crippled the division. With characteristic energy, Hobbs set about rebuilding his shattered command.

As Christmas approached, Lieutenant General William Birdwood, the Corps Commander, ordered all men to be given a complete rest. Birdwood realised that his exhausted Australian divisions were at risk of becoming a rabble. The troops had been given little respite since early July and even in the rear areas they were never free from the mud, cold or the risks posed by shellfire or enemy aircraft. They were reaching the end of their tether.

The entire 14th Brigade was concentrated in the village of Buire which Lieutenant Harry Williams (56th Battalion) described as:

> ... typical of those on the fringe of the Somme battlefield. Some of the houses showed signs of shelling. Its main street was churned into mud

> ... its inhabitants consisted of old men, frightened looking children and women with care-worn faces. Every second home was an estaminet dispensing vin rouge or vin blanc of dubious vintage.

Private John O'Grady was more scathing, regarding Buire as 'mud and stench. Billets – one big stink hole, mud knee deep in places.'[53]

It was in Buire that the battalion spent its first Christmas; the weather was cold and dark, reflecting the humour of the men. Writing in 1935, Private Jim Maynard captured the mood of the diggers:

> The war had lost a lot of its appeal by this time, and I'm not giving away any secrets when I say that it was a pretty dejected 55th Battalion that came out of the line ... for the Xmas of 1916. The Somme took its full toll of a man's health, even if he escaped without a scratch.

Private Harris was unhappy with life in general during the lead-up to the festive season: 'No parcels, no letters for a month and very little from the comfort funds. Also great dissatisfaction with our know-all sergeant.' But in spite of the limited time available to them, the battalion's officers managed to work some minor miracles and organise a few Christmas treats. Private Maynard wrote of the efforts of Captain Harry 'Von' Wilson, his company commander:

> Nothing was too much trouble for him if it was going to benefit his men. And this particular Xmas he excelled himself. New issues of clothing were provided, lousy shirts and pants changed for new ones, and when we found out there were going to be six suckling pigs provided for our Xmas dinner, why, we almost forgot there was a War. Big Bill Muir[54] and Bill Harvey[55] acted as slaughterers, and dressed the pork in the courtyard of the billet – not without plenty of advice from the mob ... Plenty of beer was provided, and the officers came around to have a drink with us at dinner – a very happy touch and characteristic of the comradeship in the 55th.'

It was a good Christmas in spite of the conditions. Even the gloomy Private Harris was moved to write on Boxing Day: 'Xmas comforts not so bad after all. Lollies, tobacco, cigarettes, dates, cocoa, roast beef and veal and plum pudding.'

Post-Christmas the unit commenced a training program: several hundred reinforcements had arrived and these new starters had to be integrated into the unit.[56] The training continued until 4 January 1917 when the battalion began making its way to the front. The men marched to Franvillers and, on the following day, moved to Flesselles. For the next fortnight the unit was involved in work parties and training; officers took care not to exhaust the soldiers given the physical ordeal that lay ahead.

On 10 January, Lieutenant Colonel McConaghy was given temporary command of the 14th Brigade and his adjutant, Captain Percy Woods, was appointed acting CO of the 55th.[57] Casualties, illness and attendance on courses had left the battalion with a chronic shortage of officers. The 24 available officers were too few to complete all the work required and there was no immediate prospect of augmenting their number except by slow monthly promotion from the officer training schools in the United Kingdom (UK).

The deficiency in officer numbers was just a small part of the battalion's leadership woes in early 1917. Senior officers in the division clearly harboured reservations about the leadership abilities of many of the 55th's remaining officers, resulting in a string of temporary appointments to the role of CO. This revolving door did little to assist in rebuilding the battalion's vacillating spirits. The progression of COs comprised: Lieutenant Colonel David McConaghy, 1–9 January; Captain Percy Woods, 10–25 January; Lieutenant Colonel Charles Robert Davies (15th Brigade), 26 January–6 February; Lieutenant Colonel Frederick Toll (31st Battalion), 7–18 February; Lieutenant Colonel Charles Stewart Davies (32nd Battalion), 19–22 February; Major Selwyn Holland, 23 February–14 March; Captain (Temporary Major) Percy Woods, 15 March–4 April; Major Robert Cowey, 5 April–11 May.

On 16 January the men moved to Montauban Camp closer to the battlefield. Here they found conditions had improved immeasurably. Not only were there now adequate huts to protect every man from the elements, but an arctic blast had frozen the muddy ground.

After a further week on fatigues, the unit moved to Trones Wood Camp on 27 January. Two days later A, B and C companies made their way to the support lines between Lesboeufs and Gueudecourt. D Company and the rear elements of the battalion remained in Bernafay Wood Camp to provide work details. Private John O'Grady described the dress of the infantryman in the forward areas in the wintry weather:

> Battle dress very heavy, you had to dress to keep warm. Long wool underwear, wool singlet, flannel shirt, corduroy knee breeches, putties,[58] long gum boots to the thigh, next tunic then sheep skin vest and last great coat, all wool. Battle equipment, canvas webbing, brass buckles, shoulder brace held by shoulder epaulette on the great coat. Ammunition pouches in front of abdomen, water bottle right side and trenching tool

in front to protect the groin if a bullet hit this area. Balaclava, tin hat on top, just nose and eyes showing. How we moved about I don't know.

The War Diarist noted that the freezing temperatures improved more than just the conditions in the rear areas: 'The frosting cold has made a great difference in the life in the line. The mud and water previously encountered has now frozen hard. Sick evacuations for the past two days from the line amount to one man only. Men's health is good.' Corporal Hunt also appreciated the changed environment of the forward trenches:

> Taking everything into consideration the frosts were not so bad as one might imagine. The feet could be kept dry and, by constant movement, even warm, then, in place of a foot of mud, we trod on a surface as firm as a rock. (Faith, it felt like a rock when we came to hammer it with our army picks!). One could keep clean, and, by sleeping together at night, fairly comfortable. On the whole, I rather liked the frosts or rather the continuous freezing.

The battalion, still minus D Company, moved to the front line proper on the evening of 31 January. Every night, patrols ventured out to prowl no man's land. Corporal Ed Bourke was Mentioned in Despatches for his work on the night of 1 February when he entered the German trenches returning with, according to his citation, unspecified 'valuable booty'.[59] There were numerous patrol clashes; on one occasion three men encountered an enemy patrol and, in the resulting firefight, managed to capture a German.

German artillery posed a constant danger: on 3 February, 2nd Lieutenant Arthur Vogan, the acting Adjutant, and Sergeant Frank Hocking were making their way along a trench towards Battalion Headquarters when a shell burst nearby, killing Hocking and mortally wounding Vogan. On another occasion a shell hit the lip of a trench, spraying jagged splinters onto the troops below, killing Private Albert Pettit,[60] Private Bill Pryce[61] and Private Percy Fenwick.[62] Five others were also wounded in the blast.

The frozen terrain also allowed work parties carrying hot stew or soup to reliably reach the front at the end of each day. This was a most welcome change from the conditions in autumn and led to a vast improvement in the men's wellbeing.

The battalion was replaced on 3 February and made its way along the duckboard tracks to Trones Wood Camp. The War Diarist commented that it had been a good tour: 'During the time in the line the health of all has been good and no cases of trench feet occurred.' The 55th remained in huts at Trones Wood and Montauban resting and providing fatigue parties until 9 February when it moved again to the support lines. On this occasion only B, C and D companies went forward — A Company remained behind to provide work details.

The weather continued bitterly cold. To obtain water, troops in the forward areas were forced to break through the thick ice in the bottom of shell-craters. Private Harris was sickened when he heard of a particular 'water hole' where 'one of Fritzy's shells lobbed right in the shell-hole and unearthed two dead Tommies and one dead German.[63] How's that for horror? Imagine drinking that water.'

On 12 February the battalion, now joined by A Company, moved into the front-line trenches it had occupied during the previous tour. During this stint, the 55th killed one of its own. One night, Private 'Billy' Quinlivan was making his way towards an outpost in no man's land when a sentry challenged him; he made no reply and continued walking towards the Australian position.[64] Several men opened fire, hitting him with five bullets and sending his lifeless body tumbling into a shell-hole. Quinlivan's brother Jack, also a member of the unit, had the unenviable task of burying Billy the next morning in a makeshift grave at the back of a trench.[65] The unit was relieved on 15 February and returned to camps at Trones Wood and Montauban.

The battalion received a further intake of reinforcements including Private Eddie Street, who wrote extensively of his wartime experiences in the battalion.[66] He described his introduction to the soldiers of the 55th:

> On entering my appointed hut, I was met with a volume of acrid smoke which filled my eyes and drove me gasping out the door. By degrees, I got into the body of the hut and peering through the smoke cloud, I found a tiny little fire made of green wood and a cluster of men around it. One man, by means of his bayonet, was chipping splinters from a piece of wet wood and putting them in the brazier. Along the walls [and] on the floor were blankets, packs, gear, food, mess tins etc. Most of the men were in bed reading Aussie letters and papers. Some were chatting [removing lice] from their shirts with thumbnails, lighted cigarettes or matches. A few were having a rough shave. Much laughter and light talk were flying around and everything looked good …
>
> Most of the men in this platoon and in the Battalion too were from the Kangaroo and Snowy River route marches.[67] Fine big countrymen they were, typical of the Monaro and the Riverina.

While the unit rested, the icy weather broke. Snow turned to slush, and mud made an unwelcome reappearance on the battlefield. On 21 February the battalion, this time minus B Company, made its way to the support lines. For a nervous Private Street this was his baptism of fire:

> The hour came for our parade beside the hut to move off [to the front]. As we lined up the men were talking and laughing, and officers had difficulty in attempting to enforce silence. My feet quaked in their sandbag wrappings.[68] ... A line of duckboards stretched up the slope of a hill and disappeared over the crest. It led towards the line ... The leading company filed off and mine followed ... Over the hill a dull crash was heard, a shell bursting in the mud ... Gritting my teeth, I placed each foot ahead of me alternately. I managed to keep going. Aussie was far away. In Indian file we followed the duckboard and the top of the ridge was reached. Our Padre, the best of the best, ran a hot cocoa joint beside this track. He and his assistants dished out, day and night, hot cocoa, coffee and pea soup and smokes to men coming out of the line and where possible to men going up. The Padre was a good game man liked by all – Hunter of Balmain.[69]
>
> It was not dark enough to proceed, we rested for a while. Alex Cameron was the man in front of me, and he being an old soldier, gave me some good advice as to bursting shells and at what distance they are reasonably safe.[70]
>
> Darkness was well on us, so we crossed the ridge and toddled on our way. The dark became intense and I could not see the muddy walk beneath my feet ... I followed Alex closely and the man behind me kept close to my heels. As the leader came to a shell smashed duckboard he would say 'broken board' and the man following, on coming to the place, would repeat the remark ... In spite of all the precautions men would occasionally slip or trip off and plunge into a shell-hole full of water or dive into the mud. These men would need to be hauled out and placed on the track again, generally with an outburst of vivid language.

There were abundant work parties in the support lines and Private Street also recorded his introduction to front-line fatigues:

> Word came to fall in for [work party] parade ... We slipped away over the mud down to Ginchy [and here] we loaded ourselves with a duckboard ... It was painful in the extreme carrying those duckboards in the inky darkness. The boards we carried were heavy and awkward and the boards

> we walked along were slippery. One man fell and broke his leg. We left him for the stretcher-bearers, congratulating himself and cursing in turn ... At length we reached our destination. The leading files were dumping their boards and there was a fair amount of noise, consequently we did not hear the shell until it was right on us. It burst with an ear-splitting scream right amongst a crowd of us, throwing the nearest ones down. One chap being very near the burst, was hurled some yards through the air, finishing up with a big splash in a shell-hole. His head remained out of the water and rested on the muddy slope of the crater. A few groans came from him, and then he cried in a terrible voice 'I'm dead, I'm dead'. He was unhurt however, beyond considerable shock ... The blast had caused him to believe with absolute certainty that he was dead and well on his way to wherever soldiers go. During our second carry we had to traverse a trench thigh deep in mud. It was a miserable journey. A barbed wire tear in my breeches 9" above the knee let in a stream of liquid mud that filled me from the puttees up.

The unit moved to the front line into the same sector as the previous tour on 25 February. To reduce casualties from enemy fire and weather, half-strength companies were sent to the firing line — those men not required remained at Trones Wood Camp.

Private Street was suffering the effects of an over-active imagination: 'The front line was a fairly wide and almost straight trench. It looked like many other trenches but seemed pregnant with death. I could visualise sudden raids and quick death.'

During this tour there was a welcome addition to the trenches — a kitchen was established in a front-line dugout and each man received hot soup or chocolate at least once a day. The freezing weather returned, heralded by a heavy snowfall; the mud again hardened.

While the ground reverted to a frozen condition, the strategic situation was beginning to thaw. On 23 February, far to the north of where the battalion was located, several German divisions retired 30 kilometres to the Hindenburg Line, an unbroken series of fortified trenches. On receiving word of the German movement, Major General Hobbs ordered an increase in patrolling to remain in contact with enemy activities in the 5th Division's sector.

Patrols from the 55th Battalion discovered that the Germans continued to garrison their front line and that their wire entanglements remained thick and impenetrable. Night after night, Sergeant 'Dally' Neville, nicknamed the 'patrol king', led patrols into no man's land; he was subsequently awarded a Distinguished Conduct Medal. The recently promoted Sergeant Henry Anson, an expert bomber, also made nocturnal journeys into no man's land. While Neville sought to avoid contact with German patrols, Anson was the opposite, hunting and killing with single-minded purpose.[71]

Private John O'Grady was a regular participant in raiding and patrolling parties:

> When doing raids on German line duty at Le Transloy, the dress was a white calico suit over all battle dress – splits in the side of the suit so you had access to Mills bombs which you carried, no rifles – NCOs and officers carried revolvers. One night, when in No Man's Land, we were pinned down for about 4 hours – hard snow and frost and perhaps well below zero. My bowels operated, everything was perfect regarding hygiene – the motion froze hard. All I had to do was take the ice lump from the seat of my pants and throw it away. [It was] the joke of the night. We did not get close enough to get any information so crawled back to our trench through German barb wire and our own barb wire defence.

The unit was relieved on 1 March and moved to Trones Wood and Montauban camps. In a letter to his mother, Captain John Marshall described improvements in conditions in the rear areas:

> It is snowing again here today and has been bitterly cold for the past two or three days. The arrangements here for keeping clean are good. You could hardly believe that I had a hot bath this morning within range of the enemy guns and a change of underclothes so I am feeling quite clean and fit. The men are treated to a hot bath each time they come out of the firing line and it bucks them up wonderfully and keeps them fit.

But Private O'Grady had a tough trip out of the line:

> I got dysentery … Not in the least surprised as the shell-hole we got water from … exposed about four bodies – German soldiers in state of decay. On midnight. … I was very bad with dysentery and was sent out, no stretcher-bearers about, so I walked. Very thirsty so broke through a big shell-hole of ice. The centre was always the weakest point. I finished up in mud and water up to the armpits … too weak to get out – no foot hold. [Thought] it was the end. Must have held on for about 2 hours. No movement around, at last heard men talking, and started to sing out – no result. At last hell let loose. A battery of howitzers heavy guns, Tommies, about five rounds each. I could feel the heat from the gun barrels. In ceasefire … I knew I must be heard so yelled out …

> Four gunners, Pommies (English), pulled me out. They took me to their officer who treated me to a dixie of porridge and half pannikin of rum. I explained my case and he directed one of his men to take me to the casualty station (Aust) at Delville Wood.

The 55th returned to the support lines on 6 March. Private Street described the move:

> Our short spell over we made [our way] forward again, this time following a shell shocked road as far as Delville Wood to our left flank ... Having reached Delville Wood ... we were issued with a pair of thigh-high gumboots. A duckboard track led through the wood and this we followed. On either [side] were smashed trenches, shattered and tangled trees. I was glad to be thru the place, for it reeked of the dead ... Skirting to the right of the heavy guns ... we followed the track up a steep hill to Switch Trench, just below the crest out of the way of German observation.

Private George Farrell was fatally wounded in the support lines, a victim of shellfire.[72] Aged only 20 years, George was the youngest of three brothers from Brewarrina, NSW, all of whom were in B Company. Ted[73] and Tim, the remaining brothers, helped to bury George in a makeshift grave and were then 'sent back for a rest'.[74] Private Fred Greathead[75] was also killed by artillery and his brother, Henry, died three weeks later during the battalion's assault on Doignies.[76] A single shell-burst killed another three men engaged in a work party: Private Arthur Oakes,[77] Private Fred Schafer[78] and Private Joe Sewell.[79]

The battalion made its way to the front line late on 10 March. C Company relieved the 54th Battalion in a section of Sunray Trench recently seized from the Germans; A Company was just behind. By now, the first glimpses of spring were apparent and, in the milder temperatures; a 'big thaw' set in. Private Street wrote of his journey to Sunray Trench:

> [With the thaw] the trench became almost impassable now, the mud reaching our knees. Price, a fair-haired youngster who was behind me, looked as if he would founder at any moment and sink into the mud.[80] I grabbed him by the waist and half dragged, half carried him along ... Bullets were flaying the mud walls around us. Pineapple bombs and 77[mm] guns were pelting the trenches in front and the one we were in.[81] Captain Palmer was heard threatening to shoot the runner who was guiding us in.[82] A block had occurred because the runner had lost his way, a thing easily done in a newly captured maze of trench. At last the runner found the track and we were soon following a shallow trench

> across No Man's Land to the newly captured Sunray trench. Whizz bangs [shells from 77mm field guns] screamed across us, bursting near at hand, the pieces whining and phetting [sic] into the mud ... Still we pushed on ... Having reached our appointed position in Sunray, we prepared for anything that might eventuate. The post did not look promising to say the least. It was very muddy and there were a number of fresh shell and Minnie [*minenwerfer*] holes.

Corporal Hunt also made the trek to Sunray, describing what he found:

> In the trench area every unsupported bank collapsed to the depth of the frost. Trenches broke away and the loose dirt half filled them with slimy mud ... We set out to make the relief as a Company but bogged on the track and in the end came in by twos and threes, ten per cent having lost their boots on the track. These we sent back next day. All the Fritz' dugouts were blown in. What was once a trench had become as wide as a sunken road with shellfire. If by chance we slipped off the banks we quietly slid into an apparently bottomless sea of soft mud.

Rumours were flying around, fuelled by reports from prisoners, of an imminent enemy withdrawal from the Lesboeufs–Gueudecourt sector. The battalion responded by pushing advanced posts and patrols as far forward as possible; the enemy were found still very much at home and alert.

Enterprising individuals often undertook informal patrols as Private Street, now a company runner, discovered one night:

> I was sent around to platoon officers to send men to company headquarters [to collect] rations. [The acting platoon commander Sergeant Neville] was out in No Man's Land inspecting a small galvanised iron hut. [I] went out to Neville and told him about the rations. He said I could have left the message with the corporal. However, he decided to come in himself. With a remark that he would 'Put the wind up Fritz', he shook the loose sheets of iron loudly. It put the wind up me. Fritz opened up [with] several machine guns and a stream of bullets thudded into the iron. Burst after burst and the bullets raised a merry din. Neville walked back to the trench casually, so I had to do the same.

The liveliness of the enemy infantry was shared by their artillery: for two days and nights the Germans poured artillery and *minenwerfer* fire into the Australian positions. Private Street was caught in one of these bombardments:

> About 3 a.m. Fritz livened up and things became uncomfortable. A heavy barrage was put on and we lay in the mud listening to the hell around

> us. A pineapple bomb picked a fight with our home. The bomb landed with a sickening crash. The broken boards and sandbags crashed on to our shivering forms. We were pulled out unhurt but severely shaken. A Minnie mortar of fearful size came down out of the blackness and burst amongst us with an earth shattering crash and we were blown around like thistle down ... The bottom of the trench heaved in a mighty convulsion. Great masses of mud and earth heaved skywards and then descended on us heavily. Those of us unhurt looked round to find the disabled. We found one man with his leg blown off above the knee and another man dead.

Captain Palmer was one of the casualties of the sustained bombardment; he was climbing up the steps of his dugout when a *minenwerfer* bomb landed on its roof, dumping three metres of earth on the officer and killing him. Private Street wrote of Palmer: 'We lost a gallant officer by his death.'

On hearing of Palmer's loss, the acting CO, Major Holland, sent the newly promoted Captain 'Bob' Chapman to take command of C Company. On his arrival at Sunray Trench in the early hours of 12 March, Chapman began checking the placement of his men. At one post, the location of a *minenwerfer* bombarding the Australian positions was pointed out to him. This was the last occasion Chapman was seen alive; he simply vanished. The War Diarist recorded: 'No trace of him can be found ... He is believed to be killed.'

Patrols searched no man's land over the succeeding nights but found no sign of the missing officer. Eventually his corpse, with a hand still holding a Colt revolver and a leg blown off by a shell, was discovered in the enemy trenches after the Germans commenced their withdrawal.[83] He was found near an abandoned *minenwerfer* position containing three dead Germans.[84] How Chapman ended up in the enemy lines can only be surmised. Perhaps it was reckless bravery; possibly he became disoriented in the unfamiliar trenches and stumbled into hostile territory. Another explanation is that Chapman, a sensitive man of undoubted courage, was traumatised by his wartime experiences and wanted to end his suffering while taking a few Germans with him.[85]

Sunray Trench contained its share of horrors. One night Private Street was making his way along the trench with several of his mates, when:

> Having traversed the trench for a short distance, we turned to the right and up a short step. As I put my foot down on the step I felt it give slightly and I almost slipped. I thought it must have been a sandbag filled with liquid mud ... During a run next morning I discovered that the step was in reality a dead Fritz.

On another occasion, Street undertook a 'run' to Battalion Headquarters, later painting a picture of the conditions the troops were enduring:

> Started off on my journey to Rose Trench ... the walk down the trench revealed terrible traces of shellfire. Here and there a dead man, walls knocked in and haggard looking men ... I found my way down the steps leading to the forward HQ dugout. My appearance caused evident surprise, and no wonder. Barefooted, plastered in mud, wet thru, bleeding from a wounded heel, wild eyes and bloody hands, I felt more miserable than heroic. I drew the dispatch from my pocket, a small splinter of shell in its folds ... The blood on my hand smeared the paper as I handed it to the Major.

The Germans continued to pound the battalion with high explosive and shrapnel. Corporal Hunt noted that, by the time the unit moved back to the support lines on the evening of 14 March: 'The men were so knocked up that twenty-five per cent came out without any equipment.' On 16 March, the resting troops discerned a welcome change, the War Diarist remarking: 'Enemy has evidently withdrawn his heavy artillery further back as he has ceased to shell [us].'

The next day the War Diarist recorded some electrifying news: 'It is reported that the enemy is evacuating his positions in front of us and retiring. All ranks held in readiness to follow him up.' The reason for the protracted bombardment of Sunray Trench over the previous week now became apparent: the Germans had used massed artillery to conceal the retirement of their forces.

Next morning the battalion advanced over the evacuated enemy trenches and took up defensive positions around Beaulencourt. For the first time in months, there was no sign of the enemy and no artillery fire; the German guns had moved rearwards and were yet to re-establish themselves. The men relished the firm ground and unbroken fields, but a few found the change unsettling. Corporal Hunt, along with two mates, took the opportunity offered by the pause in hostilities to:

> Survey the ruins of Bowlancourt [Beaulencourt] and to search for souvenirs in the shape of helmets or badges of which the Germans, in their model retirement from the Somme, left remarkably few. How strange it all seemed; to be able to walk freely over ground; to look at which a few hours previously was to court death. How strange to be able to gaze upon the wide stretches of grassland behind which was once the German line.

Private Brown also wrote of the surreal feeling in a letter home: 'When last in the line at a pretty hot corner Fritz gave us particular hell, and then cleaned out leaving little behind him with the exception of soda water and old bombs. Tis quite a treat to walk around places today [that] were three days ago a regular death trap.'

British High Command was cautious in its response to the German withdrawal. Fearing a strategic ambush, and with the majority of their artillery still having to traverse the morass of the former battlefield, the British moved forward warily. The Australians did not share their timidity; the 5th Division's advance guard, based on the 15th Brigade, rushed to regain contact with the Germans and a gap soon opened up between this formation and the remainder of the division.

On 20 March, the battalion was ordered to take up positions near Haplincourt and Villers-au-Flos. The main body of Germans had retired, but as the men approached the latter settlement there was a fleeting skirmish with an enemy patrol. Private Street described the action:

> The 55th had more unique fighting amongst the tombstones and graves in the cemetery of Villers au Flos. Our Lewis gunners and the German gunners played hide and seek as they shot at one another. As our chaps advance firing at all available targets, the targets gradually worked their way east and were finally ejected from the cemetery. Bullets whizzed and cracked ricochets off the stone monuments.

The unit remained in this location for a week. The surrounding countryside was patrolled to prevent German encroachment, but there was no contact. The 15th Brigade halted after seizing Beaumetz-lès-Cambrai and deployed a screen of outposts that maintained contact with the Germans. The 55th, along with the other battalions of the 8th and 14th Brigades, formed a defensive line in case the advance guard was driven back.

On 27 March, the 14th Brigade received orders to replace the 15th Brigade in the role of advance guard. The 55th began advancing the following day and, by late afternoon, one platoon from D Company and two from C Company had taken over the outpost line in the woods outside of Vélu. The remainder of the battalion occupied positions at Bancourt. Private Harris wrote of the devastation left by the retreating Germans:

> Got to Bancourt soaked. Rained like hell all the way. This place same as all the others. All houses blown up. The roads blown up and the wells filled in and their own [the German] dugouts either burnt or blown up.

> Trees felled across the road but most of them fell the other way. Not much good with an axe them Bosches. Ran up against a fresh horror. A well that was not filled in and where we got water for tea and drinking was found to have a dead man in it. Parts of the body were drawn up in a bucket.

The men settled down for a waiting game. By this stage the artillery on both sides had redeployed and Corporal Hunt observed: 'Soon [it] was the same old roar which had been with our sleeping and waking for a half year.' The 55th's outposts in Vélu were regularly bombarded.

The men relished their new surroundings, regardless of the shelling. On the last day of March, the War Diarist observed: 'Men's health is good due to new surroundings, in higher ground and they are in cheerful spirit.' Private Street saw a similar change: 'The green grass and firm ground cured many a sick man and the spirits of the men rose. Sunray and all those terrible trenches of mud and death were left far behind.'

The 'Somme winter' of 1916/17 was the most protracted trial of physical and mental endurance the unit was to experience. The travail had a profound effect on those who had lived through it. Corporal Hunt confided: 'The days of that terrible winter are burned into my very soul with the torture of sleepless nights, freezing cold, physical exhaustion and the loss of comrades.'

Foe his part, Private Maynard reflected on the mateship emerging from the ordeal: 'I think there must have been some quality in the Somme mud that made us stick together as well as sticking to us. There was certainly a very strong bond between those of us who had the good (or bad) fortune to go through that time.'

This bond would be put to the test in the coming weeks.

Chapter 6:
Doignies – 'what a mess-up it all was'

The Germans left numerous rearguard detachments, typically anchored in fortified villages, as an outpost line to screen their retirement to the Hindenburg Line. The mission of the enemy garrisons in the outpost line was to delay the British and Australian advance, inflict casualties and buy time for the Hindenburg Line defences to be further strengthened.

The recently promoted Field Marshal Douglas Haig tasked the Fifth Army, to which the 55th Battalion was attached, with attacking and fracturing the Hindenburg Line. However, in the Fifth Army's sector, the strength of the enemy detachments was such that they could not be bypassed. Before assaulting the Hindenburg Line, therefore, the fortified villages had to be reduced in a series of deliberate attacks, each piggybacking on the success of its predecessor. Historian Charles Bean described in general terms the nature of the topography and defences of the outpost line:

> Open country clear of all fortifications except where wire entanglements and trenches bent around the outskirts of villages, a few strong entrenched posts between the villages and a sprinkling of sentry posts dug in advance of the main posts. Most of the posts in the open were on the roads, which in this part of France were largely sunken. The Germans … largely bivouacked in these roads since the banks … not only gave protection from fire (except in enfilade) but could be quickly scooped into little one-man niches in which men [were] fairly safe against shell fire. At important posts, when they had time, they tunnelled deep dugouts in the road banks.[1]

By mid-March the 15th Australian Brigade, as the 5th Division's advance guard, had begun rolling back the outpost line by seizing Beaumetz-lès-Cambrai in a series of battalion attacks. The enemy withdrew further east, taking up positions based on the fortified villages of Hermies, Doignies and Louverval.

Hermies, the southernmost, was the largest of the three settlements. It was situated close to the 5th Division's boundary and could not prudently be assaulted until the British advance on the right flank caught up with the Australians and there was no indication when this might occur. Louverval, the northernmost, was a small community situated to the north of the Bapaume–Cambrai road, one of the main routes being used by the Germans for their retirement to the Hindenburg Line. Between Louverval and Hermies lay Doignies. A far more substantive settlement than Louverval, it was situated about one kilometre south of the Bapaume–Cambrai road.

The terrain around the three villages was gently undulating. With the exception of some woods around the chateau at Louverval, the countryside was devoid of trees. The winter snow had melted leaving only the stubble from the previous year's crops covering the ground; hedges or sunken roads bordered the fields.

The Germans had constructed strongpoints covering the gap between the three villages; these were linked by a series of almost continuous entrenchments. One of these strongpoints — a large, ruined building near the Bapaume–Cambrai road used in peaceful times for the production of sugar from sugar beets — was on high ground between Doignies and Louverval and was connected to them by defensive works. According to Ellis, the beetroot factory was 'altogether a strong position'.[2]

After seizing Beaumetz-lès-Cambrai, Brigadier General 'Pompey' Elliot of the 15th Brigade devised a plan for the capture of Doignies and Louverval. His scheme for Doignies called for a battalion attack from the south to exploit through a gap in the wire entanglements protecting the south and west of the village, and then encircle it. Louverval was to be taken by a battalion-sized hook from the north of the village. Both flanks of the assault were to link up behind the villages, trapping the German defenders.

On 29 March the 14th Brigade, commanded by Brigadier General Clarence Hobkirk, took over the responsibilities of the 5th Division's advance guard from Elliot's brigade. Hobkirk received orders from General Talbot Hobbs, commander of the 5th Division, to prepare for an assault on Louverval and Doignies. Without hesitation, Hobkirk adopted Elliot's proposed attack scheme, allocating Doignies to the 55th and Louverval to the 56th.

By the afternoon of 31 March, however, Elliot's plan had become unworkable. Observers from the 53rd Battalion, then holding the front line opposite Doignies, concluded that the Germans had formed a continuous defensive line between that village and Hermies. These discoveries, combined with information from patrols and an analysis of aerial photographs, indicated that there were now extensive wire entanglements around Doignies and the gap to the south through which the attack would move had now been blocked.

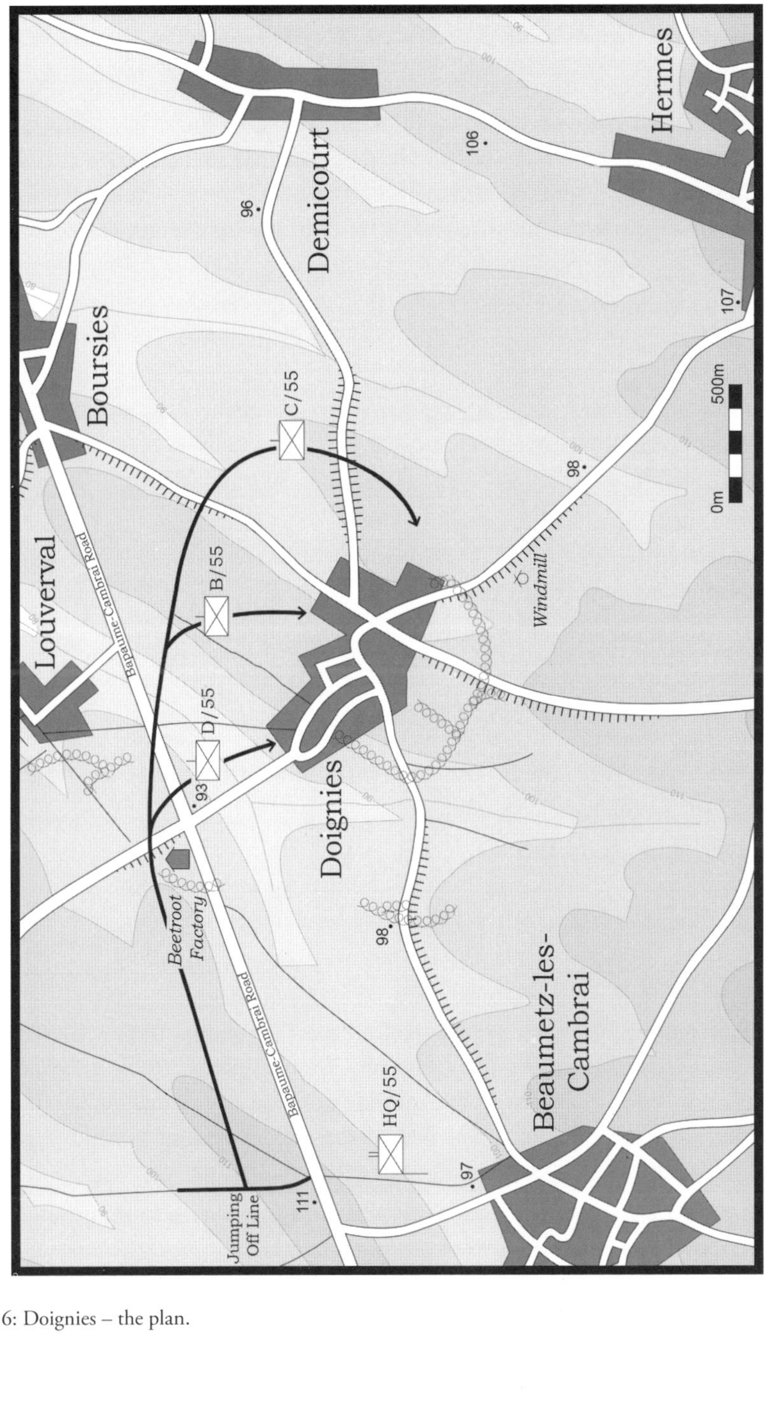

Map 6: Doignies – the plan.

The changes caused consternation to both Brigadier General Hobkirk and Major Percy Woods, acting CO of the battalion, as any attempt by the 55th to move around Doignies on the south might see men hung up on the barbed wire and exposed to a counter-attack in the flank. Hobkirk spoke to Major General Hobbs, informing him that the enemy defences 'Might make this attack very costly and, as the objective is a mile and a half – and over – from our lines the question of getting back out wounded, if we don't get in, will be very great.'[3]

A new plan was needed. General Hobbs, Brigadier General Hobkirk and Major Woods came up with a far more audacious scheme for capturing Doignies than had been proposed by Brigadier General Elliot. Both the 55th and 56th battalions were to concentrate north of the Bapaume–Cambrai road. From this location, the 56th, with the 55th behind it, was to advance parallel to the road, capture the beetroot factory, and then swing left into Louverval.

As soon as the factory was secure, the 55th was to pass through the 56th and turn right, assaulting Doignies in three company-sized columns. The intention was to envelop the village. B, C and D companies were to conduct the operation, with A Company in reserve.

Brigadier General Hobkirk added another bold stroke to his plan; he persuaded General Hobbs that the undertaking had a better chance of success without a supporting barrage. Both men agreed that artillery would only provide assistance if the infantry met difficulties during the attack. The gun batteries, however, were to be ready to deal with the massed German counter-attacks expected in response to the seizure of the two villages.

The day selected for the assault was 2 April, chosen to coincide with a major attack by the 4th Division on Noreuil, several kilometres to the north. While the 14th Brigade's attack was primarily undertaken to capture the villages, the alignment of timings suggests a secondary objective of providing a diversion in support of the more operationally important assault on Noreuil.

The battalion spent the day prior to the attack gathering in the village of Vélu and the small wood outside it. The troops were briefed on the coming engagement, drew ammunition and other supplies, and conducted a few small-scale rehearsals. Orders stressed the different objectives of the two assaulting battalions and the imperative to avoid becoming involved in the 56th's fight.

At various intervals on 1 April, the 78th and 88th Siege Batteries used their large-calibre howitzers to shell the beetroot factory, some of the known German posts, and the roads beyond the objectives. One of the 88th's 6-inch shells scored a direct hit on the factory, killing 15 German soldiers and wounding 25 others.

At twilight the 55th moved out of Vélu on their way to the designated assembly area for the assault, a sunken road north of Beaumetz-lès-Cambrai. The men's steel-shod boots struck sparks off the cobblestone roads as they travelled north through the ruined village of Lebucquiere and crossed the Bapaume–Cambrai road. The column made a sharp right turn at Chamfours Wood and continued east moving cross-country until they arrived at the assembly area around midnight; once there, they dispersed along the sunken road. The right flank of the battalion lay on the junction of the sunken road and the Bapaume–Cambrai road. The 56th Battalion, moving from Frémicourt, had already assembled along the Beaumetz–Quéant road, 200 metres east of the position occupied by the 55th.

The night was clear and frosty. A waxing moon provided some illumination as the diggers waited in the assembly area, shivering from a combination of cold, excitement and fear. Private John O'Grady recalled his time in the sunken road:

> My mate Johnny Webb[4] and I dug a hole in the side of the road embankment and settled down for the call at zero hour to advance, this was not a protected position and the Germans knew where we were so the shells peppered us all night. I went to sleep and woke up about midnight. No mate, cold and nearly frozen stiff, I saw Johnny Webb walking up and down the road, he would not take shelter and rest. He gave me his great coat so I would be a little warmer. Coming towards dawn and zero hour Johnny Webb came over to me and said, we must have a talk, he stated that he was sure it was his last day on earth.

At 4.00 am, Lieutenant Colonel Humphrey Scott, CO of the 56th, led his battalion from its assembly area to a jumping-off point along the top of a rise 200 metres further east; from here Louverval could be seen. Conforming to the plan, the 55th moved forward in artillery formation to occupy the area alongside the Beaumetz–Quéant road vacated by the 56th.[5]

Free from the scheduling restraints imposed by a supporting artillery barrage, Brigadier General Hobkirk had left Lieutenant Colonel Scott, as the commander of the leading battalion, to determine the exact timing of H-hour. Scott had earlier stated his intention to launch the assault at 4.30 am. However, while waiting in the assembly area he had been made aware of some coordination issues with patrols from the 57th Battalion tasked with securing his left flank. To allow sufficient time to address these concerns, Scott now settled on 5.00 am as H-hour. For some reason, this change was not communicated to the 55th.

To avoid intermingling, Brigadier General Hobkirk's orders had specified that a gap of 100 yards was to be maintained between the two battalions. Despite Hobkirk's direction, Major Woods and Lieutenant Colonel Scott reached an understanding that the separation was to be by time, and not an arbitrary distance. As such, the 55th was to wait for 15 minutes after H-hour before commencing its advance.[6]

Therefore, at 4.45 am, 15 minutes after the original time set for H-hour, the unit commenced its advance, unaware that the 56th was still waiting for the revised H-hour timing of 5.00 am to arrive. It was dark; the moon had set three hours earlier and the first tinge of dawn had yet to colour the horizon. On the left of the 55th's advance, B and C companies stumbled on the waiting lines of the 56th, taking up positions to the rear of them. On the right flank, D Company unwittingly bypassed the 56th and continued advancing alongside the Bapaume–Cambrai road, using the trees lining the road to assist them to maintain direction.

D Company, under command of Captain Eric Stutchbury, pushed on in silence, covering a distance of 1500 metres, unaware that they were the only Australians pressing forward. As they approached a barbed wire thicket protecting the beetroot factory, a dog about the size of a kelpie emerged from the fields, barked once then fell in beside 2nd Lieutenant Arthur Duprez, the officer leading D Company's right-hand platoon.[7]

A few minutes later figures, assumed to be from the 56th Battalion, were seen moving in front of D Company. The dog left 2nd Lieutenant Duprez's side and ran towards these figures, barking and growling. Suddenly, a single shot, then a ragged volley of rifle fire, opened up on the Australians, followed by the explosions of German hand grenades.

Despite the shock of finding the enemy less than 50 metres away, Captain Stutchbury acted swiftly. Expecting the 56th Battalion ahead of him, Stutchbury had kept his company in artillery formation. He moved his men into an extended line, shouted for five rounds rapid fire from every rifleman and then ordered a charge. It took D Company moments to reach the enemy positions.

The Germans were as surprised by the encounter as the Australians; they were spread out along a sunken road just to the rear of the beetroot factory. In the face of such an aggressive and unexpected assault, most of the Germans fled in disarray towards Boursies. Some, however, stayed and fought. In the hand-to-hand fighting which followed, several Germans were bayoneted and those enemy troops who had remained in dugouts tunnelled into the side of the sunken road were killed as Australians rolled grenades into their shelters.

Enemy machine-guns opened up from the vicinity of Louverval and the beetroot factory, but the confused tactical situation meant that the gunners were uncertain where their own forces were and the resulting fire was ineffectual. In the pre-dawn light, men from D Company could see another enemy party a stone's throw to the north in the direction of Louverval. To Captain Stutchbury's right, a German flare shot up from the vicinity of the beetroot factory, emphasising to him that this strongpoint remained a menace to his other flank.

Realising that he was ahead of the main Australian force, Captain Stutchbury resisted the temptation to attack the enemy to his north. Instead, he ordered an assault on the beetroot factory. His company responded. Second Lieutenant Duprez and Lieutenant William Morgan led their platoons in a wild charge towards the factory; both officers were awarded the Military Cross for their actions.[8] Approaching from an unexpected direction, the Australians were able to break into the strongpoint; once inside, they shot, bombed and bayoneted the 20 or so Germans defending the ruins.

Leaving a platoon to secure the factory and the captured parts of the sunken road, Captain Stutchbury positioned the remainder of his company along the Bapaume–Cambrai road facing Doignies and reorganised. It was 5.20 am, the sun had just broached the horizon and the men of D Company realised that, despite their exertions, the day's main task, the seizure of Doignies, had yet to commence. The troops settled down behind the low earthen banks alongside the road and prepared for the arrival of B and C companies.

At 5.00 am the 56th commenced its advance. The 55th waited for its sister battalion to move some 100 metres ahead of it before starting its own move forward. D Company's actions had ensured that every German in the vicinity of Louverval was either wide awake or dead. Despite this loss of surprise, the 56th's assault commenced with only minimal enemy opposition.

As they neared Louverval, the German resistance grew heavier and it took strong leadership by the 55th's officers to keep their excited men from joining in the fight. The company of the 56th detailed to capture the beetroot factory relieved D Company of responsibility for the security of this vital point.

As the fight intensified the 56th's rate of the advance slowed then stalled. B and C companies were forced to halt in a field north of the road to wait for the right time to move past the factory and behind D Company before turning to encircle Doignies. While pausing, they came under long-range machine-gun fire from Louverval; one bullet smashed into 2nd Lieutenant Charles Woolley's head, mortally wounding him.[9] The enemy fire also injured several other men.

With each passing minute, the light improved and the foolishness of moving B and C companies across open ground exposed to the unsuppressed fire of the numerous machine-guns around Louverval became apparent. Leaving the troops propped in the middle of an open field under enemy fire was also not an option.

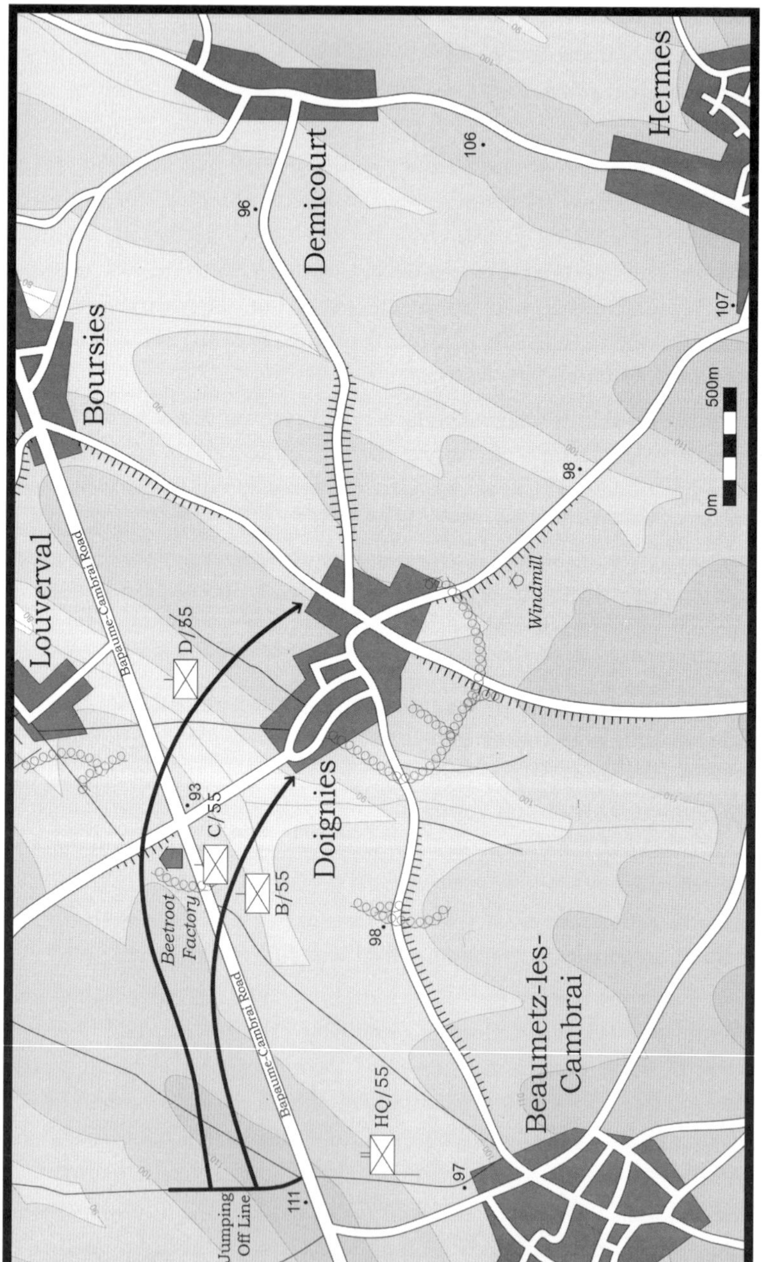

Map 7: Doignies – the attack.

With admirable flexibility, the company commanders involved in the assault developed a new plan. They decided that B and C companies would attack from their present location, aiming to enter Doignies from the north-west and avoid the belts of barbed wire protecting the western side of the village. D Company was to assault from its positions along the main road near the beetroot factory and enter Doignies from the north-east.

Captain Stutchbury moved his company into extended line and advanced across the stubble towards the village over a kilometre away. A few fleeing Germans were seen and several speculative shots were fired at them. Some distance away on the left, a German machine-gun on the high ground near Boursies commenced firing, its bullets falling among and in front of D Company causing one or two minor wounds.

This fire was annoying, but a more immediate danger came from a party of Germans taking refuge behind a hedge who now opened fire on the advancing Australians. Stutchbury's men charged and the defenders, having only the flimsy protection offered by the vegetation, dashed away, joining more of their comrades in a sounder defensive position along Rue de Boursies, a hedge-lined sunken road linking Doignies and Boursies. From here, the Germans again engaged the attackers.

Meanwhile, B and C companies had passed through a patchy German barrage along the Bapaume–Cambrai road and, in three extended lines, had commenced advancing on Doignies, visible 700 metres away across the fields. The Australians were required to move downhill from the main road, cross a small creek and then proceed up a gently rising slope to the village. There was no cover.

After crossing the road, the men strode into a hail of machine-gun and rifle fire from Doignies. Lance Corporal Archie Winter recalled: 'There was an open field of fire that Fritz took full advantage of.' Private George Seaman was one of the first to fall, his body riddled with bullets.[10] Nearby, Private Apsley Bradford was also killed, the steel bulletproof vest he was wearing providing no protection from the machine-gun bullets that perforated his abdomen.[11]

At first, the uninjured attempted to pull the wounded into the shelter offered by shell-holes, but as they closed on the objective, this work was left to stretcher-bearers. Two of these bearers, Privates Archie Rosborough[12] and William Dobson, moved among the wounded, applying dressings and placing the casualties in safe positions where follow-up bearer parties could evacuate them to the RAP.[13] Here, the RMO, Captain Hugh Wyllie, worked to stabilise the wounded and to ease the passing of the dying.[14]

The two companies worked in conjunction, small groups of men using fire and movement to advance. [15] The longer serving members of the battalion had practised this tactic many times in Egypt. Private Eddie Street described the passage across the fields:

> With short sharp rushes the men crossed the flat country between them and the village. A storm of machine gun fire sprayed the advancing companies and men went down in all directions. There was no cover for the troops and sheer guts was needed to keep advancing.

Private O'Grady recalled: 'My job was runner, C Coy. So had no choice to take cover, if I had a message from another Coy. Had to keep going. I was not hit but had my gas respirator shot off from the cross fire from German guns.'

As they neared their objective, the Australian fire became more accurate. Lance Corporal Arthur Hardy used his Lewis gun to kill several German riflemen.[16] O'Grady recalled: 'Johnny Webb shot a German sniper out of a Belfry tower in the cemetery. He sang out to me "I am taking this one with me." … I was alongside Johnnie when he was killed. His throat was cut by a piece of shell and another piece through his chest.'

With one last rush the Australians found themselves in the village. Men stormed down the streets, bayoneting, bombing or capturing the Germans wherever they found them. The enemy offered feeble resistance; the lack of fight was surprising given that the garrison was alert and in a fortified locale. Charles Bean wrote of the push through Doignies:

> In the village, when the leading party reached it, breakfast was standing ready in the dugouts and the billets. There was tinned meat, and a bottle that looked like cognac. The wary party left that cognac alone. The men have learned to mistrust the German and his belongings, however innocent they may look. Instead, someone threw a bomb into the place; the next instant the dugout was blown up – fortunately at another entrance. In another cellar an officer's breakfast was ready all laid out, and the coffee-pot steaming on the fire. The breakfast was finished then and there.[17]

Private Street offered his own opinion on the captured German provisions:

> Fritz had just received his rations. Our men were in time to get the untouched food. The coffee was the only thing that could be enjoyed; the black bread and stew being awful tack. A fair lot of tobacco was captured but was of very poor quality.

Attacked from two sides and fearing encirclement, the Germans in and around Doignies commenced a disorderly withdrawal. The retirement permitted D Company to also enter the village and assist in clearing the buildings. Bean described the impact of several booby-traps left by the Germans:

> As they pushed through the village an Australian private tripped on a wire across the street near the small village bridge.[18] A fountain of water went up from the stream along the bridge, as if a mine had exploded there, but the

> bridge was not broken. Another wire was pulled, and a house, rather better than most, was suddenly blown down. The explosion scattered amongst the brick dust a shower of papers. It was a headquarters. Four or five mines were set off as the troops entered the place, either by trip wires or by the enemy, but as far as I can hear not a man was injured, and they had not the slightest influence on the attack.[19]

Private O'Grady was close to one of the exploding mines:

> I had a lucky escape. I was given a note from our Captain Wilson to deliver to [Captain] Stutchbury, D Coy ... On my way through the village and on entering a square connecting four roads, the whole of the square was blown up, I was under the upward blast but in the vacuum caused by the explosion and finally ended up down the crater, received a crack on the back of the neck by a great cobblestone, bleeding from mouth and ears. I completed my run to contact Stutchbury. After shaking off the blast effect, but felt very sick on the belly. Stutchbury wanted me to lie down for a while but [I] had a small whiskey from his bottle, he was a tea totaller [sic].

In the space of ten minutes between seven and eight mines exploded in the village. Other tripwires were found stretched across the streets in various locations but the explosive charges at the end of these were disarmed.

The last of the Germans to leave Doignies belonged to a machine-gun crew situated near a windmill just on the southern outskirts of the village. The entire crew was killed except one or two men who escaped with the gun. The majority of the German garrison escaped the attempted envelopment and large enemy parties were seen fleeing towards Hermies and Demicourt. A few were brought down by long-range shots.

Just after 6.00 am the unit began to consolidate. Realising that the Germans were sure to counter-attack, the men formed a perimeter outside Doignies on the slight ridges that swept around from the east through to the south; the village itself was bound to be the focus of a retaliatory bombardment and was left unoccupied.

Captain Stutchbury, the overall leader of the attacking force, established a series of Lewis gun posts where they were able to achieve good fields of fire.[20] A Lewis gun was positioned at the windmill where it could shoot across the ground between the 55th and the advance posts of the 31st Battalion located east of Beaumetz-lès-Cambrai. To the north, there was a worrisome thousand-metre gap between the 55th and the 56th Battalion. Accurate and sustained fire from German machine-guns prevented the Australians from digging in, forcing them to use whatever cover they could find. The battalion settled in and waited for the German response.

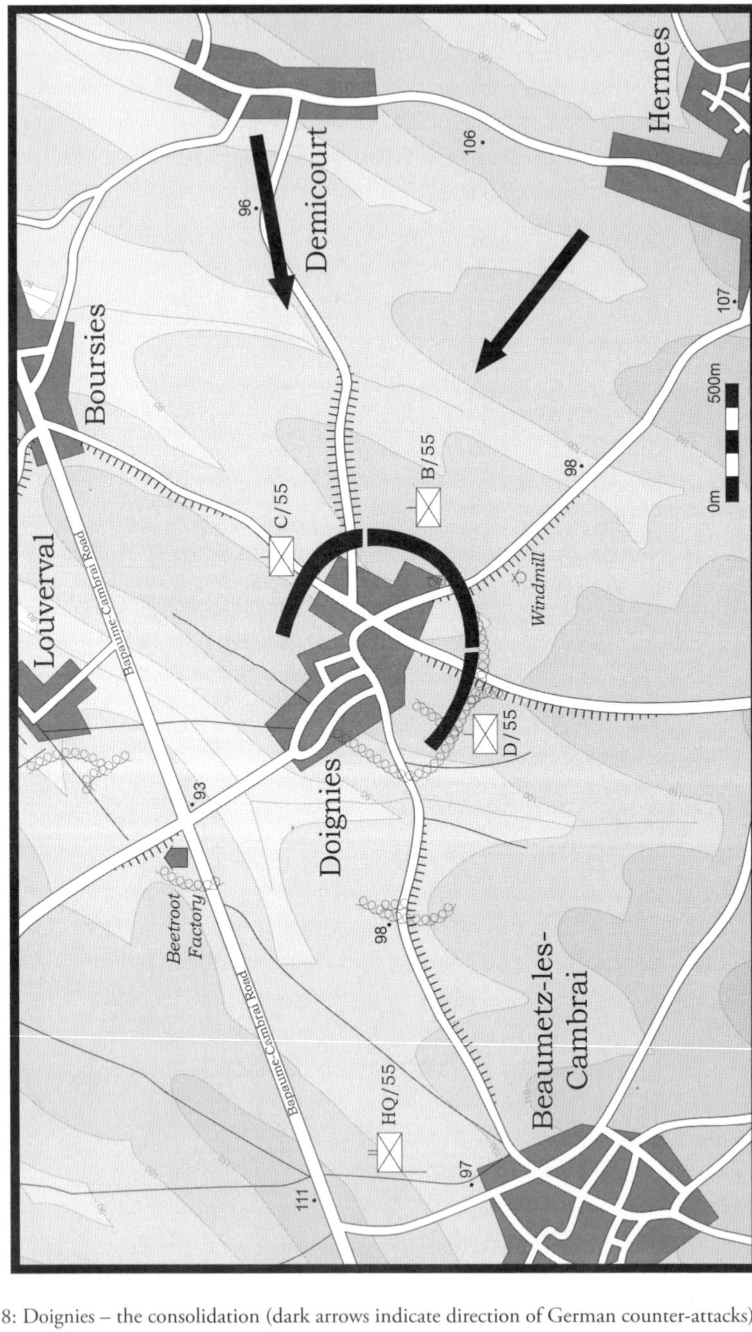

Map 8: Doignies – the consolidation (dark arrows indicate direction of German counter-attacks).

Major Woods, who had been badly gassed the night before but had refused evacuation, directed the overall conduct of the Doignies operation from his headquarters near the battalion's assembly area.[21] Captain Stutchbury told Woods of the tenuous nature of the village's defences and, in response, the latter sent two Vickers guns and two platoons from A Company to reinforce the perimeter. Soon after their arrival, an enemy shell destroyed one of the Vickers guns and caused minor casualties among its crew.

Just after 7.00 am an aerial patrol from No. 3 Squadron, Royal Flying Corps, detected a number of Germans concentrating near Boursies and moving towards Doignies. The aviators called down artillery fire that dispersed the enemy. A far more serious threat emerged at 7.25 am when reports provided by aircraft reached Major Woods of '500 enemy advancing on Doignies from east'.[22] Again the artillery, this time supported by fire from Lewis and Vickers guns, broke up the German counter-attack before it threatened the Australian positions.

Throughout the day seven counter-attacks were launched against Doignies, including a heavy assault late in the morning and another large attack at around 4.00 pm, both from the direction of Hermies. The last serious counter-attack of the day came at dusk. In Ellis' opinion, 'The infantry were not seriously embarrassed by the counter-attacks due to the fine work by the artillery, the Forward Observation Officers and the 14th Machine Gun Company.'[23] Ellis should also have mentioned the importance of the aircraft from No. 3 Squadron in their role as spotters for the artillery. The Germans shelled Doignies and the surrounding area between the counter-attacks and also kept up a sustained machine-gun barrage. Of special concern was the persistent fire from a German machine-gun near Boursies, probably the same one that had troubled D Company earlier in the day. It was possibly this gun that killed Private Jack Kellond[24] and left his partner on the Lewis gun, Private Frank Catterall, with three machine-gun bullets in his arm.[25]

Although enemy artillery and machine-gun activity was unrelenting, telephone lines were kept open between Doignies, Battalion Headquarters and Headquarters 14th Brigade. These connections enabled supporting defensive fires to be called whenever and wherever necessary. Two of the 55th's signallers, Lance Corporal Edgar Wolrige[26] and Private John Buckeridge, received Military Medals for their role in maintaining the telephone lines.[27]

Major Woods continued to fret about the tenuous nature of his hold on Doignies. He despatched the three platoons of the 53rd Battalion, placed at his disposal for use as carrying parties, to augment the defences around the village.

The arrival of these reinforcements in mid-afternoon coincided with a turn for the worse in the weather. To add to the misery caused by enemy retribution, snow began to fall and the temperature plummeted. The diggers had no option but to endure what nature and the Germans could throw at them.

Eventually night fell. In the darkness, patrols from the 56th connected with those of the 55th, closing the gap between the two battalions. Patrols weren't the only work being undertaken — Private O'Grady found himself tasked with a most unpleasant duty:

> After the stunt on the night of the 2nd a call was made to collect the dead as there were too many for the stretcher bearers. I volunteered and worked well into the morning of the 3rd April. Four worked together but still very hard to handle. It was moonlight and shells from the enemy were falling frequently and we worked all night. Near daylight we finished stacking the dead in heaps of 12 cross ways and straightened them as much as possible. At daybreak, behind the village, all those that worked all night were given the usual tablespoonful of rum and a dixie full of steaming hot bread and milk, made with condensed milk. The best meal I ever have tasted.

The 56th Battalion's War Diary described the conditions that night: 'It continued to snow and rain throughout the whole night and together with a strong freezing wind made conditions extremely trying. The men had no shelter and it was very hard to do a great deal of digging.' General Hobbs' diary entry of 2 April also lamented the appalling conditions: '[The capture of the villages] certainly has been a fitting finish to our … continuous fighting since January 18. I only hope we can hang on until relieved. Tonight is a terrible night [and with] snow on the ground our poor fellows I fear will suffer badly.'[28]

Dawn on 3 April brought no respite from the shellfire or weather although, thankfully, the enemy did not repeat their counter-attacks of the previous day. By now, the men were in a pitiful state: freezing, wet through, hungry, exhausted. Throughout the day the steady flow of casualties continued. Many soldiers had niggling wounds and yet refused evacuation. One or two men suffered trench foot, requiring their evacuation to the RAP.

Relief came after sunset when the 53rd Battalion made its way to Doignies and took over the 55th's posts. By midnight the operation was complete and the weary men marched three kilometres to their rest area at Vélu Wood.

Despite the annoyance of an occasional German shell, the unit spent the day of 4 April resting and calculating the blood-price for capturing Doignies.

Private O'Grady was in a sombre mood: 'Our casualties were not light. I saw most of my football mates from my hometown nearly all dead lying in the snow.' About one quarter of the battalion had become casualties: 47 killed, up to 208 wounded and two missing. Several of the wounded later succumbed to their injuries. Stretcher-bearers carried the fallen to Lebucquiere, where they were buried in a mass grave. A single cross with the names of the deceased was placed atop the burial mound.

The exact number of German casualties will never be known, but they are unlikely to have been too numerous. Most of the enemy chose to flee, not fight. The War Diarist claimed that the battalion had captured: 'two pineapple mortars, two minenwerfers' and a small number of German prisoners.[29] The paucity of captured materiel and human 'trophies' offered few bragging rights.

Although successful, the capture of Doignies was by no means a textbook operation. The failure by Lieutenant Colonel Scott to clarify and communicate the timing of H-hour had led to Captain Stutchbury's pre-emptive move. In turn, this had compromised the element of surprise on which the entire operation was predicated. Furthermore, the inability of B and C companies to swing around the rear of Doignies meant that most of the Germans were able to avoid being surrounded. Private Herbert Harris confided in his diary: 'Our boys took their village ... but what a mess-up it all was.'

With the benefit of hindsight, the attack plan was far too ambitious 'and had in it all the elements of a terrible disaster'.[30] The safe movement of the 55th past the fortified village of Louverval was reliant on the 56th Battalion achieving all of its objectives in accordance with a detailed timetable. As it was, the 56th took until mid-morning to seize the village. Had the beetroot factory been the machine-gun infested 'man-trap' Brigadier General Elliot feared, the entire operation could have been a debacle.[31] Despite the grand plans for encirclement, the 55th's attack was ultimately little more than an unsupported frontal attack in daylight across open ground against an alerted, fortified position. Had the enemy exploited the full strength of their tactical situation rather than meekly capitulating, the Australian ranks would have been far more extensively culled.

However, the unit's luck was 'in' and the major objective of the operation, to seize and hold Doignies, was achieved with minimum casualties by the standards of the time. After almost a month of combat operations, culminating in a successful attack, the troops were in need of a rest and a bath. In the coming days, the battalion was to be given both.

Chapter 7: Bullecourt – 'that horror of a death-trap'

On 5 April the 55th Battalion made its way to a camp comprising tents and improvised shelters set among the ruins of the village of Le Barque. Along with the remainder of the 5th Division, the 55th had been moved out of the forward areas and placed in the Corps reserve. Lance Corporal Archie Winter described his accommodation:

> The place was a complete wreck anyhow like all Auseys [Aussies] we soon made ourselves comfortable. We had a sentry over our doorway in the shape of a dead Fritz and judging by the stench there must have been some more dead ones in the cellar, but that's a mere trifle in the life of a soldier.

While the battalion was resting at Le Barque, General Hubert Gough's Fifth Army launched its Spring Offensive on a broad front between Vimy and Bullecourt. Gough's objectives were modest: capturing the German-held high ground that dominated the plain of Douai and tying down enemy forces in the north of France. The timing of Gough's push coincided with the Nivelle Offensive, a massive French operation occurring 80 kilometres to the south. The Fifth Army's first blow was delivered on 9 April when Canadian troops captured Vimy Ridge.

Two days later, the 4th Division attacked the Hindenburg Line just south-east of the fortified village of Bullecourt. In what became known as the First Battle of Bullecourt, this division, unassisted by artillery fire, was able to break through thick belts of barbed wire and into the German front and support trench-lines. However, the Australians were forced to withdraw, unable to withstand the ferocious enemy counter-attacks and intense artillery fire.

CHAPTER 7

During the morning of 15 April, the 55th Battalion was placed on standby to assist the 1st and 2nd divisions near Bullecourt. A German counter-attack threatened to sunder the Australian front. At one stage, the unit had commenced moving to the firing line when news came that the enemy attack had faltered and the Germans had been thrown back to the Hindenburg Line.

The unit left Le Barque on 20 April and made its way to a large and comfortable hutted camp built near Bécourt among the shell-holes and ruined trenches of the old Somme battlefield.[1] The weather was fine and warm and the diggers passed the next two weeks relaxing. Private Eddie Street recorded the popular pastimes of the battalion:

> Although the weather was very hot, we managed a few games of football – rugby league was the game played in the 5th Division ... Two-up schools were doing well all through the camp – Douglas[2] in 12 Platoon won about £50 in English money.[3]

As the men enjoyed the spring sunshine, the Fifth Army's push was losing momentum, becoming yet another British offensive that resulted in massive casualties and inconclusive tactical results. In the face of this reversal, the British High Command transferred the focus of military operations to Flanders. The Fifth Army forces were directed to secure a good defensive front and undertake localised operations to keep the Germans in place.

In line with the policy of limited attacks, General Gough ordered another effort to break the Hindenburg Line at Bullecourt. Accordingly, the 2nd Division, along with the British 62nd Division, attacked on 3 May. The Australians reached the enemy trenches and, despite incessant shellfire and counter-attacks, held the captured ground. While the locality around Bullecourt was no longer of military value, the fighting took on a life of its own, drawing more and more troops into the conflagration. The 1st Division replaced the battered 2nd Division and began to suffer the heavy losses that characterised this battle. On 6 May, the 5th Division, still resting near Albert, was told to prepare for the relief of the 1st Division. The Second Battle of Bullecourt, intended to involve one Australian division, was now to draw in its third.

The 55th Battalion received the order to prepare for a move to the forward areas just before midnight on 6 May. Private Herbert Harris recorded in his diary: 'Another surprise. Off to Bapaume by train then to god knows where. A hell of a battle going on. Guns going for all they are worth. We are into it this time with a vengeance. I wonder if this is to be the end.'

Despite the limited notice, the next morning (7 May) the unit marched to Albert and then made its way by train to Bapaume. From here, the battalion marched the short distance to Favreuil, arriving tired and dusty. Before leaving Albert, the battalion nucleus, consisting of seven officers and 128 other ranks, had moved to the nearby Corps Reinforcement Area.[4] This was the first time the nucleus system had been used by the battalion. Comprising 10 to 20% of a battalion's strength, including many of its most experienced officers and men, the policy of providing a nucleus arose from the need to provide a cadre around which infantry battalions leaving the front after a bloody engagement could be quickly reconstituted.

At Favreuil the men spent the day of 8 May resting, if one could call it that: the weather changed and rain teemed down, flooding their tents. What was dust the previous day now turned to sticky mud. To add to the discomfort, a nearby battery of heavy calibre guns fired non-stop. The hardship was bearable, for the troops realised that far greater tribulations lay ahead of them. At 6.00 pm came the orders the unit had been expecting and dreading: the battalion was to move to the front line near Noreuil. Two hours later the 55th was making its way to the trenches.

On either side of the narrow salient blasted in the Hindenburg Line by the 2nd Division, the Germans continued to defend their positions. Flimsy sandbag bomb-blocks were all that separated the two warring parties. The hundreds of metres of Hindenburg Line trenches taken from the enemy had been incorporated into the Australian front line. Like all salients, this was a tenuous position; German guns were able to bring fire to bear from three directions.

The 5th Division took over the captured enemy trenches and the pre-battle Australian front line to the right of the salient. The German portion of the Hindenburg Line ran in front of Quéant and Pronville. Five hundred metres separated the opposing forces.

By the time the 5th Division troops began entering their new positions, the cadence of operations had decreased from the frenetic level of the previous days. This change made it possible to hold the front line with fewer troops. Accordingly, the 14th Brigade relieved not only the 3rd Brigade in the Hindenburg Line proper, but also the adjoining 7th Brigade situated on its right one kilometre in front of Noreuil. Indeed, the 3rd and 7th brigades were so weakened that it was possible for the full-strength 53rd Battalion to replace the entire 3rd Brigade and the 55th to relieve both the 25th and 26th battalions (7th Brigade) in the forward trenches.

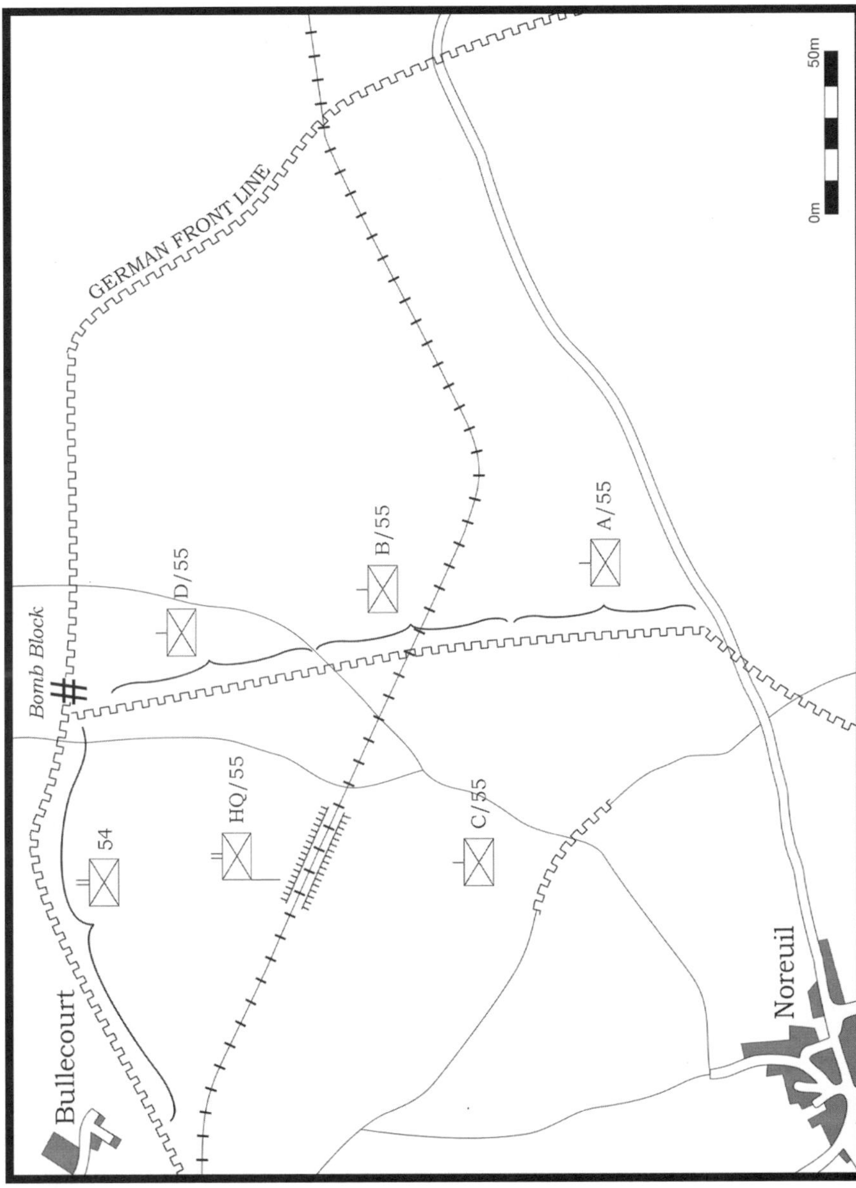

Map 9: Bullecourt – May 1917.

In the Bullecourt sector, any decrease in combat intensity was subjective. As Ellis notes, the Germans had lost none of their artillery nor the will to use it: 'Not only was the front pummelled day and night but every approach to them, especially from Vaulx to Noreuil (through the well-named Death Valley) ... was swept with a fine impartiality by an almost constant stream of high explosive, shrapnel and gas shells.'[5] As well as continual bombardment, the front-line infantry was forced to endure the constant expectation of counter-attacks as the salient 'was a menace to the enemy too grave to be endurable'.[6]

It was up Death Valley that the 55th Battalion now marched. Private Harris recorded his experiences during his journey forward:

> Was nearly a goner twice. Once [a shell burst] in a ruined house alongside the road and a piece of brick came down on my helmet. Good old tin hat. The other was here [he is writing after his arrival at the front] when a shell landed three yards away on the other side of the sunken lane. Thought we were all gone but I only got a small splinter in the side of my knee and my face seared with phosphorous. One man wounded and my mate got hit in the neck and face. A very close shave and am looking at the shell-hole while I write this and am wondering how we escaped. Had a hot time all the way in and the noise was frightful. Hundreds of guns firing from our side and also Fritzes ... Shells still passing over us. Owing to the rain the roads were awful and the mud like glue. Hope we get out alright. Just before getting here I fell down a 15 foot trench and the old tin hat saved me from smashing my head. 'Oh it was a bosker night' [Harris' emphasis].[7]

Lance Corporal George Gill was rather more matter-of-fact in his diary entry:

> Once again in the trenches, this time near Bullecourt. We arrived here late last night. We passed through our own artillery [positions] and at the time there was a heavy bombardment our guns going off everywhere. We got a few shells back. One man shot himself in the foot while we were going forwards.[8]

The relief of the 25th and 26th battalions was completed with only three men wounded which, for this time and place, could be described as uneventful. By 2.00 am the 55th was in the front line and patrols were sent out to maintain contact with the flanking units and to keep any adventurous Germans at arm's length.

Major Robert Cowey, recently returned from a course in the UK, was acting CO. He positioned D Company on the left of the battalion's front with orders to maintain close contact with the 53rd Battalion occupying the Hindenburg

Line on its left. B Company he placed in the centre and A Company on the right, adjacent to the 8th Brigade. C Company was in support, some 600 metres to the rear. Battalion Headquarters was situated in dugouts constructed in the sides of a railway cutting, often referred to as the 'sunken road' by the troops.

As dawn arrived, the men caught a glimpse of their new surroundings from their roughly hewn trenches. A kilometre away across no man's land was the ruined village of Quéant. Between the Australian lines and Quéant, the dense German wire entanglements of the Hindenburg Line were visible. As far as the eye could see the ground was nothing but pockmarked mounds of dirt, heaped haphazardly and re-arranged by the all-too-frequent shelling. Nothing green grew here; spring had withered under the fiery breath of the artillery. Stumps of trees poked up from the earth and the detritus of conflict was everywhere.

The Bullecourt battlefield was not just an assault on the eyes and ears as Captain Kennedy (53rd Battalion) relates: 'The weather had become very warm, and the stench from the vast number of unburied corpses lying about was so revolting it was almost unbearable.'[9]

The German shelling on 9 May, the unit's first day in the trenches, was relentless. The Germans primarily used 5.9-inch guns and sprinkled in gas shells among those containing high explosive or shrapnel. One witness to the shelling was Private John O'Grady, who had discharged himself from hospital that morning and made his own way to the firing line, bringing with him two litres of rum:

> Joined my platoon, had my water bottle full of rum and gave the rest away. That night we were subjected to one of the heaviest barrages of gunfire ... one of the best displays of fireworks I have ever witnessed. Very pleased I had a drop of rum, the nerves were in such a state of turmoil a person could have drank [sic] half gallon and I do not think it would have had any more effect than a Vincent powder.[10]

The battalion suffered many losses in this bombardment. At dusk, a shell burst among a group of B Company men gathered around Archie Gardner, their company commander, who had been promoted to the rank of captain the previous day.[11] Shrapnel riddled Gardner's body; his head was nearly torn off by the blast. The same shell blew Private Les Shephard[12] to pieces and severed one of Private Les Holloway's legs, mortally wounding him.[13] The shell injured six other men; one of them, Private Patrick Harpur, died of gangrene several days later.[14]

The battalion settled into a routine. During the days the snipers were kept busy as targets presented themselves and at night those men not on patrol spent their time wiring, digging trenches, improving shelters and constructing latrines. The local policy was to consolidate the existing front line while maintaining activities suggestive of future offensive operations. These latter activities were undertaken purely to deceive the enemy, for British High Command had no intention of launching another phase of the Bullecourt campaign.

The Germans continued their sporadic artillery bombardments day and night. Several dugouts suffered direct hits, the subsequent blast either killing the occupants or leaving them buried to die a slower death from asphyxiation. Lance Corporal Gill wrote:

> Second day in the trenches. Last night was a night of experiences. There was heavy spasmodic shelling, very often uncomfortably close and we had three gas alarms. The first was lachrymatory gas [tear gas]. I could smell it for a long time before I put on the [gas] helmet which I tried not to do until my eyes began to smart ... I do not think I slept last night.

Private Harris had a close encounter with death:

> Had an awful night and a wonderful escape. A shell burst in exactly the same place as the night before and a piece of it set a bag of bombs off that were on top of our little dugout. The din was fearful and the smoke also – thought it was all up this time and expected D.O. [dugout] to fall in. However, nothing happened but don't want to see any as close as that.

The fallen were given a hasty burial in a mass grave situated in the railway cutting near Battalion Headquarters.

In a largely featureless landscape, the railway cutting was one of the few identifiable places offering some protection from shellfire. Realising this, the German gunners paid constant attention to this feature. With their higher trajectory, *minenwerfers* were the weapon of choice, making life unpleasant for the Australians sheltering there. As a battalion signaller, Lance Corporal Gill's dugout was one of those situated in the railway cutting. His diary records: 'The [railway cutting] is like a part of hell. Shells mostly 5.9 [inch] bursting all around all day long. Men being constantly wounded. Yesterday afternoon Fritz sent over a load of pineapples [*minenwerfer* shells]. Some landed right in the middle of the road.'

Major Cowey was among the many victims claimed by the *minenwerfers*. A piece of shrapnel seriously wounded him in the left foot, necessitating his evacuation. Major Percy Woods was again given temporary command of the battalion, but due to his absence on duty, Captain Eric Stutchbury assumed responsibility for leading the unit in the field.

Private Harris described the lot of a work party under heavy shellfire:

> We went for rations and it was hell for we couldn't hear the shells coming for the awful din so when we go to the next trench we stayed there until it lulled a bit then on and got lost in the village [Noreuil] for three-quarters of an hour – too dark to get your bearings. Got them at last – loaded up, double load this time, and started off. Was just turning into a ruined building when four Fritzie shells came in one after the other. 20 seconds later or sooner and the lot of us would have gone up.

On 13 May the men received the welcome news that they were to be relieved that evening. This day was characterised by continued heavy shelling of the front line with high explosive and shrapnel; at 7.00 pm the Germans bombarded the front for 90 minutes without respite. Despite the enemy interference, the relief went ahead as planned and, by 2.00 am A and C companies, along with Battalion Headquarters, were in reserve positions around Noreuil. B and D companies, who had suffered the most at the hands of the German gunners, went to billets in Vaulx.

Although none knew it, the unit had moved from the frying pan into the fire.

The next day was marked by periodic shellfire but, as evening fell, the shelling of Noreuil intensified into what Lance Corporal Gill described as: 'The very hottest I have ever seen ... it was another night of hell. His bombardment started at about 8 p.m. and lasted all night ... As for us we were under gas for about seven hours. He sent over about three thousand gas shells into the village.' Private Harris was also caught in the bombardment:

> I can hardly believe I am alive this morning, after the awful time last night. Commencing at 9 p.m. and finished almost 6 a.m. today. All ideas of Fritz being short of ammunition was rudely dispelled for the rain and hail of shells during that time beggars description and of course they took their toll. Among the first being our corporal (Burch of Goulburn[15]) and Percy Peacey my mate.[16] Burch killed outright and Peacey badly wounded. A third of the shells were gas shells and I had my [gas] helmet on and off hundreds of times. It out did Fromelles that's all there is to say. Peacey got almost 20 wounds (small ones). Corporal Burch had his throat cut by a piece of shell.

The Germans used vast quantities of gas with the objective of silencing the British gun batteries around the village. However, the toxic vapours did not discriminate between gunner and infantryman. Private James Groutsch won a Military Medal for repeatedly entering gas-filled trenches to rescue men who had been buried by shell blasts or overcome by fumes.[17]

At 4.00 am on 15 May, the reason for the diabolical shelling became apparent. A massive enemy counter-stroke crashed against the 54th Battalion which was holding the right-hand section of the Hindenburg Line. Although the Germans were able to seize a toehold in the Australian positions, in the ferocious close-quarter fighting that followed the 54th was able to recapture and retain its original positions. Victory came at a heavy cost, the 54th suffering enormous casualties during the bombardment and subsequent infantry assault.

To reinforce the weakened battalion, at 8.15 am Captain Stutchbury ordered A Company to move forward and report to Lieutenant Colonel Stephen Midgley, CO of the 54th. On arrival, A Company was placed in the first line of trenches and commenced rebuilding the shattered defensive positions while keeping a sharp watch for any follow-up assaults.

The 54th's casualties were so overwhelming that, at 9.00 am, volunteers with a knowledge of first aid were called for from C Company and 30 men were despatched to act as stretcher-bearers. For the next 17 hours these men worked without respite, often under fire and sometimes up to the enemy wire, to evacuate the wounded. Four of them, Private Alex Thompson,[18] Private Ernie Corey,[19] Private Les Jackson[20] and Private John 'Jack' Buckley, were awarded the Military Medal for their bravery.[21] Author Darryl Kelly describes an episode involving Corey:

> On one of his forays [to collect wounded] he encountered two Germans who were treating an Australian casualty. The startled Germans stood back in amazement as the tough little digger approached them without a hint of fear or hesitation. Then, under their very noses, he gathered up the wounded Australian and headed back towards his own lines, while the confounded Germans simply looked on in bewilderment.[22]

Having depleted his forces in the reserve trenches, Captain Stutchbury ordered B and D companies to move to Noreuil from Vaulx; they were in position by noon. At dusk, two platoons from C Company were placed under command of the 54th Battalion and despatched to support positions around the railway cutting to augment the overall defence of the sector.

The precautions against a further German attack proved unnecessary. After the storm of 15 May passed, life in the reserve trenches reverted to its familiar pattern. Positions continued to be heavily bombarded at intervals and the British artillery positions around Noreuil were regularly saturated with gas. On the morning of 16 May, Private Harris wrote of several near misses during the previous night:

> I joined some other fellows in another part of the ruined house and then I sat all night on the floor expecting every second to be blown to atoms. But although they came alongside of us, over us, none [hit us]. I got up and walked into the next room. Had just got into it when a shell came into the room where I had been, [I] walked out of this room and the same thing happened, a shell came into the room I had just left amongst a lot of men, wounding some. I then went to the place where I had taken my things to see if it had been blown down and no sooner had I got there when another shell came through at the corner. That was enough. I went out for an early morning stroll up to the road away from the shells and after getting my nerves a bit steady made to come back when crossing what was once a railway line a whizz bang just missed my head by ducking and landed about 10 yards away. Some of the boys say I am not born to be killed by shells.

Later that day Harris made another diary entry, reflecting on the cumulative impact of the German bombardments: 'My nerves are getting jumpy under this terrible shellfire. This morning I was thoroughly unnerved and fit for nothing, as were all of us.' The stress proved too much for some — Private Clarence Gosling shot himself in the foot, necessitating his evacuation.[23]

Every day the battalion's casualty list grew longer. So far 23 men had been killed on this tour, and four times this number had been wounded. To address the slow decrease in the battalion's combat effectiveness caused by this attrition, on 17 May the majority of the nucleus were sent to Noreuil as reinforcements.

On the evening of 18 May, under moderate shellfire, the 55th relieved the 56th Battalion in the front line. The 54th was replaced at the same time and A Company, plus the two platoons from C Company attached to this battalion, also came out of the line and went to Vaulx. The 30 men from C Company engaged on stretcher-bearer duties had completed their work by the morning of 16 May. They were so physically and mentally exhausted that they had also been sent to Vaulx for a respite and were not called forward again during the remainder of the 55th's time in the Bullecourt sector.

B and D companies occupied the same stretch of trench line as they had during the previous tour. The reinforcements from the nucleus were sent to

C Company and, together with the remaining platoon of C Company, they occupied the positions around Battalion Headquarters in the railway cutting and on the right originally held by A Company.

Over the next few days, the ad hoc shelling of the front continued and a single *minenwerfer* kept up a regular bombardment of the railway cutting and the surrounding area. The battalion suffered another 13 fatalities. One of these, Sergeant John Nelson, had his right shoulder severed by a shell splinter.[24] On the way to the dressing station, another shell burst nearby, further traumatising the wounded man; he succumbed to his injuries shortly after.

Patrolling on both sides became more active. On the night of 19 May, an Australian patrol came under German machine-gun fire and was showered with grenades; the men reciprocated with grenades of their own and then withdrew. The following night a strong German party approached an advance post manned by Private Arthur Gillett and one or two other men. Concerned he might engage a friendly patrol, Gillett refrained from firing his Lewis gun until the enemy were within bombing distance. In the sharp engagement that followed, Gillett was wounded in the thigh by a grenade fragment but continued operating his gun until the Germans were repulsed.[25] A subsequent reconnaissance found a quantity of grenades, rifles and blood, indicating that the Germans had suffered several casualties but had taken them away.

Lance Corporal Gill was singled out for some personal attention from the enemy during one trip to the forward areas:

> Sniping has got very bad there now, both from machine guns and rifle. Had one experience with a sniper on going forward to the [railway] cutting about 8 a.m. There is a piece of road about 100 yards to get along. This sniper snipes over the last 50 yards. The corporal the day before was sniped in this place. I was expecting it, but never the less got a shock when I heard the ping of a bullet going past me. I ducked, looked for a place to hide for a while, then decided to run. Got a few yards only when another caused me to run the faster. I was thankful when I reached the sap. I was not in much danger for the bullets went high, but I think it is the first time I have been deliberately sniped.

With each passing day, the venom was drawn from hostilities. The Second Battle of Bullecourt was drawing to an inconclusive close and this part of the Western Front was reverting to 'normal' trench warfare. The Germans had retired from the village of Bullecourt and had apparently come to terms with the presence of an Australian salient in the Hindenburg Line. They adapted to the circumstances by constructing a new line to the north of Bullecourt, linking this to their defences around Quéant.

On the night of 22 May, again under shellfire, the 55th was relieved and returned to the reserve positions around Noreuil. Over the next few days the unit was kept busy improving defences or conducting battlefield salvage. The shelling had decreased to represent nothing more than a dangerous nuisance.

The battalion's tour ended on 25 May. Replaced by the 12th King's Royal Rifle Corps, the troops made their weary way to Favreuil and from there to Beaulencourt, situated just to the south of Bapaume. The next day, Private Harris recorded: 'Thank God we are out of that horror of a death trap at last. Headed straight up Death Valley with full packs and a hot night to cheer us on with sweat poring [sic] out of us. Got clear away to Vaulx ... Had tea and cake at the YMCA then on again.'

The 55th had been spared the massive casualties suffered by most units that served at Bullecourt. The unit was fortunate that the battle's intensity had begun to diminish before they were committed to it. Nonetheless, 33 men had died and over 100 had been wounded defending the gains made by the 1st and 2nd divisions.

As usual, the men spent their first few days out of the line resting and cleaning. There was plenty of hard work to prevent the diggers thinking too deeply about the events of the previous fortnight; a second line of defence had to be constructed and work commenced on this task. Second Lieutenant James Murray[26] described the area around Beaulencourt:

> I am at present on the fringe of the storm in comfortable quarters just behind the line. The country around was not very long ago the scene of one of our most terrible battles and its traces lie very evident to the eye and nose ... The country is intersected with trenches running in all directions – dugouts, shell-holes, barbed wire, empty shells as well a loaded ones, bombs, ammunition, equipment, graves scratched just sufficiently deep to hold their burdens, dead horses and dead men, British and German, that have not yet been buried. One needs strong tobacco and a good stomach.

Private Street wrote of the treatment meted out to several newly arrived officers, his story illustrating the assertiveness of the soldiers in choosing their own leaders and their confidence that their actions would be supported by those senior within the battalion:

> Two new officers joined our company, they had just been promoted to commissioned rank from the Gingerbeers [engineers]. Having retained the snobbishness and conceit learned in their old units, they bumped

> trouble with our men. The Company put up with them good humouredly for a week, but the two one-pippers [2nd lieutenants] became really unbearable. After a particularly trying day in the heat, we refused to obey the drill commands and counted the two officers out.[27] ... A week later they were transferred to other units.

It was now over seven months since the 5th Division had arrived on the Somme, and of this time five months had been spent in the forward areas. Indeed the division had received no sustained break from the fighting since arriving in France. The troops were physically exhausted and their nerves stretched taut. There were muttered complaints that the Australians were doing more than their fair share of the heavy lifting. All this was about to change.

Major General Talbot Hobbs, commander of the 5th Division, made strong representations concerning the deteriorating condition of his men. Subsequently, the 55th Battalion, along with all other units in I Anzac Corps, began to enjoy what Bean described as 'Probably the longest, most complete, and most pleasant rest ever given to British infantry in France.'[28]

The 55th spent a fortnight at Beaulencourt before moving further back. Over the next four months the battalion spent significant slices of time in billets at Millencourt (where, according to Private Street, 'the bedding consisted of straw a foot deep and chats [lice] six inches deep'), Mailly-Maillet, Rubempré and Lynde. Local and UK leave was freely given, and the beautiful French summer days were spent training and relaxing. The War Diarist for June 1917 recorded:

> The time spent in billets at Millencourt ... has been most beneficial to all ranks. In addition to the training – cricket and other sports have been indulged in after parade hours. 10% leave [i.e. 10% of the battalion strength] to Amiens has been granted daily. The rest in barracks has been more appreciated on account of the Battalion not having been out in back area billets since January.

War news told of an attack on Messines Ridge but few of the soldiers were interested in these events. They knew their time for war would come once more. For now, they enjoyed not hearing the growling of distant artillery nor seeing the flash of guns flickering across the night sky.

CHAPTER 8

Polygon Wood – 'not worth the life of even one man'

The 55th's prolonged rest and training period ended on 17 September when the battalion marched from Lynde to Steenvoorde, close to the French–Belgian border. Private Eddie Street wrote of the move: 'On the march … we rolled through hop-fields, past villages, the stubble fields and the haystacks. The dull continuous booming of the line grew softly louder [sic]. At times the air quivered beneath the distant blasts.' The following day the unit crossed into Belgium, marching to Reninghelst, a trying undertaking owing to the dust and poor condition of the road.

The battalion left its nucleus at Reninghelst and, at dusk on 21 September, tramped to the staging camp at Chateau Segard, near Dickebusch. Private Street and his mates 'tramped over the cobblestones of Reninghelst ... following a broad tree-lined road. Darkness had descended and the horizon was a quivering mass of flashes from firing guns.' At Chateau Segard, final preparations were made for their introduction to the Ypres salient.

At the start of the Great War the German Army had reached Nieuport near the North Sea before being halted by deliberate flooding of the terrain. Anchoring their left flank on the inundated areas, British and Belgian troops established a hasty defensive line that looped around the Belgian city of Ypres before continuing south to France.[1] These trench lines solidified following the German defeats in the autumn of 1914, leaving the British holding a considerable salient of low-lying ground circling around Ypres. After the failure of initial German efforts to pinch out the salient, the battlefield around the city had remained relatively benign during the latter half of 1915, the whole of 1916 and into 1917.

The Anglo-French offensives of 1916 had not achieved anything tangible in terms of ending the war. The Allied commanders now began looking for a new arena in which to stage a strategically decisive engagement. Over time a consensus emerged that a successful offensive around Ypres would loosen the German grip on the North Sea coast and force them to abandon their submarine bases in Ostend and Zeebrugge; it would also permit the supply lines from the UK to be shortened. In the longer term, an Allied breakthrough at Ypres offered the prospect of striking Germany through Belgium, rendering irrelevant the German defences running the length of France.

Field Marshal Douglas Haig selected Passchendaele Ridge as the point at which the breakthrough was to be made. Passchendaele Ridge is an almost imperceptible crest of land that runs in a crescent shape some eight kilometres east of Ypres. German control of the ridge allowed them to observe the entire salient and direct artillery fire with great accuracy.

Instead of a massive offensive of the type that had failed so disastrously on the Somme, the predominately British forces settled on a 'bite and hold' approach to capturing the ridge. This called for small, tactically important pieces of ground to be taken by overwhelming force and then used as a springboard for subsequent undertakings. The first of these 'bite and hold' operations commenced on 7 June 1917 with the capture of the Messines–Wytschaete Ridge, eliminating the German threat to the left flank of the Ypres salient. The success of this attack enabled the Allied commanders to turn to the main game, the capture of Passchendaele Ridge itself.

General Herbert Plumer's Second Army, to which I Anzac Corps (including the 55th Battalion) was attached, was given the task of seizing the long southern branch of the Passchendaele Ridge between the villages of Broodseinde and Becelaere. Plumer decided to approach Becelaere via the spur that ran from this village in a south-westerly direction towards the British lines. It was astride this spur that the bulk of I Anzac Corps operations in late 1917 were to be conducted.

While the British had been refining their offensive strategy, the Germans had adjusted their approach to holding ground. As 'fixed' line defences based on strong trenches and deep dugouts had proved ineffective in stemming British advances, the Germans adopted a policy of 'defence-in-depth', fortifying whole areas rather than just trench lines.

This policy most visibly manifested itself in the construction of thousands of reinforced concrete shelters called 'pillboxes' spread thickly over the forward area and extending to a great depth, often several kilometres. Each pillbox accommodated between a dozen and 40 men and contained at least one machine-gun.

In most cases the pillboxes were built so that the garrison fought outside the shelter, but a number had embrasures allowing a machine-gun to be operated from inside. The pillboxes were sited so that their garrisons could fire across the fronts and flanks of other pillboxes in the area. The defensive hypothesis was that, as an attacking force advanced, it began by crossing virtually undefended front-line trenches. However, as the force proceeded it would meet increasing resistance from the pillbox garrisons until, exhausted and bloodied, the attackers were forced to halt or withdraw.

The tactics devised to attack pillboxes involved an infantry advance close behind a creeping barrage.[2] Thus the pillboxes would be overwhelmed before their garrison was able to exit. Where this could not be achieved, the infantry was trained to use Lewis guns and rifle grenades to suppress the front of the pillbox and those members of the garrison who managed to get out. Under the protection of this covering fire, other parties would work around the flanks of the pillbox and assault it with bomb and bayonet, making further resistance futile or suicidal for the garrison. The 55th Battalion had practised these new tactics tirelessly during the summer months of 1917.

On 20 September, the 1st and 2nd divisions, also elements of I Anzac Corps, used a 'bite and hold' operation to seize Nonne Boschen Wood, Glencorse Wood and the western edge of Polygon Wood. The 5th Division was the Corps Reserve for this operation but was not required as all facets of the attack were successful.

The next 'bite and hold' was intended to sweep the enemy from Polygon Wood and the high ground to the north of it, establishing a base from which subsequent operations could strike up Passchendaele Ridge towards Broodseinde. The 4th and 5th divisions were to form the centre of this assault. The 5th Division, which was better rested than the 4th, was given the toughest assignment — seizing Polygon Wood.

Polygon Wood was shaped like an arrowhead, with the tip pointing in the direction of Broodseinde. At its base, the wood was 700 metres wide, narrowing to a few dozen metres at the apex. From base to apex the wood extended for a distance of approximately one kilometre. The ground was flat, with a slight tendency to fall on the right flank.

The wood had long been the site of an artillery range where generations of Belgian gunners were trained; *polygone* is the French term for 'firing range' and the surrounding forest had become known as Polygon Wood. In the centre of the wood was an oval-shaped horse-riding track used by the artillerymen to train and race their horses.

About 100 metres from the apex lay the wood's most distinctive feature, a large earthen mound, some ten metres high, known as the 'Butte'. Its original purpose was to act as a stop-butt for artillery firing practices in the days when the range of cannons was restricted to a few hundred metres.[3] In 1917, when the range of artillery was measured in kilometres, the Butte's height provided the Germans with a vantage point offering sweeping observation over this part of the salient.

Polygon Wood had been the scene of fighting during the First Battle of Ypres in late October 1914. The Germans had captured the wood in May 1915 and it had been shelled at varying intensities almost daily thereafter. By September 1917, Sinclair Hunt, recently promoted to sergeant, observed that:

> Polygon Wood was ... now totally devoid of any life, and so much squashed by shell fire that no sign of an original sapling was left. The whole [wood] appeared like a forest of charred and splintered stumps standing about three or four feet high. From our position in the more marshy part of the wood, the ground gradually rose to a ridge, the objective, which dominated our position ... Between us and the ridge stood a huge mound of earth [the Butte].

The trees had been replaced by thick, gnarly undergrowth that grew to about one metre high. In all directions through this tangle of brambles and tree stumps ran the remains of old trench systems and scattered barbed-wire entanglements, interspersed with the ubiquitous shell craters, a legacy of years of bombardment.

The Germans had fortified the wood to make use of any minor features offering a local tactical advantage. Their defences were based on pillboxes but the now superseded trench systems were still in evidence and formed a useful adjunct to the overall defence of the wood. By September 1917 the Germans had taken a piece of ground offering significant natural defensive properties and fortified it into a position of enormous strength and depth.

The 5th Division's plan for seizing Polygon Wood involved a two-phase operation: Phase 1 was a push of 700 metres to the 'Red' Line; Phase 2 was a further advance of 300 metres to the 'Blue' Line. In the sector allocated to the 14th Brigade, the Red Line ran along the axis of the Butte and the Blue Line followed the line of Jetty Trench which ran along the top of a low ridge in a roughly north-south direction.[4]

The 5th Division's attack frontage was 1000 metres. Major General Hobbs backed his two most experienced brigade commanders and thus split the frontage between Brigadier General Hobkirk's 14th Brigade and 'Pompey' Elliot's 15th Brigade. Of the two, he gave Elliot the more demanding assignment on the right. The 4th Division was to advance on the left of the 14th Brigade.

Phase 1 of Major General Hobbs' plan for seizing Polygon Wood called for one battalion from each brigade, preceded by a creeping barrage, to secure the Red Line. Once the Red Line was captured, the creeping barrage would lift, although artillery fire would continue for an hour to allow the Australian infantry time to reorganise.

At the start of Phase 2, two battalions from each brigade would pass through the Red Line and advance to the Blue Line, again supported by a creeping barrage. While the infantry consolidated on the Blue Line, the barrage was to continue creeping to a point some 300 metres beyond the Blue Line; once it reached this point it was to halt and keep firing for a further three hours to prevent enemy interference with the consolidation of the captured ground.

Brigadier General Hobkirk gave the job of capturing the Red Line, including the Butte, to the 53rd Battalion. Hobkirk expected the assault to the Blue Line to be a tougher proposition and he allotted this task to his most reliable battalions — the 55th on the left and the 56th on the right.

To reduce the risk of casualties from German retaliatory barrages, Hobbs designed the assembly area to be as compact as possible and as close to the enemy front line as was prudent. The entire 53rd Battalion was to be squeezed into an area 20 metres deep with a frontage of 500 metres. The 55th and 56th battalions were even more compressed. Five metres from the rear men of the 53rd, the four infantry companies of the 55th (approximately 760 men) crowded into a frontage of 250 metres at a depth of 20 metres.

The plan required the 14th Brigade's assembly area to hold approximately 2500 men along a front of 500 metres to a depth of around 45 metres. Had the Germans caught wind of the impending operation and brought their artillery to bear on this concentration of troops, the 14th Brigade would literally have been blown out of existence. As they had done at Doignies, Hobbs and Hobkirk took a gamble to obtain significant tactical advantage and reduce the risk to their men's lives.

Characteristically, Hobbs had meticulously organised the fire plan. In support of the 14th Brigade, he dedicated twenty-four 18-pounder field guns and eight 4.5-inch howitzers.[5] In addition to this firepower, the I Anzac Corps commander, Lieutenant General William Birdwood, had allocated up to 205 pieces of heavy artillery to be distributed over the 2000-metre frontage of the I Anzac Corps attack.[6] This equated to an artillery piece every ten metres, not including the 18-pounders, 4.5-inch howitzers and trench mortars supporting the respective infantry brigades.

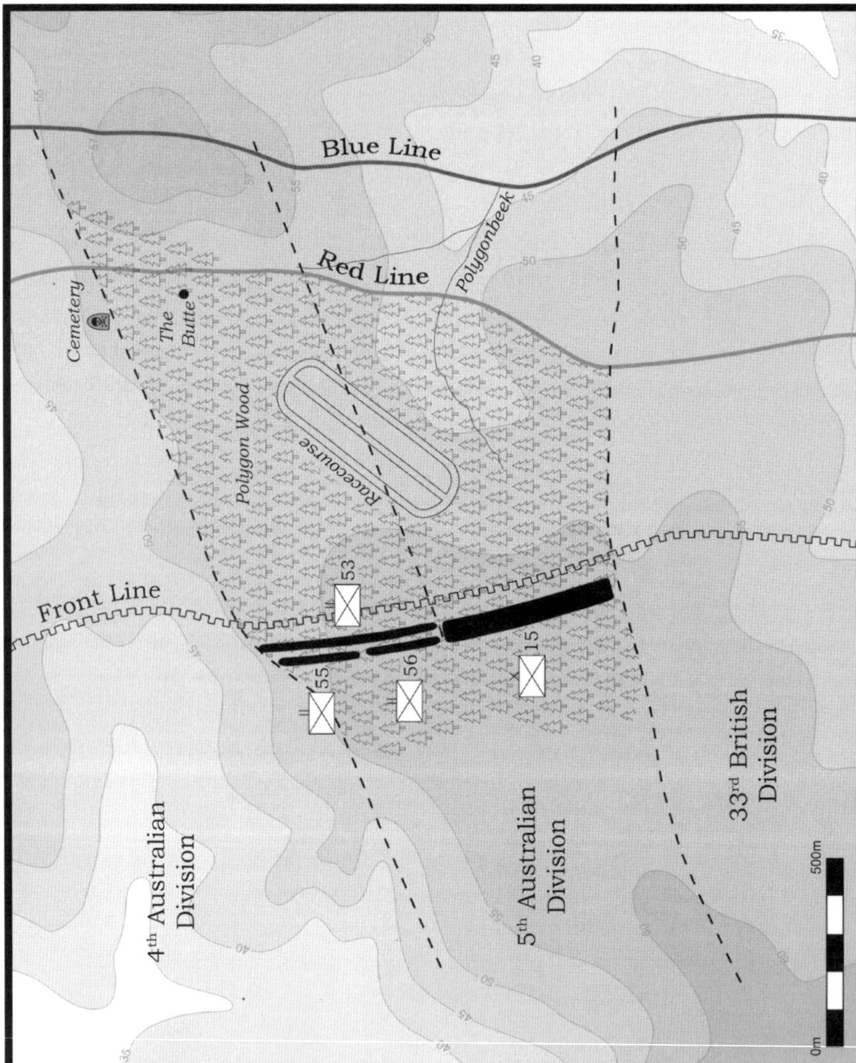

Map 10: Polygon Wood – starting locations.

In addition to the artillery, Major General Hobbs ordered all 64 Vickers guns of his division centralised under command of the division's Machine Gun Officer. Half of these guns were tasked to prepare a barrage ready to drop in front of the assaulting troops in the event of a counter-attack by German infantry. Of the remaining 32 Vickers guns, 24 were to add their fire to the creeping artillery barrage. The remaining guns were provided to each of the attacking brigades for use as they saw fit.

In addition to the support from the artillery and machine-guns, Hobkirk used his own light trench mortars and was allocated six medium trench mortars by Hobbs. He placed two of the latter at the disposal of the COs of each of the three attacking battalions.

While his staff were planning the attack on Polygon Wood, Major General Hobbs saw advantages in giving his division an opportunity to view the ground over which they were to assault. As such, on the morning of 22 September, the 55th was informed that it was to occupy the front line that night; despite being well rested, the men had seen too much fighting to relish the prospect of more. Sergeant Hunt described the pre-deployment preparations: 'All morning we spent packing ourselves up with rations, flares of all kinds, extra ammunition, bombs and in the afternoon the human camels set out to take over various sectors.'

Private Eddie Street wrote of the pre-deployment jitters affecting some of the men:

> Several chaps went to the Quack [RMO] and told him they had a premonition of death and they wished to be left out of the line. The Quack gave them permission. A couple more men told the Quack that they would in all probability go to pieces in the line. These were left out also. A few more men with the above complaints did not go to the Quack, but kept out of sight ... Before any stunt there were a few men who find it impossible to get a grip on their nerves. Not the same men by any means. It might be me one day, it might be the Sergeant the next ... The effectiveness of the Battalion was not impeded; on the contrary the fighting strength and spirit was stronger without the timid men.

As darkness fell, the battalion relieved the 1st and 2nd battalions in the forward trenches on the western edge of Polygon Wood and the support trenches in Glencorse Wood. Private Bert Bishop wrote of the unit's journey:

> We reached Ypres at dusk, the shattered city looking uncanny and ghost-like in the starlight. Out through the Menin Gate, onto Menin Road, then over corduroy tracks heading linewards. It was a depressing march, all the horror and heartbreak of war came back with a bang as we entered that indescribable morass. The earth had been pulverised, completely

> blown and blasted out of all semblance of what it once was. Not a blade of grass, not a tree or bush showed the slightest life. The area was a jumble of tremendous shell craters and shell-holes. Smashed wagons and limbers littered the two sides of the track. Piles of swollen bodies of horses and mules were entangled in the wheeled vehicles. There were great cavities where ammunition dumps had exploded. A wrecked tank here and there, blown out artillery, wheels and guns in grotesque positions. Filthy putrid lakes and mires. Everywhere among the shambles of death and horror were little wooden crosses or an up-ended rifle denoted that a human body, English or German, was below the mud, and sometimes the body was partly above the mud. Overall was the sickening stink of high explosive, cordite, gas and the rotting bodies of human beings and animals. Screaming, blasting shells were falling about us. We ran, we gasped our way through this hell of abomination.[7]

The relief was performed under searching artillery fire that Sergeant Hunt noticed 'made some of the new hands look anxious'. The troops discovered that there was no front-line trench but a series of shell-holes loosely connected. Bert Bishop reflected the sombre mood: 'Nobody laughed … We split into groups of twos and threes and tried to find some shelter in the least repellent of the shell-holes.'[8]

Next morning, Private Fred Farrall was forced to endure heavy shellfire: 'We were subject to a merciless barrage. The worse that I'd experienced. With little or no protection, only a shell-hole. It went on and on. Even an hour seemed like an eternity.' Sergeant Hunt noted that this first barrage set the tone for the remainder of the tour:

> For two days we spent the usual time in supports, lying low by day, plenty of work by night, and enough shells to go around with plenty to spare at times ... We could depend on three good 'strafes' a day, morning, afternoon and night. When the very earth and air seemed torn asunder and we poor impotent mortals, squeezed against the sides of the trenches and endured it as best we were able. Some lay down and shivered, others sat quietly and smoked, all ducked as one came very near, except the sentry who stood head above the parapet, hands in pockets, and looked out into the darkness.

Private Street had a close call during one bombardment:

> A battery of 5.9 [inch] shells rained down on our short trench. Shell after shell screamed down on us out of the sky and the trench quivered and

> broke under the massive blows ... the screaming death filled me with horror ... We two runners hugged the parados ... two light pieces of wood and a few sheets of galvanised iron were our sole head covering. We sat in the mud shivering with fear and cold, each burst rocking our hearts as they throbbed at high pressure ... A shell burst nearer, the next still nearer then a rushing roar and a crash and the earth heaved up and then descended on my head. I was engulfed in a smothering pressing darkness. I was buried in a sitting position, my arms by my side, crushed and helpless. My grave became most uncomfortable as the weight of the earth increasingly pushed my head down into my shoulders. My chin was pressed into my throat, just about suffocating me. The one free space was between my chin and the brim of my tin hat. It needed a [considerable] effort to get a gasp of air from this space. Practically no air could reach my lungs, except by shrinking my body and straining the head upwards but every movement packed the earth more tightly around me. I was sinking into unconsciousness ... my back was nearly at breaking point with the weight of the earth ... At last I realised that weight was off my head. A pick clanked against my tin hat. Fresh air reached me ... a stretcher-bearer said 'Here he is'.

In the quiet time between barrages, the officers attempted to point out the features of Polygon Wood, notably the Butte, to their men. This was largely a futile exercise as most of the soldiers were located in shell-hole 'possies' far removed from their officers and where, as Bert Bishop complained, 'in the daylight hours it was impossible to make the slightest movement, it brought a hail of bullets from the German concrete pillboxes'.[9]

Once night fell, it was a little safer for the soldiers to stretch and move about. A number of officers took advantage of the darkness to conduct a personal reconnaissance. It is indicative of the dispersal of the German pillboxes in the forward areas that one of these patrols, led by Captain Fred Cotterell (Officer Commanding A Company), made it as far as the Butte. Cotterell wanted to enter the tunnels that honeycombed the position, but the two sergeants with him refused to participate in the officer's hare-brained scheme.

After 48 hours in the trenches, the tired and shaken men were withdrawn to allow them a few hours of rest before the looming assault on Polygon Wood. Private Street described the march from the firing line to Halfway House, the massive underground shelter that was to accommodate the unit:

> Our crowd went through a wild night on the way out to Halfway House ... The shelling was very scary. Transports on the road were being blown

> to pieces, riderless horses, overturned and smashed wagons, dead horses, dead men were all mingled with one another ... Across the mud and shell-holes we paddled, dodging shells and deep water. At last we reached our new home in the midst of a forest of guns. The vivid flashes of the cordite dazzled our eyes and we blundered on our way half-blind, groping through the dark, down into the depths of the earth we went ... Wearily we threw off our gear but revivers in the shape of rum, tea, soup, stew, bread, butter, jam and bacon cheered us up.

By the time the battalion was relieved it had suffered a total of ten killed and dozens wounded. Reflecting on these two days in the trenches, Ellis wrote that, given the unformed nature of the front line occupied by the unit, the casualties suffered during the tour were 'lighter than may have been expected'.[10] The 14th Brigade War Diarist noted that: 'the battalions which came out of the line had a moderately good time.' The final word on the value of this 'moderately good time' should be left to Private Bishop: 'The forty-eight hours were to give us an idea of what our job would be. If we had just rammed mud into our eyes we'd have learned as much.'[11]

The attack on Polygon Wood was to commence on 26 September. The men of the battalion spent the daylight hours of 25 September resting in Halfway House and readying themselves for the battle ahead. Private Street used some of his time to explore the underground city:

> I took a walk ... The entrance shafts sloped down from an angle of 45° and were stepped all the way down. Stout timbers supporting the roof. Entrances were numerous ... 50 or 60 feet below were the long galleries, heavily timbered [the galleries] varied in height and width and as a general rule they were about eight feet wide and about nine feet high. The whole affair was a vast network of these galleries and it was very easy to become lost. Bunks lined the walls ... The beds were in tiers of three and consisted of wooden frames with wire netting for mattresses. A pumping plant kept the whole place free from water. The dugout was lighted with electricity generated in one of the galleries ... Men were everywhere, resting, eating, talking and preparing equipment.

Despite the foul smell and abundance of lice and other vermin, Bert Bishop found his time in Halfway House tolerable: 'A bunk was available for every man and we were given good food. The [time] we spent there gave us a surcease from the horror we'd been in and we all felt thankful.'[12] Still the thought of what was to come was never far away; Sergeant Hunt 'stewed in the smoky atmosphere and waited for the next morning.' By dusk every man was ready. The War Diarist chronicled optimistically: 'All men are in high spirits and eagerly awaiting for a chance to get to close quarters with the Hun.'

Every leader had been issued with a sketch of Polygon Wood and each man was familiar with his role in the attack. Equipment was made ready; in addition to his helmet, rifle, bayonet, entrenching tool, gas mask and water bottle, every rifleman carried at least 170 rounds of ammunition, two sandbags, two grenades and a field dressing. On top of their individual burdens, the troops were loaded up with at least two shovels per section, spare magazines for the Lewis guns and a host of other miscellanea.

While the 55th rested and prepared, trouble was brewing on the right. It was not difficult for the Germans to work out that the next instalment of 'bite and hold' was likely to occur in the vicinity of Polygon Wood. To disrupt British planning, they launched a counter-attack on the section of front held by the 15th Brigade and a collection of British units. All day the battle raged, the 15th Brigade eventually prevailing over the German offensive, but at a terrible cost in lives on both sides.

That evening, Major General Hobbs faced a dilemma. The brigade to which he had entrusted the most difficult task in a major assault the following morning had spent the day engaged in a pitched battle.

Hobbs made an audacious call.

Instead of postponing the operation, he decided to continue with the set-piece assault as planned, allotting Brigadier General Elliot two battalions from the 8th Brigade to reinforce his battered command. All night long Elliot scrambled to get his brigade ready for the attack. By sheer willpower and determination, Elliot and his commanders succeeded in getting the 15th Brigade into position for the attack by Zero Hour. If the Germans believed the ferocity of the day's fighting had thwarted the impending Australian attack, they were gravely mistaken.

After receiving their rum ration and confirmation from Major General Hobbs that the operation was 'on', the 55th Battalion began tramping to the assembly area on the western edge of Polygon Wood at 11.45 pm, the last men underway by 12.15 am. Not every man was present, however, as Sergeant Hunt fumed: 'Three of the platoon reinforcements "squibbed it" when marching out time came … The rough handling by the Germans … and the lack of sleep had made them for a time nervous wrecks and quite useless.'

In small groups the battalion made its way to the assembly area. As usual, there was sporadic shelling along the Menin Road, but nothing to cause undue concern. The fierce bombardment of the day died away and the move to the assembly area was conducted in complete silence; any noise would have brought a German barrage certain to cause carnage among the packed troops. Private Street recorded his impressions of the journey:

> The night would have been black as pitch under ordinary circumstances. Now it was lit up by a million splashes of light, with no space between splashes ... We followed the duckboard track through the forest of raging guns to Birr Crossroads and on the duck-walk crossed the shell-torn ground finishing at Corduroy Road. The road gave better footing to march on but was being heavily attacked by crowds of shells ... The road was splattered with dead men, mules and horses. We walked over blood and entrails. The shell-holes amongst the dead still smelt of acrid fumes ... Coming to the end of this road of horror, we struck out overland across the mud and shell-holes but [it was] quiet around our sector.

The assembly area was situated in no man's land just in front of the forward line of trenches. The actual location had been surveyed and staked out the previous night by parties from each of the assaulting battalions under the supervision of Major Henry Bachtold (14th Field Company).[13] At dusk on 25 September white tapes were laid between the stakes oriented to show the direction of the assault. Even this mundane activity was not without its dangers; a random shell killed Sergeant George McRae as he was rolling out the tape.[14]

Once in the assembly area, the troops sat down in the nearest shell-hole and commenced the nervous wait for Zero Hour. By 3.00 am the entire battalion was in the assembly area, the War Diarist commenting: 'Excellent discipline was maintained and the Battalion moved into position without casualties.' Private Bishop recalled that the assembly area was 'Cold and miserable and all the time the German shells landed here there and everywhere, just as though the gunners were amusing themselves with pot shots.'[15]

Judging by the large number of flares fired, the Germans were nervous and, as dawn approached, Sergeant Hunt noted: 'A light barrage began to fall on us. It caused a few casualties and was very uncomfortable.' C Company copped the worst of the bombardment and Private Street witnessed some of the results:

> There were men killed during the last hour of waiting and these deaths were caused by a fearsome strafe ... The members of a Lewis gun section were sitting in a big shell-hole to my right when a particular shell out of

> the storm picked out their possy and crashed fearfully into the centre of the old hole ... The smoke blew away and one man could be seen looking dazedly at the mangled remains of his four mates.

One of those killed by this shell was Lance Corporal Alfred 'Mick' Carmichael, an original 3rd Battalion man who was doubtlessly proud to show off the scars from being shot in the backside at Gallipoli.[16] Private Varney Davis[17] was badly wounded by the same blast — Sergeant Hunt, his platoon sergeant, described the man's condition: 'Great lumps had been chopped out by pieces of shell and one leg was smashed. He bore up like a true man, would insist on seeing every wound as we hastily dressed it and quietly weighted up the chances of life, which as he said didn't look over bright.' Davis' self-diagnosis was correct: two days later he died of his wounds.

The placement of the assembly area a little in front of the actual front line spared the battalion heavier casualties as the Germans concentrated their barrage on Glencorse Wood just to the rear. Private Jim Maynard, also in C Company, 'just hugged the earth as closely as possible, and hoped for the best.' After dealing with the casualties, Sergeant Hunt settled down again to wait for Zero Hour at 5.50 am:

> Half an hour to go. A fog had fallen and we could see Fritz flares only hazily through it … Ten minutes, a man rose here and there to tighten a belt or stretch his cramped limbs. Three minutes – the fog was more dense, and the sections became very restless as they quietly fixed bayonets and prepared to advance.

Private Maynard recalled his last moments in the assembly area: 'Presently a whisper went around there was only a few seconds to go, and we started to get up on our feet, look to our bombs, equipment etc. and get ready for what was to follow.' Sergeant Hunt marked that, at the stroke of Zero Hour,

> A gun boomed louder than the rest, suddenly the whole earth seemed to burst into a seething, bubbling roaring centre of eruption and as at the touch of an enchantress wand out of the ground sprang a mass of men in little wormlike columns, each wriggling its way forward to a sparkling, shouting seething line of earth, fire and smoke in front of them.

Years later, Jim Maynard remembered:

> Suddenly, like a gigantic clap of thunder, our barrage opened. No one who was present that September morning, if he lives to be a hundred, can ever forget that tumult of sound ... Over our heads rushed hundreds of shells of all calibres mingled with an endless stream of machine gun bullets as the gunners further back were 'lobbing 'em.

Bert Bishop was awe-struck by the spectacle:

> The very earth itself seemed to be exploding. Throbbing, quivering, the noise grew in venomous murdering intensity … The rushing, screaming shells above us, the great steady swishing of countless machine gun bullets covered our advance … The ground before us was being churned up, blasted, chewn and blown about in terrific fashion … I found myself pitying the poor wretches of the Germans who were catching it.[18]

Given the certainty that a German bombardment would fall on the assembly area within three to four minutes of Zero Hour, it was of the utmost importance that the assaulting troops were clear of the danger zone. Therefore, as the barrage commenced, the 53rd Battalion ran forward some hundred metres and then propped, still 60 metres short of the exploding shells. A churning cocktail of dust, smoke and mist limited visibility to about 15 metres.

The 55th adopted a more leisurely approach, almost every man lighting a pipe or cigarette before leaving the assembly area. Captain Cotterell 'led the Battalion's advance, walking casually, cigarette in mouth, map in hand, behind him the thick line of 'worm columns' each led by an NCO.'[19]

Disoriented by excitement, fear and the noise of the supporting barrage, Private Fred Farrall left his rifle behind in the assembly area: 'I'd lost my rifle, as a matter of fact I'd nearly lost my senses … so I just started off across No-Man's Land in a bit of a daze. I discovered that I didn't have a rifle, but it wasn't long before I could pick one up easily enough.'

With the slow pace the 55th soon caught up with the rear elements of the 53rd and pressed up behind them. At Zero plus three minutes the bombardment began to creep forward and the infantry assault commenced in earnest. Private Maynard described this moment:

> The smoke drifting back towards us from the barrage, and the early hour of the morning made visibility difficult, as with rifles at the high port and cluttered up with all sorts of odd ironmongery, usual in these type of affairs, we moved slowly after the barrage in the direction of the Butte.

Ellis recounted the plight of the defenders:

> Somewhere behind the line of destruction lay [the Germans], shuddering in pillboxes, staggered by the sudden commotion, dazed by the concussion of shells, petrified by the terror of the gleaming line of bayonets which they knew came grimly behind the line of fire.[20]

As the Australians began their slow advance, the anticipated enemy protective barrage began to fall to the rear. Here and there a few shells landed amid the packed ranks; Sergeant Will Cook, the Company Sergeant Major of D Company, was eviscerated by one of these shells.[21]

The 53rd did a thorough job during its advance to the Red Line; every shell-hole was carefully searched. However, Private Maynard, and his mates still had mopping-up to complete:

> Here and there a pillbox was rushed and prisoners extracted. For the most part there was little resistance. [I remember] Lieutenant Slater (with a small party of about a dozen of us) swinging a big 'Colt' revolver working around to the entrance of a pillbox shouting out, 'Out of it you B____s.' The boys collected a few souvenirs, and away went Fritz to find his way back to the rear.

Private Bishop had a narrow escape:

> Nugget suddenly stopped with a rifle about to fire. He had stumbled on three Germans with a machine gun, and by a fraction of a second their hands went up before he fired. They could have got us both before we saw them. They had caught the first part of our barrage, they were dazed and stupid. Nugget motioned them to the rear and went on.[22]

In the reduced visibility and disoriented by the swirl of battle, the battalion drifted to the left. The left-most group, A Company, found itself in the middle of a cemetery. Consisting mostly of German graves, the cemetery was in the 4th Division's sector, not the zone allocated to the 55th. To add to the surprise of being in the wrong place, A Company came under concentrated fire from a number of enemy strongpoints in the area. These positions were rushed and quickly subdued.

Most men did not engage the Germans. Lieutenant Colonel Percy Woods, the 55th's CO noted: 'Section Commanders everywhere led their sections forward in good order in worm formations as recently trained and followed the barrage closely ... The only difficulty in moving to the first line [the Red Line] was to hold the men in check.' Even for the leading troops of the 53rd Battalion, the 500-metre advance was largely uncontested. The majority of Germans simply surrendered. Private Bishop observed their capitulation:

> Most Germans came out with their hands up. Where none came out we tossed Mills bombs through the trapdoor. The German trenches had completely disappeared under our bombardment. We came upon quite a few [German soldiers] some of them were crawling about on hands and knees, blubbering like babies. Most Germans had fled, it seemed ... We found their water bottles all contained cold coffee. We knew how to handle that.[23]

In a post-battle report, one of the unit's officers wrote of the surrendering enemy:

> He [the Germans] streamed out into our ranks holding his hands high filled with souvenirs of every description as a peace offering. Some of the prisoners taken were whimpering boys, and drew pity rather than the fire of our men, who accepted their souvenirs and let them pass to civilisation and safety.

On the right of the battalion, through the fog, smoke and dust, Sergeant Hunt found: 'Presently ... the big mound [the Butte] loomed strangely, and in a few seconds the men [of the 53rd Battalion] were swarming over it like ants, rooting the scared Germans out of their dugouts.'

In line with the general experience, German resistance at the Butte was feeble and the garrison, which consisted of administrative and medical personnel, surrendered. As planned, the artillery fired smoke shells to indicate that the Red Line had been reached. This smoke further reduced visibility, adding to the complexity of reorganisation, but also screening the activities of the Australians from the German reconnaissance aircraft circling overhead. It was 6.25 am and the advance to the Red Line had taken exactly 35 minutes. Everything was going according to schedule.

The plan of attack called for a pause of approximately an hour at the Red Line. Despite the tight control kept on the men, poor visibility and enemy actions had resulted in considerable intermingling of the battalions. The 55th's slight slide to the left also needed rectification. The noise of the protective barrage made speech almost impossible but slowly order was re-established. The troops sat in shell-holes smoking and trying to talk while they waited their turn to attack. Lieutenant Colonel Woods informed Brigadier General Hobkirk that his battalion had contacted the friendly units on both flanks and that casualties were 'very slight'.

The German artillery observers kept in close touch with the progress of the battle and positioned their retaliatory barrage just behind the Australian advance. The assaulting troops were in the 'safe' area between the Australian and German barrages but runners, stretcher-bearers and carrying parties suffered casualties from the enemy shellfire as they made their way to and from the rear.

It was not just the enemy artillery that was active. A small counter-attack by German infantry was launched against the Butte, but this was repulsed. The garrisons of pillboxes on the fringe of the barrage also fired on the Australians and these were dealt with by fighting patrols despatched for that purpose. One enemy rifleman took up a position in a tree and unwisely opened fire; four Lewis guns were turned on him and he was killed before he could fire a second shot.

At 7.30 am the Australian shellfire increased in intensity and commenced creeping forward. At this signal, the 55th passed through the 53rd Battalion and advanced. The mist had burned off and visibility was now slightly better. Sergeant Hunt caught up with the barrage: 'The boys hugged it to a yard … now and then some too venturesome spirits could be seen right in it, and realising their danger back they would come for their lives.'

For a short time the men remained in their section-sized 'worm' formations. However, once they met the enemy this structure broke down and 'the area was covered with little parties of eight or nine men deployed each looking for its shell-hole full or pillbox full of Boche.'[24]

Private Street wrote of the close-quarter battles in and around the pillboxes where the enemy garrisons, whether better led or less shaken up by the barrage, chose to fight:

> A big pillbox nearby attracted us. We rushed it right off. Another chap and I were first in the race. My cobber went to the rear entrance and I went to the side one. As I tumbled in through the doorway a German gunner rose from where he had been firing his gun. He fumbled for his 'squirt'[25] but my bayonet was within an inch of his heart and he was mine to do with as I liked. Our eyes met and I saw he was ready to meet his fate. I always had a horror of taking life, so I grinned and pointed to the door for him to get. I took his squirt and he departed giving me a smile of thanks … I hastened to the sleeping quarters, my rifle and bayonet in the 'on guard' position. The place was dark and I could barely distinguish objects for a moment. From behind some hanging blankets which were used to screen off the bunks came a sudden blinding flash, and the narrow confines of the pillbox rang with the crack of an automatic. A Fritz … stepped out from behind the blankets evidently expecting to see me. 'Na Poo'.[26] I jumped forward and with the whole weight of my body behind the thrust I drove my bayonet deep into his right shoulder … the automatic dropped … Fritz put up his hands and got outside quick and lively.

Private John O'Grady shared a similar experience:

> We had to get behind the stronghold and blow the door in to get them out which we did … this is where I came [in] physical contact with a German soldier in action. He was a Bavarian officer and he opened fire at a few yards with a revolver, missed me. I was going to bayonet him but realised he would hit me first so I pulled the trigger of the .303. He was hit. It was many years before I could forget this incident.

Sergeant Hunt conveyed the uncompromising nature of the fighting:

> Before the last shell fell on a pillbox it was swarming with 'Aussies' who scrambled over it looking for 'flues' or ventilators through which to drop a bomb. Fritz however did not want any coaxing in most cases but ran out with hands up at the first call and shewed [sic] no signs of fight ... Rumour has it that 'the boys' are awful with the bayonet and terribly bloodthirsty, don't believe it ... if a Fritz fights he gets fight with interest but the moment he throws up his hands the fight finishes ... Of course there are cases and will always be cases, while war is war, when men wrought to a pitch of momentary insanity by the terrific strain of a barrage endured for hours perhaps, or by the loss of mates killed at their sides, act as they would never do in saner moments. Such cases however are rare, and certainly do not conform to the Australian idea of a clean fight.

Private Eddie Street witnessed one such instance of 'momentary insanity':

> There remained but one pillbox to be taken and then our Blue Line was reached. A crowd of us rushed it. Poole getting there first ... he met a crowd of Germans coming out with their hands up.[27] Savagely he rushed the Fritzies, bloodcurdling inarticulate sounds coming from his mouth ... One by one he bayoneted the Germans. Quick savage thrusts to the heart, throat or stomach and there were six dead Germans before we could interfere. I do not know what he was thinking.[28] ... The surviving Germans horror-struck, stood like statues and then moved off to the rear.

Two large pillboxes between them provided over 70 prisoners. Sergeant Hunt was amused at the antics of the soldiers around him: 'No sooner had a bunch of prisoners issued from a "pillbox" then they were roughly searched and passed to the rear. Watches, pouches, cigarette cases, revolvers and field glasses were the most common trophy.' One or two men were left behind to ensure that each pillbox was 'mopped-up' and to escort the captives.

The heavily cratered area between the Red Line and Blue Line was densely sown with pillboxes and other forms of German strongpoints — one soldier counted 18. The tirelessly practised tactics for assaulting pillboxes proved so effective that the enemy garrisons were quickly overwhelmed and the Australian troops were able to keep in touch with the creeping barrage. In the middle of the melee, several men distinguished themselves in attacks on enemy defences. Private Fred Howell,[29] Private Abraham Smith[30] and Lance Corporal Philip Kelly were awarded Military Medals for their heroism in attacking pillboxes.[31]

After his close-quarter action in the pillbox, Private John O'Grady:

> witnessed two sights I will never forget. One of our Battalion boys in front of me was hit on the back of the neck. It must have been one of our 18lb shells. It took his head completely off. He walked two or three steps, blood spurting in the air like an ornamental fountain. Not long after I run into two German soldiers in a fox hole only about ten feet away. I was going to give them the bayonet and one draws a revolver and fired a shot – it missed me. Not waiting for a second, I pulled the trigger of the .303. I hit him in the centre of the forehead right between the eyes. I looked at the other German and found he was sitting up, a hole in his chest, dead, hit by shrapnel or a stray bullet. I also had a look at the German I shot, very sorry I did, a neat hole between the eyes but no back in head, that is what a .303 bullet can do.

Eventually smoke shells told that the Blue Line had been reached. It was 8.30 am and the objectives had been captured within a minute of the planned time. Sergeant Hunt reflected: 'The advance itself was the finest we had ever experienced. The artillery barrage was so perfect and we followed it so close, that it was simply a matter of walking into the position and commencing to dig in.'

Private Street chronicled the destruction wrought by the bombardment: 'Jetty Trench was reached and it proved to be a shambles. The trench was flattened and pieces of men strewed the ground.' The battalion's casualties were light, mainly caused by men walking into the creeping barrage or the occasional artillery round falling short.

The Germans, however, were far from beaten. Consolidation activities were hampered by machine-gun fire from pillboxes located in the 'sweet spot' between the Australian barrage and newly captured Jetty Trench. Two pillboxes, one on the right and the other on the left flank, were particularly troublesome. The fire from these emplacements inflicted a number of casualties, one of these Private Stan Whipp who was hit by a bullet in the back of the head that exited under his left eye, killing him instantly.[32]

Fighting patrols were despatched to take care of these German strongpoints. When attacked by A Company, the pillbox on the left was quickly overrun and its garrison of 20 men and two machine-guns captured. The strongpoint on the right flank, close to the boundary with the 56th Battalion, offered more formidable resistance than any of the other pillboxes encountered during the assault. The fighting patrol sent by C Company to subdue this pillbox comprised

Lieutenant Henry Slater, Sergeant Harry Mortlock,[33] Corporal George Peters and Private Ron Dowling.[34]

Lieutenant Slater and his three men advanced with little in the way of covering fire from their colleagues but with plenty of attention from the alert German garrison. In short rushes the Australians worked their way around to the flank of the pillbox. Once within throwing range, they showered the position with grenades and, after putting its machine-gun out of action, stormed and seized the pillbox. The party killed several members of the garrison and captured 30 prisoners and the machine-gun. While Slater's clothing and equipment were riddled with holes caused by near misses from machine-gun bullets, none of the patrol was injured in this audacious attack. Slater was recommended for the Victoria Cross but had to be content with a Distinguished Service Order — a prestigious award for such a junior officer.[35] Sergeant Mortlock, Corporal Peters and Private Dowling were each awarded Distinguished Conduct Medals for their part in the assault.

Lieutenant Slater's actions, and those of the A Company patrol on the left, allowed the consolidation to continue without further disruption and casualties. Under the protection of the barrage, Private Maynard spent the remainder of the morning 'Mostly digging and feeding on what we had and watching the artillery plastering the area in front of us.'

The focus now moved to linking up shell-holes into a primitive trench system, constructing strongpoints and rolling out telephone lines to the Butte and forward companies. When the covering artillery stopped firing just after 11.00 am the battalion was well prepared for the counter-attacks that followed.

Patrols were pushed out in no man's land to 'mop-up' German laggards hiding in pillboxes and other shelters. The lifting of the barrage and fine weather also gave the enemy machine-guns located several hundred metres away the opportunity to fire on the troops as they were digging in the dry, sandy soil, an annoyance that was to plague the Australians for their remaining time in the line.

During this time Private Street saw what he called 'the last scene in the tragedy of the Poole Brothers' — Private Fred Poole was the man Street had witnessed bayoneting prisoners: 'Poole was still avenging his brother's death. He shot a sniper away to the left and then commenced a duel with another. Shot after shot rang out, bullets were hitting the parapet beside Poole. [Suddenly] Poole sank back into the trench gasping. The bullet had entered his throat.'

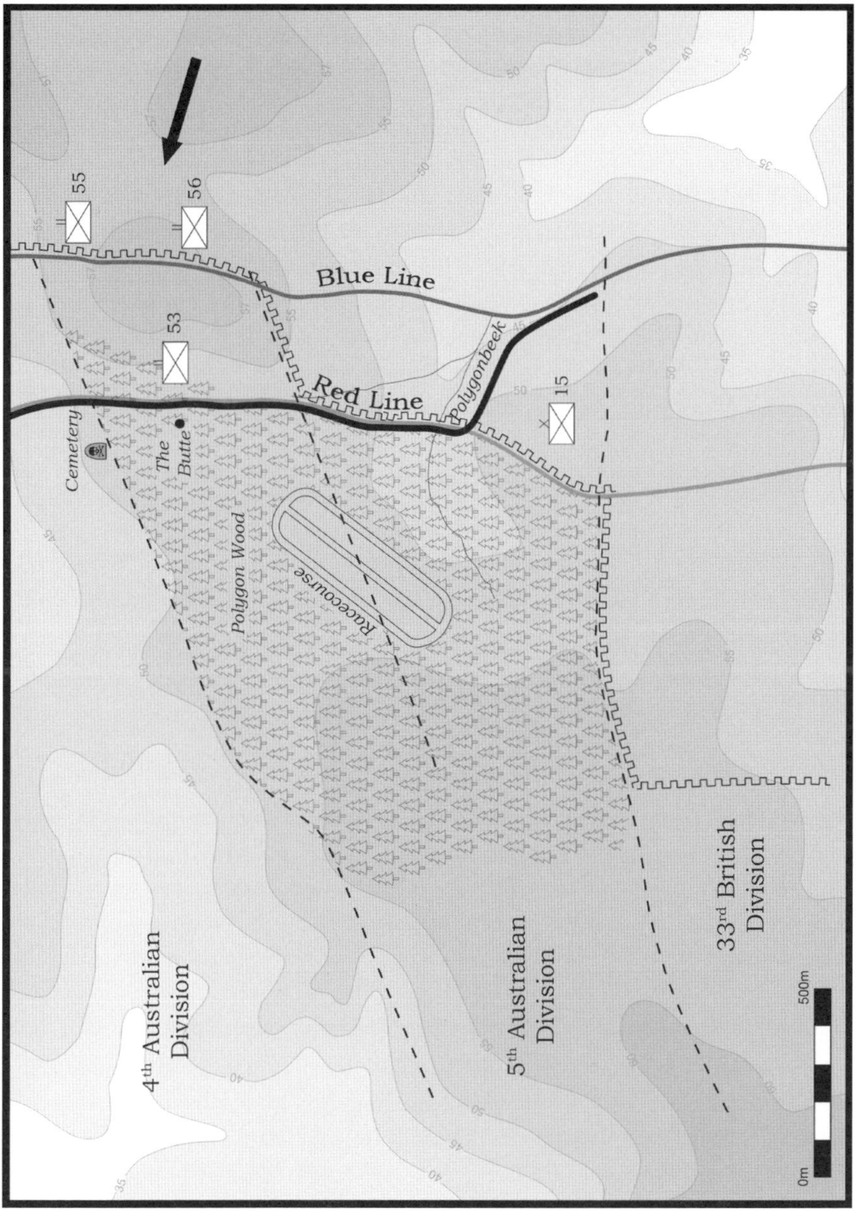

Map 11: Polygon Wood – the consolidation (dark arrow indicates direction of German counter-attacks).

At midday, intelligence sources advised of an approaching counter-attack. Sergeant Hunt didn't need warnings from 'intelligence' to tell him that the Germans were coming, for by 1.00 pm:

> From a sunken road about a mile away Fritz was seen debouching and advancing up to our position. Soon thousands were scattered over the open fields in no apparent order, bobbing, jumping, diving, like a flock of playful sheep in the distance. Then a sudden transformation took place ... A loud crackling broke out behind and a million bullets 'swished' in an endless stream over our heads. Then the roar of the guns increased into deafening noise as the barrage again descended. As we looked the ground beyond seemed to break into convulsions, and this and the enemy disappeared in the rising smoke.

German stretcher-bearers were later seen carrying a large number of dead and wounded to the rear.

At around 4.00 pm, enemy infantry were again observed starting to mass on the right, flitting from pillbox to pillbox. An hour later observers on the Butte could see dense masses of Germans beginning their counter-attack 'like chaps coming from a football match'.[36] Hell descended on the Germans; Vickers guns located on the Butte, several 6-inch howitzers and a few 18-pounders scythed through the enemy. The counter-attack broke up without gaining any momentum.

At sunset, approximately 7.00 pm, hundreds of German infantry were observed coming forward in small groups and again forming up for an attack. A heavy bombardment of the Australian positions began and 15 minutes later the Germans launched their assault, advancing in three waves. The protective barrage descended with devastating effect; enemy troops not hit by the supporting artillery and machine-guns were picked off by fire from Lewis guns and rifles. The Germans got no closer than 50 metres from the Australian lines. There were no further counter-attacks that day. Private Maynard wrote of the indifference he felt towards the German efforts: 'Though we had plenty of odd shooting, nothing serious developed.'

The battalion's losses during the consolidation had, by now, exceeded those suffered during the attack. Casualties were not always a result of enemy activity — Jim Maynard complained: 'Our chief trouble was caused by one of our guns firing short and dropping 'em right into our trench'. Bert Bishop also wrote of the casualties caused by 'friendly' fire:

> To add to our woes a battery of our own eight-inch howitzers began dropping shells short ... There was nothing that we could do about it ... Then one landed in the middle of our platoon. It made a huge crater,

> almost demolishing the trench. Bill[37] was dead, another man would soon join him. Several were wounded.[38]

Lieutenant Slater was another man injured by poorly directed artillery fire:

> My own departure from C Coy ... was occasioned by a 'short one' from our own artillery bursting right in our trench after the counter attack; partially burying me and badly wounding two other Diggers who were alongside me at the time. The force of the explosion blew my steel helmet clean from my head and over the parapet and I was stone deaf for about a week. I went to Bttn. HQ [Battalion Headquarters – located in the Butte] where I remained till we moved out.

There were accidental deaths as well: Private Norman Dark[39] and Driver John Babbage were killed in their dugout when a safety pin was accidentally removed from a hand grenade.[40]

C Company also lost Captain Cotterell; a shrapnel ball struck him in the left shoulder and deflected to his chest, necessitating his evacuation.[41] His place as Officer Commanding C Company was taken by Lieutenant William 'Big Bill' Clark, described by Private Maynard as: 'A tower of strength and worth his weight in gold for the way in which he kept things going. Cheery, game and blasphemous, he put new life into us.'[42] Clark spent the afternoon devising and implementing the defensive scheme for the right flank of the battalion, including linking up with elements of the 56th Battalion.

Fifty metres behind the front line in the support positions, D Company took a pasting from the ferocious artillery barrages that accompanied the German counter-attacks. Private John Quinlivan, one of the Company's Lewis gunners, was blown to pieces. Later that afternoon another shell landed in the trench occupied by six D Company men — only one of them survived.

Despite the distractions of shellfire and counter-attacks, the work of consolidation continued and by nightfall the whole front line was 'wired-in', a number of strongpoints had been constructed and a series of loosely linked trenches had been dug. The darkness provided an opportunity for the removal of the wounded from the battalion's aid post at the Butte and for additional supplies to be brought up.[43] It had proven very difficult to evacuate the wounded during the day due to the German artillery barrage and sniping at the stretcher parties and the legendary bravery of the stretcher-bearers had again been on display. Private Ernie Corey was in the thick of the action, tending the wounded regardless of the German fire and winning a Bar to the Military Medal he had been awarded for his work at Bullecourt.[44]

Throughout the night, which was quiet by the standards of the Ypres salient, both sides patrolled no man's land, probing the location and strength of the other's defences.[45] Private Frank West and his Lewis gun were in a post plugging the gap between the two forward companies (A Company on the left and C Company on the right). A German patrol consisting of an officer and five men stumbled into West and was promptly captured.[46]

At daybreak the Germans once again massed for a counter-attack and, following the pattern established the day before, artillery and machine-gun fire again broke up the enemy concentrations before any real threat emerged. While German aircraft had been above the battlefield the previous day, on 27 September they made their existence felt in earnest, cruising low up and down the Australian line.

Sergeant Hunt had no illusions about what the increased presence of German aircraft meant to the infantry: 'The result we all foresaw. On the second day the enemy artillery had all our positions ranged to a nicety, and bombarded them no longer in an erratic manner but systematically and heavily.' To add to the soldiers' discomfort, the Germans mixed in gas shells. The battalion's padre, Chaplain David Hunter, was killed by a shell as he made his way rearwards after a visit to the front line.

The enemy aircraft were not just the precursor for more accurate artillery fire; they also took the opportunity to rake the trenches with machine-gun fire. Friendly aircraft were conspicuously absent; the 55th's after-action report noting the 'numerous' enemy planes before observing: 'Our aircraft were not quite as numerous.'

There were other dangers to contend with. During the night a number of German snipers had infiltrated various nooks and crannies in no man's land. The fire from these marksmen added to the danger and discomfort of the Australian troops. Throughout that day, and in the days that followed, the 55th's own snipers engaged in a deadly contest with their adversaries.

'Friendly fire' from several guns of the supporting artillery saw heavy shells routinely dropping in the support trenches and on the front line. This fire caused around 20 casualties. The exasperation felt by Lieutenant Colonel Woods at the ongoing attrition of his battalion prompted him to send a terse message to Headquarters 14th Brigade in the early hours of 28 September: 'Still shells falling short (10th message).'

Sergeant Hunt wrote of the utter exhaustion associated with the post-assault period:

> Then came the hanging-on, the hard part which tests the endurance and discipline of all troops ... The excitement of the first days had now worn away, leaving in its place just like a heavy drug, a feeling of acute fatigue and listlessness which it took all our latent energies and will power to conquer ... A few parties pushed out into 'No-Man's Land' to collect stragglers from the enemy's line and to gain information regarding his dispositions. Other than those engaged, men not thus engaged trudged hour after hour backward and forward over ground beaten by long range machine gun fire, or through barrages of varying degrees of intensity carrying food, ammunition and a thousand and one articles needed for consolidation.'

The 14th Brigade's War Diarist noted in his entry for 30 September that the weather was 'Dry, bright and cold [and] the 55th and 56th Battalions are tolerably comfortable in the line and have the prospect of being relieved tonight to cheer them up.' The day passed quietly and, by 9.30 pm the battalion had been replaced, without casualties, by the 20th Manchester Regiment.

In dribs and drabs, the men made their way to a rest area 14 kilometres away at Dickebusch. Leaving the front did not alleviate all their trials as Private Bishop recounted:

> It was a pitch black night, there was no track to follow and, despite our guide, we became helplessly lost ... We walked, scrambled, fell into shell-holes full of water, accumulated more and more mud and ran out all of the cusswords we knew ... Three hours of tramping, questioning and arguing at last brought us to our bivouac ground.[47]

Private Street described the unit's reception at Dickebusch:

> German bombers were up but the cooks had got a great dinner ready, and crowds of men were eating around the cookers ... Men were straggling in all night, utterly exhausted. The night bombers played havoc with the surrounding bivouacs. We were too dead tired to worry about getting killed and sleep, the great refresher, held sway and unheeded went the thunder and flashes of the battlefield surrounding us.

The battle was over. Lieutenant Colonel Woods was in no doubt as to what the battalion had achieved, his after-action report stating:

> Taken together it will be seen that the 14th Brigade attack both in design and execution was well nigh flawless. So careful had been the

> preparation so gallantly had all ranks fought, and so free from any untoward happenings that a success complete to the smallest detail had been achieved within a few hours ... The number of enemy dead left on the field was very great.

The battalion had captured approximately 150 prisoners, seized 12 machine-guns and killed a significant number of enemy combatants.

As always, however, victory had come at a cost: 48 men killed and at least 145 wounded, around 20% of the battalion's strength. Eddie Street recalled that, after the battle:

> The roll call was not a spectacular event. Seven, including the Sergeant, were paraded in No. 12 Platoon lines. The Sergeant jotted down the names of the six of us. There were seven of us left out of a platoon which had been fairly strong before the stunt. The majority of absentees were wounded.

The Battle of Polygon Wood was, arguably, the battalion's finest hour. All of the unit's objectives had been attained — Jetty Trench was seized and held. Over the coming weeks other 'bite and hold' operations would exploit the cleft in the enemy defences made by the battalion and other units of the 5th Division. Private Bishop, however, perhaps with the benefit of hindsight, was less upbeat at the 55th's achievements: 'We'd had a battle and a victory, and we knew it meant nothing as far as winning the war went. We gained a few hundred yards of abominable, filthy, stinking countryside, not worth the life of even one man.'[48]

David McConaghy, the courageous CO of the 55th Battalion from its formation, the action at Fromelles and through the dark times of the Somme winter of 1916/17 (AWM H19209).

Officers at Ferry Post 1916. Left to right: unknown (possibly John Mitchell, 54th Battalion), Norman Gibbins, unknown, Norman Pinkstone (AWM PR02053).

Soldiers of an unknown Australian unit in the front line at Fleurbaix in 1916. This picture shows the breastwork defences typical in this stretch of the Western Front at the time of its occupation by the 55th Battalion (AWM P00437.017).

A section of the advanced line constructed at Fromelles during the night of 19 July 1916. This remarkable photograph, taken the next morning while the Germans were reoccupying their old positions, shows the boggy conditions and haphazard nature of the defences. Note the Australian dead in the foreground (AWM A01562).

Men of the 55th Battalion moving up to the trenches along the Montauban–Mametz road in December 1916. The soldier fourth from the left is Harold Walsh (AWM E00017).

Doignies seen from the beetroot factory on the Bapaume–Cambrai road. The 55th Battalion crossed this open area under heavy fire to capture the village (AWM E01367).

Polygon Wood photographed two days after the assault by the 55th Battalion and other elements of the 5th Division. The Butte is clearly visible (AWM E01912).

Portrait of officers of Headquarters 55th Battalion in 1917. Front row, left to right: Hugh Wyllie (RMO); Eric Stutchbury (2IC); Percy Woods (CO); Norman Pinkstone (Adjutant); Carl Gow (QM). Back row, left to right: David Hunter (Chaplain); William Campbell (Lewis Gun Officer); James Murray (Intelligence Officer); Reginald Wilkin (Asst Adjutant); Alfred Smith (Asst QM) (AWM A01429).

The corduroy track across Anzac Ridge in late October 1916 (AWM E01241).

The 55th Battalion's aid post near Wytschaete in February 1918 (AWM E04552).

Looking across the Wytschaete–Messines Ridge from the 55th Battalion's headquarters in Prince Rupert's Dugouts in February 1918 (AWME04555).

View along the railway embankment near Anvil Wood looking towards the destroyed St Denis sugar factory near Péronne. The Albert–Ham railway is visible on the right and St Denis hamlet is located among the trees in the background (right). The 14th Brigade attacked across this open ground on the afternoon of 1 September 1918 and again the following morning. Photo taken a fortnight after the assault (AWM E03752).

The shell-damaged hamlet of St Denis looking along the road towards Mont St Quentin. The ruins of the sugar factory are in the background (left) (AWM E05601).

'Sugar refinery, Péronne' by Laurence Howie (AWM ART93068).

Looking towards Cabaret Farm [building in background (middle left)] from near the tunnel embankment, Bellicourt. The 53rd and 55th battalions crossed this ground from right to left during their assault on 1 October 1918 (author photo, 2009).

John 'Jack' Ryan VC, a hero at Bellicourt, he never recovered from his wartime experiences, returning to Australia as an alcoholic and dying destitute (AWM P01383.013).

Edmund Street, a decorated 'runner' who spent the years after the war writing an account of his life in the 55th. It was never published (AWM P00376.001).

Sinclair 'Clair' Hunt, an articulate and prolific recorder of his time in the 55th Battalion until his death at Péronne on 1 September 1918, aged 26 (AWM P07969.001).

Walter 'Bert' Bishop, who wrote an evocative book of his experiences in the 55th Battalion. This photograph was taken in 1918 when Bert was aged 20 (private collection).

Ernest 'Ernie' Corey, a gallant stretcher-bearer and the only man ever to be awarded four Military Medals (AWM A05109).

Chapter 9: Anzac Ridge – 'mud, mud and more mud'

After the tribulations of the Polygon Wood operation, the unit was tired and in need of some time away from the fighting to recharge. The 55th, along with the remainder of the 14th Brigade, was ordered to Reninghelst, where the battalion was to have the break it so desperately needed. The unit's officers were concerned that the worn-out condition of the men, combined with the poor state of the roads, would result in the journey becoming a shambles. As it turned out, they had nothing to fear; the prospect of a protracted break in a well-organised camp proved an adequate incentive and the march was completed on 2 October in good order.

The accommodation in Reninghelst comprised mostly tents and not the huts characteristic of many other camps in the vicinity. The weather remained mild and the green fields surrounding the camp were appreciated by the troops. Major General Hobbs knew what the diggers had been through and, anticipating other trials to come, directed that all units were to have four days of complete rest.

While September had been unseasonably dry, October is typically the wettest month in Flanders and 1917 proved no exception. Most of the land within the Ypres salient was, in peaceful times, low-lying pasture with an elaborate drainage system to keep the ground from reverting to the marsh it once was. Shellfire and rain could quickly turn this terrain into a muddy bog and there had been no shortage of either over the preceding four years.

On 4 October, on a day distinguished by rain and a bitterly cold wind, the 1st and 2nd divisions undertook the next in the series of 'bite and hold' operations, this time accomplishing a tremendous victory on Broodseinde Ridge. The 14th Brigade's War Diarist noted: 'Men aware of big push and are looking forward to next spell in the line.' Given the deteriorating weather in the days following the seizure of Broodseinde Ridge it is unlikely that too many men were genuinely 'looking forward to their next spell in the line'.

On 7 October unofficial word was received from Headquarters 5th Division that this formation, along with units of the 4th Division, was to conduct another 'bite and hold' action, advancing north along Broodseinde Ridge towards Passchendaele. Reconnaissance parties, including selected officers from the 55th, were despatched to the front. That evening the 14th Brigade's War Diarist recorded his impression of the front: 'This district "out-Sommes" the Somme already. It will be very difficult later on if two days rain reduces it to this condition.'

The following day the battalion's nucleus moved to the division's reinforcement camp and the remainder of the unit marched to nearby Ouderdom, the first stage in their return to the forward areas. The men were accommodated in well-maintained huts, a circumstance they appreciated as rain moved in during the afternoon, accompanied by a howling gale.

While the battalion waited at Ouderdom, on the morning of 9 October the soldiers of II Anzac Corps commenced an unsuccessful attack on Passchendaele, eventually defeated by a combination of dismal weather, stubborn German resistance and mud. Another attempt by this corps to capture the village on 12 October also foundered. These twin failures ended all offensive action by Australian infantry in the battle for Passchendaele. Despite the large number of casualties it had sustained, II Anzac Corps remained in the trenches for another eight days before being relieved by the Canadian Corps.

During these operations I Anzac Corps, to which the 5th Division was attached, was given the role of protecting the right flank of II Anzac and the Canadian Corps and of defending the gains made along Broodseinde Ridge. Accordingly, on the night of 9 October, the 5th Division replaced the 1st Division on the extreme right of the I Anzac Corps sector. Major General Hobbs placed the 8th Brigade in the front line, the 15th Brigade in support and left the 14th Brigade in reserve around Ouderdom.

The 55th Battalion spent a few days at Ouderdom with training occupying most of the time, also providing work parties to carry supplies to the forward areas. The battalion suffered casualties on these fatigues: Private Stan Crane[1] was mortally wounded by a shell that blew off his left shoulder and one of his legs, Private Lawrence Halpin[2], Private Tom Wolfe[3], Private Les Pardey[4] and Private Tom Bald were also killed by shellfire.[5] The recovery of the bodies was often impractical due to the mud and ongoing shellfire. The fallen were given a field burial, placed in the nearest shell-hole and covered with a few spadefuls of mud. A makeshift cross or upturned rifle marked the interment spot until this too was swept away in a subsequent barrage.

The mud, cold and occasional rain made conditions in the forward trenches atrocious. Consequently, Major General Hobbs directed that, on the night of 13 October, the 15th Brigade would replace the 8th in the front line, with the 14th Brigade moving to support.

There were heavy downpours of rain during the day of 13 October, but these had cleared by the time the 55th commenced moving at last light. The troops used corduroy roads for the journey from Ouderdom; these were in an awful state owing to the lack of drainage and enemy shellfire. Notwithstanding these difficulties, the unit reached its billets in Halfway House by 7.30 pm. It was in this enormous underground complex that the battalion had prepared for the attack on Polygon Wood three weeks previously.

The troops spent the next few days providing work parties, carrying engineering stores, conducting burials and completing battlefield salvage. Private Bishop described the grisly work performed by some of these parties:

> For four days we toiled on the job of trying to bury rotting bodies of horses and mules and burning the remains of limbers and wagons that littered the countryside. I thought hell itself could not possibly be more heartbreaking and horrible. And every little while we would unearth or find under the debris the decomposing body of what had once been a human being.[6]

All the support areas were subject to indiscriminate enemy shelling. During the night of 14 October, the entire 5th Division sector was subjected to sustained bombardment with mustard gas. Casualties were light, as this attack did not come as a surprise; a German prisoner had disclosed details of the planned gas concentrations. Two days later the German gunners paid particular attention to the artillery batteries in the vicinity of Halfway House, the 55th suffering several losses as a collateral result of this shelling.

The next intra-brigade troop rotation occurred on 17 October. On this occasion, the 54th and 56th battalions occupied the front line on Broodseinde Ridge, and the 53rd and 55th battalions took over the support trenches on Anzac Ridge, a low rise located one kilometre west of the front line.

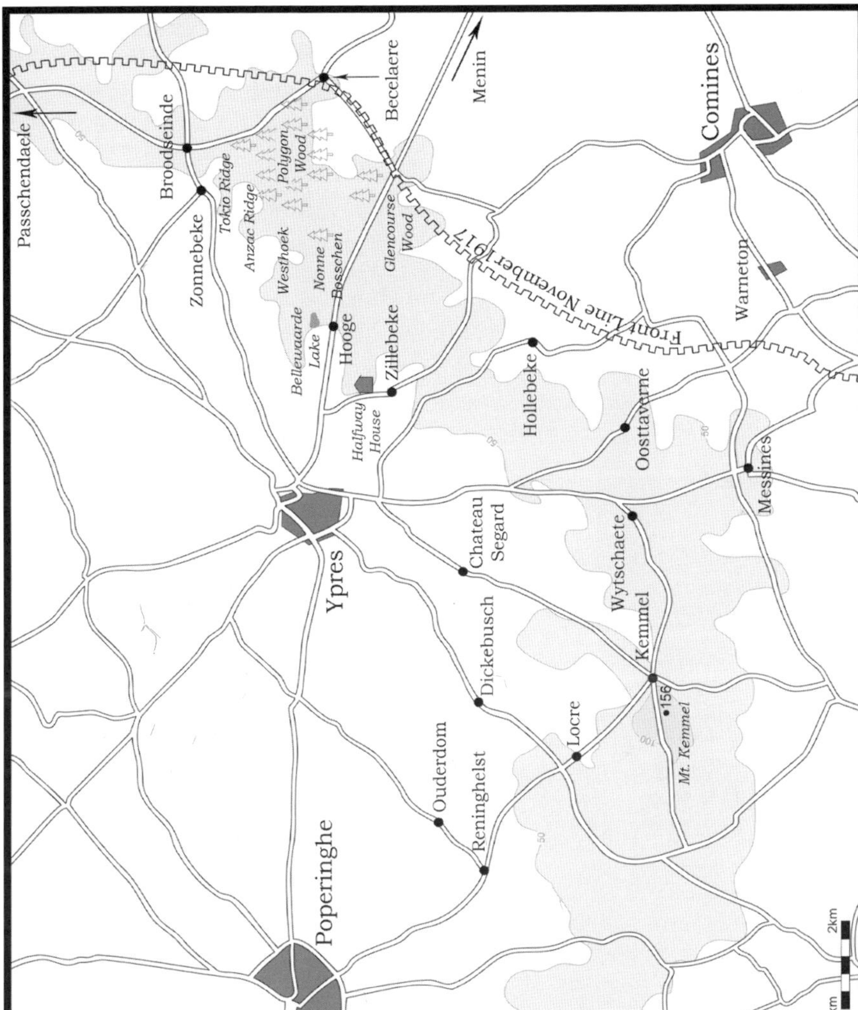

Map 12: Ypres area of operations – November 1917.

The day was clear and bright with a cold wind blowing. The corduroy and duckboard tracks up to the forward positions were, by Flanders' standards, good: no rain had fallen for the previous two days and the sun and wind had partially dried the tracks. The battalion, having passed through the stumps of Chateau Wood and the powdered brick heap that had once been Westhoek, crossed the swampy Hanebeek stream and made its way to its new positions. Private Eddie Street wrote of the journey:

> We marched past old Hellfire Corner on to Birr Crossroads where we followed the corduroy track ... It was broad day and Fritz was going hell for leather. Huge shells fell amongst the guns and our crowd ... We followed the long ANZAC duckboard 'til Anzac Ridge was reached. Pillboxes and dugouts there were in plenty ... with two others I crawled into a low sandbag shelter and this was to be our future home.

Private Bishop was unimpressed with his new circumstances: 'The country was just mud, mud, and more mud. German artillery had our position ranged perfectly and every little while a shell added to our non-joyful outlook.'[7]

By 4.00 pm the men were settled in the dugouts and abandoned pillboxes that were their new lodgings. The shellfire remained constant as darkness fell, Private Eddie Street observing:

> Fritz was playing merry hell with the fatigue parties. A crowd of new men had just joined us and a lot of them were killed the first night before we knew their names ... The drum fire[8] continued and the night was made hideous by the crashing death as it threw up fountains of mud ... We lay awake shivering, listening to the howling madness outside.

The battalion was ordered to construct a defensive position, so over the next few days many troops found themselves constructing trenches, dugouts and strongpoints. This work presented a difficult undertaking in the cold, boggy conditions and near-constant enemy bombardment. Other men were allocated a variety of work parties: carrying stores, maintaining roads, interring human and animal remains, conducting battlefield salvage, pumping out dugouts and laying duckboards.

James Murray, recently promoted to lieutenant, had been a member of the 55th before being appointed the 14th Brigade's Intelligence Officer. He accompanied Brigadier General Hobkirk on an inspection of the 55th's positions:

> Bad weather has made everything very muddy up here. I was around the Battalion this morning with the General [Hobkirk] and finished up with mud to my hips through direct sinking and sitting and with arms

> and shoulders plentifully covered from slips off the duckboard. One sees ludicrous little scenes in these tours. When the screeching buzz announced the coming of a shell in the immediate vicinity, different people are affected in different ways and the majority approach nearer the prone position according to the nearness of the shell ... I saw a shell land amongst a small party. They were all lively kinds and before he [the shell] hit the ground they were all down flat. Not one of them was hit, although I suppose the furtherest [sic] was only ten yards from the burst. They rose immediately and trotted out of the danger zone before the next arrival. One gets very wise as to the landing place of these beggars and so if you are pretty quick it is bad luck if you are hit.

The German observers, with intimate knowledge of their former positions, directed artillery fire on pillboxes and along the tracks; the difficulty of traversing the morass of the battlefield ensured a continuity of troop and transport movement along these fragile paths. The enemy artillerymen knew their trade; one morning the pillbox housing Battalion Headquarters suffered several direct hits, forcing the evacuation of the staff to another pillbox further up the ridge. Bert Bishop was caught in one of these bombardments:

> Suddenly a German barrage came down. It was heavy. It sought out all those tracks. Heavy traffic going up and coming back was caught. I stood and watched the holocaust. Horses reared and plunged everywhere, whips cracked, drivers shouted. Teams in frenzied terror shot off the tracks, dragging or leaving the smashed remains of their vehicles as the harness broke and freed them. Most of the wreckage ended in a deep shell-hole impossible to climb out of. Nearby a big shrapnel shell burst over the heap of a six-horse team tied to a wagon. The horses reared, tried to tear themselves away, and screamed in terror. The drivers were thrown away from the shambles, the horses and wagon ran over them and galloped off towards Ypres. Many men and horses were killed. Horses still tangled to their vehicles were cut free and left to do the best they could for themselves.
>
> It took only a few seconds for that heavy, regulated traffic to become a crazy, screaming, bloody inferno. The roads were clear except for the mangled heaps of horses and smashed wagons and limbers. Wounded horses, unable to move, screamed again as shells burst near them and they fought to get on their feet. The whole area was spread with desolate horror.[9]

In addition to high explosive, every night the German gunners saturated the sector with gas. The uncomfortable masks precluded sleep and made every chore difficult. Private Street was on sentry duty during one of these gas attacks:

> A temporary calm and the air became still and the battlefield grew into queer shapes. The stricken battery of guns lay in the mud ... the heroic gunners, dead now, and the faithful horses whose shapes I could discern lying in the mud ... Once more the flashes began to appear all around the horizon. The big guns were talking and the dull rumble was troubling the night ... The night quickened once more, a few white and a few coloured flares shot up along the front ... despite the mud and water there was activity in No Man's Land and death stole along the ridge ... a series of gas shells landed … A good whiff of it before getting one's helmet on would cause vomiting. I hammered into the gong[10] and its clatter resonated through the night ... The gas passed off and the company went to sleep again.

Given the incessant bombardment, a steady stream of fatalities was inevitable: Private Nathaniel Pritchard was returning from a carrying party when a shell burst next to him, blowing off a leg and leaving him to bleed to death over the next five minutes, surrounded by his mates, who were powerless to stop the profuse haemorrhaging.[11]

Four stretcher-bearers, Private Lawrence Mulvihill,[12] Private Alex Masterton,[13] Private James Simons[14] and Private William Mackenzie, were sleeping when their dugout suffered a direct hit, killing and burying them outright.[15] The next morning, several of their mates dug up the bodies to retrieve personal effects before reinterring their comrades in a field burial. This was a grim task for, as one of the party related, 'only pieces [of Masterton] could be found here and there'.

Owing to consistent and accurate shelling, C Company's commander, Captain Harry Wilson, was forced to continually adjust the positions of his troops. Finally, he succumbed to the inevitable and moved his entire company to a healthier locality further up the ridge.

The dry spell ended on 19 October with several heavy showers making, in the words of the 14th Brigade War Diarist, 'the conditions underfoot "sticky"'. However, the immense network of roads and duckboard tracks kept conditions in the trenches far less difficult than they had been on the Somme the previous winter. Herbert Harris, now a lance corporal, wrote of his surroundings: 'Raining and miserable. Nothing but mud. Our rum issue helps to keep us warm. Lot of boys got terrible colds.'

The change in weather coincided with a modification in the nature of fatigue parties. Instead of working by day, these were sent forward at night to assist the defensive preparations of the front line, particularly laying wire in no man's land and improving strongpoints.

On 21 October the men of the 55th received the welcome news that they were to be relieved that evening. To minimise the risk of casualties and to take advantage of the fine day, the CO, Lieutenant Colonel Percy Woods, ordered the battalion to immediately start thinning out troop numbers. Throughout the afternoon small parties made their way back to Dickebusch. By 6.00 pm the last elements of the battalion had handed over to the relieving battalion.

While spared the tribulation of service in the front line, the soldiers were shattered and filthy after their tour in the support trenches. A large percentage of the battalion was suffering from colds, sore throats and chest infections due to the chilly, wet weather; many had lost their voices as a result of ingesting small quantities of gas.

The 14th Brigade's War Diarist noted that, even a week after being relieved in the forward areas, 'The men still appear to be suffering from the effects of gas.' Private Street wrote that time away from the front quickly refreshed the unit: 'By afternoon [24 October] the Battalion had taken on a wonderfully smart appearance. Each man had shaved, bathed and donned his [clean clothes]. A few kicks at the football, more feeds and we were set once more.'

The 5th Division Diarist described the 14th Brigade's two-week tour in and around the Broodseinde and Anzac ridges as 'comparatively quiet'; judged against the bloodletting occurring elsewhere in the Ypres salient the sojourn was 'quiet' indeed. Nonetheless, the battalion had lost 18 men killed and dozens wounded in a month in which the troops had not engaged in any warlike operations or even been to the front line, except as members of working parties.

Despite the casualties, the battalion was still in good shape, ending October with a fighting strength of 36 officers and 840 other ranks. With the bloody battle for Passchendaele still raging, the men pondered how long their near full-strength battalion would be allowed to rest. A few days later their questions were answered.

Chapter 10: Wytschaete – 'a quiet spell in the line'

On 25 October 1917 the battalion moved from Dickebusch further to the rear in the area between Poperinge and Abele known as Wippenhoek. Here the unit commenced a program of physical training, bayonet fighting drills and short route marches. It wasn't all work: rugby league was played most afternoons and the picture and concert shows in Poperinge were popular in off-duty hours.

Private Eddie Street noted that even these leisure activities were not without risk: 'It was not very comfortable owing to the number of shells screaming over and bursting in the town … Of course in the estaminets you could procure that which made you openly defiant of screaming shells.'

While the battalion recuperated at Wippenhoek the newly formed Australian Corps, which included the 5th Division, was ordered to take over defence of the Messines–Wytschaete Ridge in the south of the Ypres salient; this was the ground that had been hard-won in battles during June. The 5th Division was allocated the role of securing the northern half of the ridge from Hollebeke to near Warneton. Frederick Cutlack, an Australian war correspondent, described the landscape he saw in late 1917:

> The western side of the [Messines–Wytschaete] ridge is in several places torn by great mine craters, but the whole scene is littered with the ruins of everything that ever stood there and so ploughed and pitted by shellfire that the craters of the mines which heralded the British attack in June 1917 will almost miss the eye … There is no shelter there save some holes in the ground and those perpetually muddy, and shattered concrete forts built by Germans in turnip fields.

On 4 November, the battalion began travelling to its new sector. The first stage was a 25-kilometre march to Outtersteene; Lance Corporal Herbert Harris described this as a tough day: 'Almost 14 miles, clean done up. My feet awful. Did another three miles looking for our packs but did not get them so here we are, no blankets or great coats and freezing.' The men were eventually reunited with their bedding and warm clothing and spent the next five days training, boxing and playing football. Once a day's formal activities were over, the soldiers were left to their own devices. Private Bert Bishop took advantage of the opportunities to leave the camp:

> In the evening twilight we wandered about the village. Civilian French people occupied the town, estaminets and shops were plentiful. We almost felt civilised again as we spent our pay on the good things in the shops.[1]

The 14th Brigade's commander, Brigadier General Hobkirk, gave notice of an impending move to the front on 7 November. Leaving the troops to their training, the battalion's CO, Lieutenant Colonel Percy Woods, joined other senior leaders from throughout the 14th Brigade in visiting the forthcoming area of operations. Initial impressions were favourable as the Brigade War Diarist observed: 'New sector promises to be quiet and comfortable.'

Three days later the 55th began its journey forward. The first march was to Locre, Lance Corporal Harris remarking: 'Got to Locre at 12. Marched all the way in the rain and wind. Very cold at first but soon got warm. Splendid camp. Best I've been in yet. Huts with fireplaces in them and fires.'

The following day the battalion arrived at Lindenhoek Camp after a miserable trek in the rain across terrain described by Harris as: 'Just like the Somme last year … nothing but a sea of mud and slush.' The wet men became more despondent when they found that they were accommodated in tents pitched over the muddy ground. At Lindenhoek, the men made their final preparations for a stint in the trenches.

The 14th Brigade was detailed to take over the portion of firing line then held by two brigades, the British 21st Brigade and the Australian 15th Brigade. This meant that the 14th Brigade would need to defend a four-kilometre segment of the front, requiring all four of the brigade's battalions to be in the forward trenches. At mid-afternoon on 13 November, the 55th began occupying its new positions, taking over from the 19th Battalion, Manchester Regiment.

In this sector, the campaigning season was over and activities were focussed on preparing for winter. The defensive layout in the Wytschaete area differed from that the soldiers had previously encountered. In the low, wet country both sides preferred to organise their advance defences in a chain of strongpoints rather than a continuous trench system.

The Australian front-line strongpoints, sometimes referred to as outposts, were based around captured pillboxes or short trenches. Usually manned by a platoon or reinforced section equipped with one or more Lewis guns, strongpoints were sited to take advantage of local tactical conditions and their ability to provide supporting fire across the front of adjacent strongpoints. These defensive locations were not designed to halt an enemy attack but to disrupt it as much as possible. In front of, or sometimes between, these strongpoints was a listening post, typically occupied by three men.

To add depth to the Australian positions, a support trench was established several hundred metres to the rear of the front-line strongpoints; half a kilometre behind this was the reserve trench. Where the terrain offered a defensive advantage, isolated 'forts', normally based around one or more Vickers guns, were constructed between and behind the trench lines.

With a large expanse of ground to defend, Lieutenant Colonel Woods was only able to allocate one company to the front-line strongpoints; the remaining companies occupied the support and reserve lines. These dispositions meant that there was some distance, often several hundred metres, between strongpoints. This left the front line porous and vulnerable to infiltration by German patrols and raiders.

While listening posts afforded some protection against surprise enemy attacks, the pressing need was for Australian patrols to dominate no man's land, preventing the Germans from getting close in the first place. Therefore, even as the relief was underway, patrols were despatched to ascertain the lie of the land in the battalion's new patch of the Western Front. Private Street was unimpressed by the relief operation and the condition of his allotted strongpoint:

> There were no duckboards between close support and the outposts. The night was very dark on which we took over. A drizzling rain had saturated our clothes, and the torn battlefield was a sea of oozing mud and water. Every one of us had two or more falls in the slippery mud and shell-holes. At last we slid into our appointed outposts. Shells did not worry us during the relief, but a few machine guns spattered [sic] at us from out of the darkness. This 'possy', which was used as a day quarters, was about as rough as any I had come across. It was an oblong excavation, fifteen feet by nine feet and about four feet deep. Curved sheets of corrugated iron was the sole roof covering. The shell-holes around us were overflowing with water and our dugout suffered by reason of this and the soakage. There was a depth of about two feet of water in the possy. Duckboards had been placed on sandbags so that they would just clear the evil smelling water.

> The mud in the sandbags and in the bottom of the hole made very poor foundations. The boards were continually sinking in the water taking us with them … A shell landed very close, severing a bank which dammed up a number of water-filled shell-holes. The water poured into our possy. Things became so hopeless that we just took no notice of it and carried on as if it were our natural state.

The first night in the Wytschaete sector was eventful; two Germans, lost in no man's land, wandered too close to an Australian strongpoint and were taken prisoner. The 14th Brigade's War Diarist wrote of the captures: '[it was] a good start and they [the 55th Battalion] expect to account for some more tonight.'

As dawn broke, the men were confronted by an archetypal Western Front landscape: shell craters, mud, and blasted trees. Trenches and wire entanglements criss-crossed the area and sprinkled throughout were pillboxes. After the inferno of the June battle, the tempo of military operations along the Messines–Wytschaete Ridge had died down. Private Street and his mates appreciated the relative tranquillity for:

> To our left we could hear the terrible rumble of drum fire at Ypres. The struggle for Passchendaele ridges was still on … Many a man amongst us gave a silent prayer to whoever was in charge of the universe, thanking him for our withdrawal from the inferno.

The troops settled into a routine. The poor condition of the positions inherited from the British meant that their prime occupation was drainage and cleaning up trenches, dugouts and pillboxes. Private Bishop described his living conditions:

> No dugouts, no duckboards, we lived in rain and mud, sleeping while leaning against the trench sides. It rained unceasingly, then came the snow, feet of it, and then, worst of all came the freeze. The mud and ground became like a solid rock to a depth of two or three feet. The heavens were black and a howling, piercing wind raged and moaned over the dreary waste. Crouched in frozen shell-holes, unable to move by day, we learned what cruel pain cold could be. Except for a sniper here and there on either side of no-man's-land, the war ceased.
>
> A partial thaw came. The ice and earth became sticky mud and clay. At night we had to bring our own rations up. We'd start off reasonably clean, each man finding a piece of timber or limb of a tree to use for a walking stick. In no time the whole party, one after another, became a mobile piece of mud. There was just no dodging, in the blackness of the night, those shell-holes full of filthy water and mud. Staggering and slipping

> along we soon were all so dirty that the sticks were no help. We became so muddy that soon no more mud could attach itself to us. Things would get so bad that they could get no worse. Someone would say something funny, we would laugh, we'd go on again.[2]

Some men were accommodated in pillboxes. The 14th Brigade's War Diarist wryly observed: 'The Boche pillbox provides good shelter in the way of battalion headquarters, company headquarters and billets for men … It is hoped that the Boche continue to build them as they are never much trouble to take and are quite useful after they are taken.'

But the concrete emplacements did not always offer the desired degree of protection. On one particular occasion the pillbox sheltering Battalion Headquarters was heavily shelled, forcing Lieutenant Colonel Woods and his staff to flee for their lives. Lance Corporal Harris described an incident in which a German shell exploded inside a pillbox: '14 out of A Coy skittled yesterday (one killed – Ted Sharkey[3]). Sgt Mansell, Tom Martin,[4] Geo Mussett wounded … It's a wonder they were not all killed.'[5]

Days were punctuated by desultory artillery activity on both sides, often including a quantity of gas shells. Patrol clashes disturbed the nights; occasionally enemy parties approached a strongpoint but most were driven off by Lewis gun and rifle fire. Private Street recounts an incident in which an undetected enemy patrol surprised a strongpoint:

> Two o'clock in the morning or thereabouts. We were sitting in our trench smoking and talking. Sentries were at their stations. The Lewis gun was in position on the parapet … The night was very dark. Our outposts were wide apart. The post to our right was about 100 yards distant. The one on our left about the same … Things were running along nicely [when] from behind the parados a loud voice commanded 'Put up your hands Australians, you are my prisoner'. We were very much surprised at this audacious order and looked to see who had the cheek to give it. We saw a Fritz standing on the parados above our heads … In one hand he held a potato masher and the other an automatic [pistol].[6] 'Come with me' he ordered. We picked up the nearest weapons … At this sudden move of ours, Fritz threw his bomb, hopped across the trench and made for his home … via the barbed wire. The barbed wire hung him up and he got a bullet.

For a time, the German patrols were intent on locating listening posts and attempting to capture their tiny garrisons. Every one of these endeavours was unsuccessful and sometimes costly in terms of enemy lives lost. Gradually, the nature of the German patrols became defensive, seeking only to keep the Australian patrols at a 'safe' distance from their own lines.

The relative tranquillity of the sector allowed a well-developed rear area infrastructure to be constructed and, as a consequence, mud was nowhere near the problem it had been the previous winter. Lessons on how to keep men healthy in cold and wet conditions had been learned; companies were rotated, rationing was carefully planned and most troops received one hot meal a day. Medical staff were successful in their efforts to prevent a re-emergence of trench foot — not a single case was reported throughout the entire winter.

On 28 November, the tour came to an end. Private Bishop made his way independently to Lindenhoek Camp and found that conditions there had improved:

> Orders were to get out as best you could, anyway you could, with the Battalion to assemble at the heap of bricks and rubbish that had once been Messines. We then marched to a camp at Lindenhoek. Midnight saw us in lovely bow-huts, each hut having a brazier fire in its centre. We sat around great dixies of steaming stew, we drank glorious hot tea with rum. And, best of all, a huge mail from Australia was distributed. In a good camp with duckboard footways, as much good food as we could eat and a huge mail from home, we were in paradise.[7]

The battalion spent the next fortnight in Lindenhoek Camp, resting and providing work parties. Every afternoon was set aside for sporting competitions. Private Street, a keen participant, wrote of the importance of football in the life of the 55th: 'We took out the footballs and had a kick about … Matches were soon organised … Football did more to rejuvenate a Battalion than anything else. Food kept us alive but footer [sic] made us enjoy living and forget death.'

Local leave was also given and Bert Bishop described the positive impact of these short excursions away from camp:

> The lovely country about us was a delight to look at. After weeks of filthy mud and water the green pastures, the neat rows of trees, the village with neat houses and bright gardens seemed to belong to another world. The fair-sized town of Dranouter was only two miles away, and when walking around it we really felt something like human beings again.[8]

A highlight of this period in Lindenhoek Camp occurred on 11 December when the men were required to cast their votes in the second conscription referendum. The conscription proposal was a divisive subject and passionate debates, often inflamed by liberal ingestion of *vin rouge* or *vin blanc*, occurred within the battalion. In the end, Private Bishop believed that the vote was 'about fifty-fifty in our platoon'.[9] Around this time the battalion received welcome news — the entire 5th Division was to be given an extended break from the fighting.

On 13 December the battalion moved by train to Desvres in France. The following day the unit marched to billets in Halinghen and Le Turne, with two companies in each village. The accommodation at Halinghen was deemed unsatisfactory and, several days later, the companies billeted there moved to better lodgings in nearby Niembourg.

Close to the seaside and the major town of Boulogne, the battalion settled in for a complete rest. At first, Private Bishop found it difficult to adjust to his new surroundings: 'Living in a world without a war around us was almost impossible to visualise, it was difficult to believe.'[10] The time was spent in a daily cycle of training and sport, but there was liberal leave to Boulogne and a general sense of being 'on holiday' prevailed.

A heavy snowfall occurred just before Christmas and a frost settled in, making the roads and paths like glass. With the exception of football, the organised afternoon sports continued despite the bitter cold. Preparations for Christmas festivities commenced: a huge barn was decorated and makeshift tables and benches constructed.

Christmas Day was far more enjoyable than the previous year. The War Diarist recorded that the men were provided with: 'Tables, seats, plates etc. and an Xmas dinner equal to that which they would normally have in their own home.' Private Bishop was enthralled by the whole celebration:

> How they did it, where they got money from, nobody seemed to know, but a couple of wagonloads from Boulogne pulled into the yard and the food was arranged about the tabletops. Poultry, sucking pig, ham, meats of many varieties, piles of vegetables, big dishes of plum pudding, custards, sauces and sweets. Tea, coffee, nuts by the bucketful, cigarettes, beer, barrels of it. Our officers waited on us, it all seemed a dream.
>
> We ate, we drank, we sang. For a few hours we forgot the war. Heavy falls of snow had completed the Christmas environment. So, feet of snow outside, warm braziers inside, we were so comfortable and happy that the Christmas dinner continued into the night.[11]

New Year's Eve was very cold, with snow and ice everywhere. The officers held a dinner to welcome 1918, leaving the soldiers to make their own fun. Private Jim Maynard recorded one highlight of the evening:

> I was in an estaminet drinking with several cobbers late in the evening

> when the old madam kept running in, shouting, 'Au feu, Au feu'[fire, fire]. We didn't comprez, but on going out in the street found our billet was blazing up to the skies. It was a fine brick barn. How it started, nobody ever knew, but there wasn't much left of it at the finish. Of course, a round of enquiries followed, and this put rather a damper on that Xmas. Part of our training had to be curtailed, as a rare lot of our rifles and equipment went west in the blaze … So ended another Xmas – on a note of comedy for the mob, but of tragedy for the owner of the barn. I'm afraid the inhabitants of Le Turne won't write any testimonials for the 55th Bn after that.

Private Bishop, as often seemed the case, was close to the action:

> New Year's … eve had been celebrated well and truly, and at midnight some men were still celebrating by their beds in the hay. It was always strictly forbidden to light a match, day or night, near hay in barns. This group, continuing their party, were well up on a pile of stacked hay. One of them, too fuddled to worry about orders, lit a match for his cigarette. The match fell from his fumbling fingers. He tried to grab it but it landed on hay …
>
> Despite the efforts of the men beside it the flame jumped like lightning, running up the side of the stacked hay. I never saw, I never thought, that fire could travel like that blaze did. In seconds the fire was leaping and jumping like Chinese fireworks. It began to roar. Men were scrambling and pushing to get out … The doors were opened, the inrushing air fanned the spreading blaze and in two or three minutes that fine barn was a blazing inferno. Nothing at all could be done to fight it, and from the freezing air outside we watched our lovely home destroyed …
>
> We realised we were cold, we wandered round searching for a sleeping place. At last we picked on the ditch that ran beside the road outside. Ordinarily the ditch would be full of filthy water and mud. Now it was solid ice, and we finished up trying to sleep on groundsheets spread over the ice, our clothes, blanket and greatcoat tangled about us.[12]

The fire destroyed C Company's billets, although other accommodation was quickly found. The War Diarist wrote of the blaze's impact on morale, noting that the 'fire set a slight gloom over it [the battalion]'.

January started bitterly cold, but as the month progressed it showed a tendency to moderate into spring-like weather. A regular training and sports program continued throughout the month, with inter-unit competitions organised to keep boredom at bay. In their spare time, the men went tobogganing and skating, much to the chagrin of the quartermaster and medical staff confronted by a steady stream of damaged clothing and bodies in the aftermath of these activities.

The battalion was now trained in defensive tactics, a novel way for the officers and men to operate. Each person was aware that over 100 additional German divisions were heading to the Western Front following the Russian capitulation in late 1917. Everyone knew that the Germans would try to smash the French and Commonwealth forces before the arrival of troops from the United States (US). If anything detrimental could be said of the protracted break, it was that it gave the diggers too much time to mull over the real and imaginary threat posed by the imminent German offensive.

In late January, preliminary orders were given for the 14th Brigade to reoccupy its former sector along the Messines–Wytschaete Ridge. Of the rest period, the brigade's War Diarist wrote:

> [It was] a particularly happy and useful spell in a comparatively comfortable area ... The mild weather, opportunity for sport and competitions, the freshness of the country and distance from forward positions all combined to make a holiday of what might have been a very cold and muddy spell of training drudgery.

The men were unaware that this was the last real break they would have until the end of hostilities 11 months later.

The 14th Brigade was to relieve the 1st Brigade as the left forward brigade on the 5th Division front. In their autumnal tour, the brigade had all four battalions in the front line. This time, to add defensive depth, the brigade did not take over the whole of its original sector but only the northern half, with two battalions manning the front line while the remaining two were kept in reserve.

On 29 January, the holiday ended. The battalion travelled by train to Kemmel in Belgium. The following evening the unit relieved the 2nd Battalion in the reserve area around Wulverghem, near the ruins of Messines. The 54th and 56th battalions occupied the front line.

For the next three weeks the diggers were kept busy with manual labour in what the RMO described as 'a morass of mud'. Despite the surroundings, the 14th Brigade Diarist observed that the troops 'Consider themselves to be lucky in this sector. It certainly is a nursery when compared with some of the sectors further north.' Private Bishop wrote of the work parties and defensive preparations:

> We worked in low sheltered places during daylight. We worked in higher open positions by night. And, day and night, we worked very, very hard. New trench lines, new outposts, new communications trenches came into being. Great masses of barbed wire entanglements grew even bigger out in front. Great new trenches were constructed behind us, new artillery positions corresponded to them. Great heaps of ammunition grew even greater beside them. We knew what all this meant … that a great German offensive was coming as the winter ended [and] the sector we held was among the top few priorities the Germans were expected to attack.[13]

On 20 February, the 55th replaced the 54th in the forward trenches. To their north, the 53rd relieved the 56th. It was a quiet relief, for the Germans had yet to throw off their winter lethargy. Bert Bishop captured the mood: 'What immediately struck us was the quietude that existed along the front. The war was completely at rest.'[14]

The unit was tasked with defending the north-east slope of the Roozebeek Spur, a branch of the Messines–Wytschaete Ridge just north of where the 55th had been in November. The Australian positions on the forward slopes of the Roozebeek Spur were several metres higher than those of the Germans, providing wonderful observation into the enemy's rear locations. There was a price to be paid; all the 55th's defences forward of the support line were under direct observation of the German gunners. Private Eddie Street described the front-line positions:

> We soon settled down into our new home. The trench was very clean and dry … The filthy posts to which we had been introduced on our first arrival at Wytschaete had been discarded and on our front every post and trench was in the best of order. In front of us stood acres and acres of barbed wire, shells landed at regular intervals … sometimes on the parapet or parados. Occasionally one would smash itself furiously, but impotently, on the pillbox behind our outpost.

Life soon settled into a pattern. Nights were spent wiring, digging, laying duckboards and patrolling. During the day, men would rest, snipe and take their turn on sentry duty. Every five to seven days the companies would rotate between the front, support and reserve lines. Battalion Headquarters was established in 'Prince Rupert's Dugouts', a series of deep, vermin-infested concrete shelters.

While life in the trenches could never be described as ideal, compared to the previous winter it was comfortable. The Battalion Quartermaster reported:

> Men and officers were in every way satisfied with the quantity, quality and

> variety of all cooked rations … The front line companies were especially cared for inasmuch that a daily supply of cocoa was always available which they prepared for themselves with the aid of trench cookers. (AIF Comfort Fund provided this issue).'[15]

From time to time the Germans would launch an artillery barrage or fire machine-guns into communications trenches, a lucky shell or bullet occasionally wounding a member of a working party. Enemy patrols usually shied away from confrontation with the Australians and seldom managed to get more than halfway across no man's land before being spotted and engaged.

The battalion's patrols ranged at will in no man's land. Several patrols went out each evening; the exception being those nights when the full moon illuminated no man's land to the extent that unobserved movement was impossible. In an effort to hamper the Australian patrols, the Germans took to randomly lobbing *minenwerfer* shells into no man's land.

Patrolling was rarely dull. The report of a patrol conducted on 26 February describes the typical experiences of the officers and men involved in these activities:

> Patrol moved out to hedge in front of strongpoint … thence past old British strongpoint … at this point three enemy were seen approaching about 200 yards away from the direction of Whiz Farm … Enemy took cover and sniped at our party, who took cover. Owing to the bright moonlight and to the uncertainty as to whether the enemy were working in conjunction with a larger party, our patrol did not advance on the enemy. After three-quarters of an hour [the] enemy party moved back in direction of Whiz Farm.

Warrant Officer Class 2 Henry Anson was a fixture on patrolling duties; on two occasions, patrols he led encountered and engaged German reconnaissance parties.[16] Private Street described Anson as 'a bombing crank and a real fire-eater, both in and out of the line. He was absolutely without fear and his daredevil actions in No Man's Land staggered us.'

February was kind to the battalion, which suffered no fatalities despite being in the front line. As the Brigade's War Diarist noted:

> [the] month in the line … has been passed as pleasantly as it is possible … The weather has been good, the sector is well organised and fairly clean and we have had few casualties ... Nothing could be more profitable than a quiet spell in the line, from the point of view of the education of all ranks, unless it be a battle which of course, has its disadvantages.

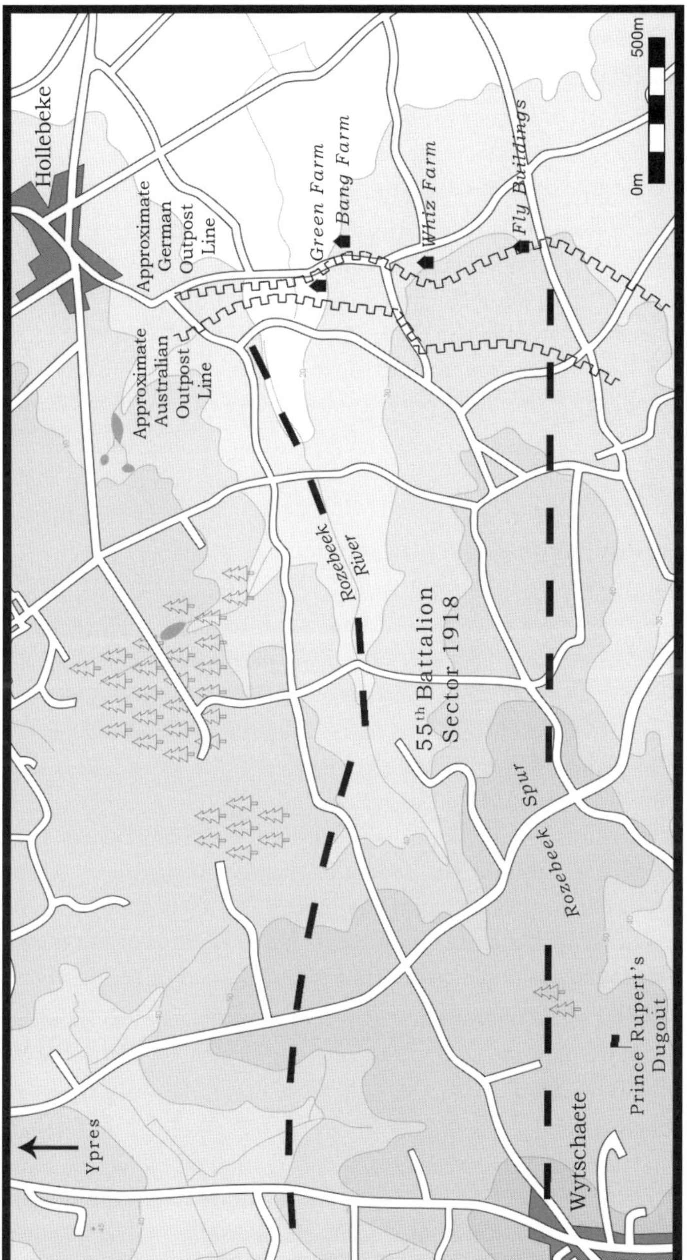

Map 13: Wytschaete sector – February/March 1918.

On 26 February, Major General Hobbs and Brigadier General Hobkirk visited the battalion. They informed the acting CO, Major Selwyn Holland, that a significant raid by the 3rd Division in the Warneton area to the south had been planned for 3 March. In support of this operation, a series of minor raids was to be conducted in the 5th Division's sector, one of which was to be undertaken by the 55th. The three officers settled on Whiz Farm as the unit's objective.

Whiz Farm was one of the 'farm forts' characteristic of enemy defences in the area. The position consisted of a ruined farmhouse, three large pillboxes and an encircling section of trench extending over a front of some 100 metres. The site was located approximately 500 metres in front of the nearest Australian strongpoint. The pillboxes were difficult to locate from the Australian positions as they rose only a little above ground level and were covered in mud and detritus.

Raids were intended to keep the men in an aggressive spirit and primed for the big battles that spring would bring. Major Holland's orders for the operation were explicit: 'To capture and kill as many of the enemy as possible; to damage and destroy materiel; to obtain identifications.'

Preparations for the operation commenced immediately. Volunteers for the raiding party were requested and a detailed reconnaissance program of Whiz Farm and its approaches began. The call for participants was over-subscribed, the War Diarist noting: '[The] final selection of raiding party was made from a very long list of volunteers.' Given the operation's limited objectives and the desire to minimise Australian casualties, the size of the raiding party was restricted to 24 troops, led by Lieutenant Stan Colless, a sergeant at Fromelles who had subsequently been commissioned. Ellis wrote of Colless: 'The Division had produced few subordinate leaders of greater enterprise or daring.'[17] Five men were chosen from each company, and a 'searcher' and German linguist were picked from the battalion's Intelligence Section. All the raiders were veteran fighters. The party was organised in four sections, each under an NCO.[18]

The members of the raiding party were sent to the rear and set about training for their forthcoming enterprise. They constructed a life-sized replica of the Whiz Farm locality and this was used day and night to develop and refine the tactics to be used.

On 28 February, the preparatory 'shoots' began. Artillery and trench mortars fired on the German forward defences and wire entanglements. Extreme care was taken in preparing the fire support plan to ensure that it did not provide the enemy with any inkling of the impending operation.

Over the nights of 1 and 2 March, the Battalion Patrol Section, consisting of Ed Bourke, now a sergeant, and eight other ranks, escorted each member of the raiding party into no man's land, providing him an opportunity to study the ground and see the objective from the assembly area.[19] These familiarisation patrols were by no means the only reconnaissance activities occurring.

On the morning of 2 March, Lieutenant Colless, Lieutenant Reg Gardiner (the battalion's Intelligence Officer) and four men took up a position in Green Farm, a series of unoccupied buildings in the middle of no man's land.[20] They spent the rest of the day here, observing Whiz Farm and other German strongpoints in the vicinity.

Dawn on 3 March was dull and rather cold, conditions that persisted throughout the day. Heavy trench mortars bombarded Whiz Farm and the nearby German post known as 'Fly Buildings' in an attempt to destroy wire entanglements. At 3.00 pm a sustained artillery barrage was laid down on Whiz Farm, Fly Buildings and 'Bang Farm', another enemy strongpoint near the raid objective.

For the rest of the afternoon and into the evening artillery fire pounded the German positions. The timings and targets of the shoots were selected to preserve a sense of normality as far as possible. The supporting fire plan had called for gas shells to be included in the bombardment of Fly Buildings to neutralise the machine-guns known to be sited there. This particular barrage was cancelled at 9.00 pm because of concerns that the light and variable winds might cause gas to drift over the raiders.

Behind the lines, the raiders completed their arrangements. Faces were blackened and equipment checked. Each raider was armed with a rifle, bayonet, two 5-round clips of cartridges, two hand grenades and a gas respirator. Metal helmets were dispensed with. To prevent the Germans gaining intelligence on the unit involved in the operation, the raiders carried no identification except a raid number, a small metal disk with a number on it. Lance Corporal Archie Winter, one of the raiders, described the last-minute preparations:

> Firstly the Padre held a Communion Service and later we indulged in a great tuck in, a rareity [sic] in the Army. At 10 o'clock, we moved from our position having a good way to go. We were given a nip of either rum or whisky. It was very dark.

The plan called for the raiders to leave the battalion's middle strongpoint and traverse no man's land to a spot just in front of some old trenches designated the assembly area. At Zero Hour, set for 11.45 pm, the attackers were to advance under protection of an artillery barrage and, two minutes later, when the shellfire lifted, they were to rush the objective. The withdrawal was to be made at Lieutenant Colless' discretion.

The raiding party left the Australian lines at 11.10 pm. The sky was overcast and the raiders were hindered by the near-complete darkness. However, the early actions were completed in time and the men made their way to the assembly area without incident. The keyed-up men didn't have long to wait; Lance Corporal Winter was 'just in position in No Man's Land when our barrage came down'. Guns from an 18-pounder battery and a 4.5-inch howitzer battery sent volleys of shells plunging into Whiz Farm. Additional artillery, supported by trench mortars and machine-guns, pounded other enemy positions in the neighbourhood. Private Street watched the start of the raid from the Australian lines:

> A single gun spoke, and then there followed a mighty crashing and smashing. The wild tornado of shells rushing through the air filled me with awe ... A minute later the air was a mass of shooting and quivering coloured lights [German flares]. The reds, the blue, the green, the white, the orange they all soared up and descended brilliantly ... The machine guns got an early start and thousands of bullets cracked across No Man's Land. The sound of rifle fire swelled the chaos.

As the first shells landed the raiders stood up and, line abreast, walked towards the objective. The German barbed wire was patchy and formed no serious obstacle. Once through the wire, the raiders wheeled slightly right so they struck all three pillboxes simultaneously. Lance Corporal Winter was just behind the barrage, and commented that 'It was good oh following it … along then we had to change direction right. I happened to be on the left flank so had furtherest [sic] to go.' Winter, with some modesty, fails to mention that he led the flank responsible for keeping the correct direction for the entire assault.

As they advanced, the raiders came under withering machine-gun fire from Whiz Farm and surrounding enemy positions. The Australians paused for a few seconds some 30 metres from the objective while they waited for the barrage to lift. Then, as the bombardment suddenly shifted to other targets, the raiders surged forward to close with the pillboxes before the Germans could organise a coherent defence. As planned, they fell on all three pillboxes simultaneously. It was furious, grim work. The flashes of exploding shells and flares illuminated the night and, in this flickering black-and-white world, it was difficult to tell friend from foe.

From the southernmost pillbox, a machine-gun placed on top of the concrete structure fired until the attackers were within a few metres of it. Private Charlie Arnett rushed forward, bayoneting both members of the machine-gun's crew.[21] Around him, the rest of his section captured the two Germans remaining in the pillbox. The prisoners were despatched under escort across no man's land.

Responsibility for the capture of the northernmost pillbox was given to Lance Corporal Winter's team. Winter was reluctant to describe the subsequent events, his diary stating: 'What transpired there I need not relate anyhow we had a win.' However, the citation for the Distinguished Conduct Medal he was awarded for his part in the raid provides some insight into Winter's activities:

> The [German] garrison defended the strongpoint with the greatest resistance and in the hand to hand struggle which ensued [Winter] accounted for two of the enemy, each of whom were in the act of firing on other members of the party.

Private Les Curran was also involved in the savage fighting around this pillbox, killing three Germans with his bayonet.[22] Two Germans were encountered in the rear of the central pillbox. Lieutenant Colless shot and killed them as they were aiming their rifles.

Using the precious seconds bought for them by the defenders of the pillboxes, the remainder of the enemy garrison in Whiz Farm chose to flee rather than fight. A party of up to 15 Germans was observed retreating along a trench leading away from the position. Lieutenant Colless ordered a limited pursuit of the withdrawing enemy. The raiders sprinted after them throwing grenades, and five or six Germans were seen to fall.

As abruptly as it started, the fighting was over. Lieutenant Colless instructed Lance Corporal Albert Turner, the intelligence specialist, to search all three pillboxes.[23] None was found to contain papers or anything else of intelligence value. As the objectives of the raid had now been achieved, Lieutenant Colless ordered the withdrawal. Just after this directive was issued, a flare exposed a hitherto unnoticed German machine-gun firing from a shell-hole around 15 metres away; three grenades were thrown, killing the gun's two-man crew. That a machine-gun could be firing in such close proximity without coming to the attention of the raiders is testament to the noise and confusion of the battle.

The retirement was undertaken as rapidly as possible. It took the enemy gunners a few minutes to respond, but as the withdrawal commenced, a heavy artillery and *minenwerfer* barrage descended on no man's land and the Australian forward positions. To bypass the retaliatory fire, Lieutenant Colless instructed the raiders to make their way towards one of the battalion's northern strongpoints. As they proceeded, a party of eight Germans was observed approaching from the direction of Bang Farm, apparently with the intention of counter-attacking. Lieutenant Colless collected the seven men nearest him and formed an impromptu rearguard. The Germans and Australians clashed in a sharp, brutal engagement in the middle of no man's land. Six of the enemy were killed, one of whom managed to land a punch on Colless before the aggrieved Australian officer slew him. The two remaining Germans fled.

The majority of raiders reached the safety of the Australian lines at 12.14 am. Lieutenant Colless and his rearguard were forced by enemy fire to make a wide detour northwards, entering the Australian lines in the 53rd Battalion's sector at around 12.45 am. Shortly after Colless' return, the supporting artillery ceased fire.

The German prisoners and their escorts had been shelled as they made their way back across no man's land and one of the captives was killed. The second prisoner was despatched to Brigade Headquarters for interrogation. Private Arnett carried the captured machine-gun back to the lines.

As the raiding party had returned in dribs and drabs, it took until 6.00 am the following morning for them to realise that one man was missing. Private Reuben 'Darky' Smith had been knocked over by a shell-burst from the supporting artillery just as he neared the objective.[24] When asked if he could find his way back to the Australian lines he responded, 'I'm alright – carry on.' It seems likely that Smith had become disoriented and stumbled further towards the German lines, where he later died. His body was never recovered despite careful searches of no man's land on the nights following the raid.

Private Smith was the only Australian fatality. In terms of other injuries to the raiding party, Private Bill Hickswas wounded by German artillery as he was escorting the prisoners back across no man's land but was able to reach the Australian lines.[25] Lieutenant Colless had been struck by a piece of spent shell but was able to remain on duty. The retaliatory shelling had wounded several men in the Australian positions, none of them mortally. While performing his duties as company runner, Private Street came across one result of the enemy barrage:

> At last I reached Number 3 post. It was in a sad state. The post itself was smashed about, one part being obliterated. Five men lay on the floor of the trench, getting their wounds dressed. The fumes of the burst explosive filled the trench with their ghastly odour.

By mid-morning on 4 March, the full extent of the raid's achievements had become apparent. At least 18 Germans, and probably several more, had been killed and a prisoner taken, all for the loss of one man. Major General Hobbs and Brigadier General Hobkirk visited Lieutenant Colless and the raiders later that day. Hobbs told the men that 'the raid was the most successful he had ever had anything to do with', a sentiment echoed by Hobkirk who 'considered the raid to be the most successful ever carried out by the 14th Brigade'. The outcome of the operation was testament to Colless' leadership, bravery and planning, as well as the raw courage and professionalism of the raiders.

While the Australians had been patrolling and strengthening their defences throughout the winter, the Germans had been relatively benign in their own endeavours. However, at the start of March, as if awakening from hibernation, the enemy attitude became more defiant and aggressive. Shelling of the Australian positions increased in intensity and the Germans implemented a vigorous patrolling policy. Large enemy parties were encountered more frequently in no man's land.

Private Frank Brown, in a letter dated 9 March 1918, commented on the noticeable change in the German tempo of operations: 'When in outpost we hold a former German pillbox and this past week Fritz has tried his utmost to blow it up with big stuff and minnies [*minenwerfers*] but it's some post I can assure you, not to be evacuated lightly by our boys.'

In every strongpoint, trench and latrine, the talk was of the anticipated German offensive. These were anxious days for the soldiers, for they knew that, if the enemy launched an offensive against their part of the front there was no hope of salvation. Private Bishop captured the nervous atmosphere:

> Men joked not, and at times we found ourselves speaking almost in whispers … It seemed so uncanny. We never parted with our rifles. When we lay down in the hope of getting some sleep, our rifles were beside us. We knew our first sight on waking could be hordes of grey-coated Germans.[26]

The battalion's officers were emphatic concerning the role of the unit in the event the blow fell on their sector. Bert Bishop's company commander told his troops:

> If the Germans are strong enough and we can't hold them, [this] trench is where our battalion dies. Other battle lines are behind us, other troops will be manning them. If the Germans reach them, our battalion has ceased to exist.[27]

On the evening of 10 March, German artillery lashed the 55th's sector, supplemented by masses of machine-gun fire. The systematic nature of the shelling raised the battalion to a fever pitch of preparedness, as it was thought the enemy were about to undertake a raid. Nothing eventuated; the Germans were playing cat and mouse with their foes.

A few nights later, on 14 March, the 15th Brigade raided two German positions. Lieutenant Harry Miller was ordered to support the raids by orchestrating a 'Chinese Attack' — a ruse to draw enemy fire using dummy figures placed in no man's land operated by pulling strings to make the figures rise and fall realistically.

Just after midnight on the morning of the raid, Lieutenant Miller and 21 soldiers placed the dummies in no man's land and ran the controlling strings back to the two closest strongpoints. Due to the volume of hostile fire the 'Chinese Attack' was expected to generate, these two strongpoints were likely to become very 'hot'. To reduce the risk of casualties, the majority of Miller's party, and the usual garrisons of the strongpoints withdrew. This left Miller and four 'pullers' as the sole occupants of one strongpoint and two 'pullers' and a Lewis gun to defend the adjacent position.

At 3.30 am the raid commenced and, for the next 30 minutes, Miller and his intrepid band manipulated the dummy figures. The deception was a complete success, drawing considerable machine-gun and trench mortar fire. Just before dawn, when the raid was over and the front had calmed, Miller was reunited with the rest of his working party and they again moved into no man's land, recovering the badly damaged dummies. Miller's party suffered no casualties during this cameo role.

While gas attacks were a routine occurrence, on 18 March the Germans launched a massive gas bombardment of the battalion's reserve positions. The War Diarist noted it as 'one of the heaviest the Battalion has yet experienced' although the gas caused few casualties. Private Street was caught in this bombardment, recording his experiences:

> Fritz put over a fearful storm of gas shells. All night they sighed and popped. Shell after shell and soon [Ostaverne] Wood was a mass of reeking fumes ... Towards morning when visiting the pillbox where the cooks, quarter master staff and company clerks etc. were I found the whole crowd badly gassed. A gas shell had burst in the doorway, the fumes entering the narrow confines of the blockhouse. The whole mob were rushed away to hospital. None died, but most of them were sent back to Aussie unfit ... The shelling night drew to a close and daybreak appeared. The order to 'Stand to' was a permanent thing now. Every morning we expected the German masses. We were 'standing to' wondering if the gas barrage of the night was the herald of the storm.

The next day the German mortars were again very active, and the evening brought another heavy gas attack on the reserve positions. This time the Australians were in a position to respond in kind, using 400 gas projectors to launch a retaliatory gas attack, after which the War Diarist reported: 'The enemy was fairly quiet for the remainder of the night.'

The frequent German barrages resulted in a steady trickle of casualties to the unit. Private Robert Keyte suffered an agonising death after being hit in the stomach by a piece of shell while on a fatigue party.[28] Private Arthur Metcalfe died when his dugout suffered a direct hit.[29] Shrapnel also killed Corporal Alex Roddan,[30] Private Merion Moriarty,[31] Private Alfred Browne, Private James Kinsela[32] and Driver Rupert 'Skinny' Hinson.[33] The same shell that killed Hinson also badly wounded his platoon commander, Lieutenant John Byrne, in the chest and legs.[34]

At dawn on 20 March the battalion's two southernmost strongpoints were subjected to sustained fire by *minenwerfers* and machine-guns. A significant number of *minenwerfer* bombs were landing close to the barbed-wire defences and a patrol under Lieutenant Miller was sent out that evening to check on the condition of the Australian wire.[35] Moving under machine-gun fire and *minenwerfer* explosions, Miller's party found a gap in front of one of the strongpoints. This was quickly patched.

Next morning the Germans ensured that the Australians were fully awake by firing a few heavy *minenwerfer* projectiles at daybreak. These were the first rounds in a bombardment that, by 7.00 am, had grown to include German artillery blasting the whole of the battalion's sector. The barrage was again focused on the two southernmost strongpoints. Enemy machine-gun fire also began converging on these posts.

At 7.25 am the barrage was lifted to the rear of the two strongpoints. The Australians in one of the posts were shocked to see 30 to 40 Germans attempting to breach the wire closest to them, having passed through the outer wire belt under cover of the bombardment and the fog that had risen just after dawn. The enemy were engaged by rifles, Lewis guns and grenades. Overwhelmed by this fire, the Germans scattered, leaving several dead and wounded men draped over the wire.

Another German force, also numbering between 30 and 40, approached the second Australian strongpoint. They did not appear to be aware of the post's existence, as they were moving to outflank the strongpoint being attacked by the other enemy party. The enfilade rifle fire of the Australians in the second post surprised the Germans and drove them off.

As soon as the enemy disappeared from sight, Sergeant Tom Buckingham and another man ventured out and conducted a reconnaissance of the ground in front of the Australian wire.[36] Two wounded Germans were discovered, one of whom subsequently died. There were also several enemy dead on and near the wire. The wounded man was brought in and evacuated. Buckingham searched the dead men closest to him, deducing from their paucity of equipment (they were carrying grenades, rifles, no bayonets and only 20 rounds of ammunition) that they were part of a raiding party rather than members of a full-blown attack.

The lifting of the fog made it impossible for Buckingham and his comrade to investigate further so they returned to the Australian lines. Sounds of groaning were heard throughout the day coming from no man's land. At dusk, the same patrol went forward and searched the dead as thoroughly as the incessant German machine-gun fire would allow. Eight enemy dead were found and searched and four additional corpses could be seen.

Over the next few days it became apparent that the raid had been coordinated with the opening of the long-expected German offensive that burst on the British forces further south. If the intention of the enemy foray had been to demoralise the 55th, it had achieved quite the opposite effect. When the battalion was relieved in the front line that evening, Lieutenant Colonel Woods found 'all men in good spirits', a situation that surprised him after such a long and strenuous period in the line.

While their morale may have been high, some of the men were worn out and tired, a point highlighted in a letter home by Sergeant William Cosgrave: 'I am still kicking around but not in the best of health … I got a mouthful of gas and it has weakened my lungs a great deal, and one catches cold very quickly and it's awfully hard to get rid of.'

This was to be the 55th's last front-line tour in Belgium. They could not have foreseen that, within three weeks, the ground they had spent the winter defending would be back in enemy hands. For now, they waited to see what their role would be in confronting the looming German juggernaut.

CHAPTER 11

Chapter 11: Villers-Bretonneux – 'altogether it was a very bad business'

The capitulation of Russia in late 1917 permitted the Germans to shift massive numbers of men from the Eastern to the Western Front. These reinforcements tilted the military balance in Germany's favour. General Erich Ludendorff who, along with Field Marshal Paul von Hindenburg, was the de facto political and military leader of Germany, realised that the American troops soon to arrive on the Western Front would completely negate this advantage. Ludendorff began planning a series of offensives to conclude the campaign on the Western Front before the arrival of the Americans became a decisive factor.

On 21 March 1918, the first and largest of these offensives, 'Operation Michael', was unleashed against the British forces south of Arras. The aim of 'Michael' was to drive a wedge between the British and French armies then wheel north-west and push the British into the sea.

German artillery and storm troopers stunned the defenders. The British Third and Fifth armies were forced into a disorderly retreat. The French civilian population fled in panic. German troops advanced across the old Somme battlefields towards Amiens, a vital rail hub and the junction of the French and British forces.

To capture Amiens the Germans needed to pass through Villers-Bretonneux, a town situated on a low plateau which provided an uninterrupted view to Amiens 15 kilometres away. Four Australian divisions in the Ypres salient were told to prepare for a move south to reinforce the depleted British units defending Villers-Bretonneux; the 5th Division was one of these.

On the day 'Operation Michael' was launched the 55th Battalion was resting near Wytschaete. Here the men received the news that, as part of the AIF's policy of replacing British officers with Australians, the highly regarded Brigadier General Clarence Hobkirk was to hand over command of the 14th Brigade to 34-year-old Brigadier General 'Cam' Stewart from Melbourne, Victoria.[1]

In the reserve areas, the unit settled into a semblance of out-of-the-trenches normality. While training and work parties kept the men occupied, their thoughts were ever on the battles raging to the south. Wild rumours circulated about the German offensive. Private Bert Bishop, on returning from leave, found 'Everyone was uneasy, upset, a strange all-up-in-the-air feeling obsessed us all.'[2]

On the evening of 23 March, confirmation came that the entire brigade was to move to an unspecified destination. Private Eddie Street had no doubt where they were going, despite the absence of official clarification:

> We knew it was the Somme for us again. Absolutely sure. Despite the catastrophe down south, the AIF were in great spirits, and morale was raising higher and higher ... We were feeling in great form, the mighty responsibility about to be thrust on us served to make us more confident.

Next morning the curtain of uncertainty was swept aside. Private Bishop related the manner in which news of the German successes were communicated:

> 'I've just come from brigade,' said our [company commander] ... We crowded in more closely. The captain's voice was calm, steady.
>
> 'To put it plainly ... the Germans have smashed through on a front of many miles. They have captured an enormous number of British troops and practically the whole of our artillery in the line of their advance. Our entire front – men, artillery, tanks, dumps, even our observation balloons, have simply been overwhelmed and captured. The enemy has advanced many miles and is still advancing. Only scattered remnants of our troops face him. He is practically marching through.
>
> There was no attempt to hide facts, it was a situation that allowed of no subterfuge. We learned of the dreadful disaster in plain, truthful terms ... As the captain's words sank in, the faces about him seemed to change. Men looked sterner, grimmer, there was no talking.
>
> 'Well, now we've got something to think about. The only news I still have to give you – and I think you're expecting it – is that the whole of the AIF, as fast as lorries and trains can handle them, are now trekking back to the Somme. We leave tonight.'[3]

The military wheels turned quickly. Orders were issued for the unit to prepare for a 'long move'. That night the battalion marched to Vierstraat and on the following day to Reninghelst. Of this latter trek the War Diarist recorded: 'The strenuous period in the line has not affected the discipline and good bearing of the men, who completed the move in perfect order, there being no stragglers. The health of the men is good.'

On 26 March, the battalion tramped to Wippenhoek; once again, no-one fell out. The next day the unit boarded a train destined for the Somme. The War Diarist wrote: 'All ranks were eager for the completion of the trip, as they knew it would bring them to the scene of the big German offensive.'

The train trip ended at Mondicourt early next morning. Such was the degree of confusion in the British High Command that Major General Hobbs was uncertain where to deploy his 8th and 14th brigades; the 15th Brigade was already engaged at Corbie. Hobbs chose to move the uncommitted brigades to an inactive sector around the villages of Louvencourt and Vauchelles-lès-Authie.[4]

After breakfast the men commenced marching, among them Private Eddie Street:

> Full packs up and we were off down the long white dusty road ... through familiar fields we wended our way. Crowds of refugees were strung along the road, sad and weary-eyed. A few smiled and an odd one shouted encouragement but the great majority seemed to be in no condition to cheer themselves hoarse ... The dust hung in the air and we were covered from head to feet with a film of it. It was a hot day and we grew thirsty ... village after village we passed and we dared not hope the next one would be our home.

The battalion reached its 'home' at Louvencourt as night fell.

Since the exact location of the Germans was unknown, the 55th's CO, Lieutenant Colonel Percy Woods, positioned a series of platoon-sized outposts around the village. He also received word that the battalion must be ready to move at short notice.

The next few days passed in a surreal manner. While some of the largest battles of the war raged, the area around Louvencourt remained tranquil, as if in the eye of a cyclone. For a week the unit rested and conducted limited training. Finally, the long-expected orders arrived for the battalion to prepare for deployment to the forward area. The actual mission of the 55th once it arrived at the front remained unclear.

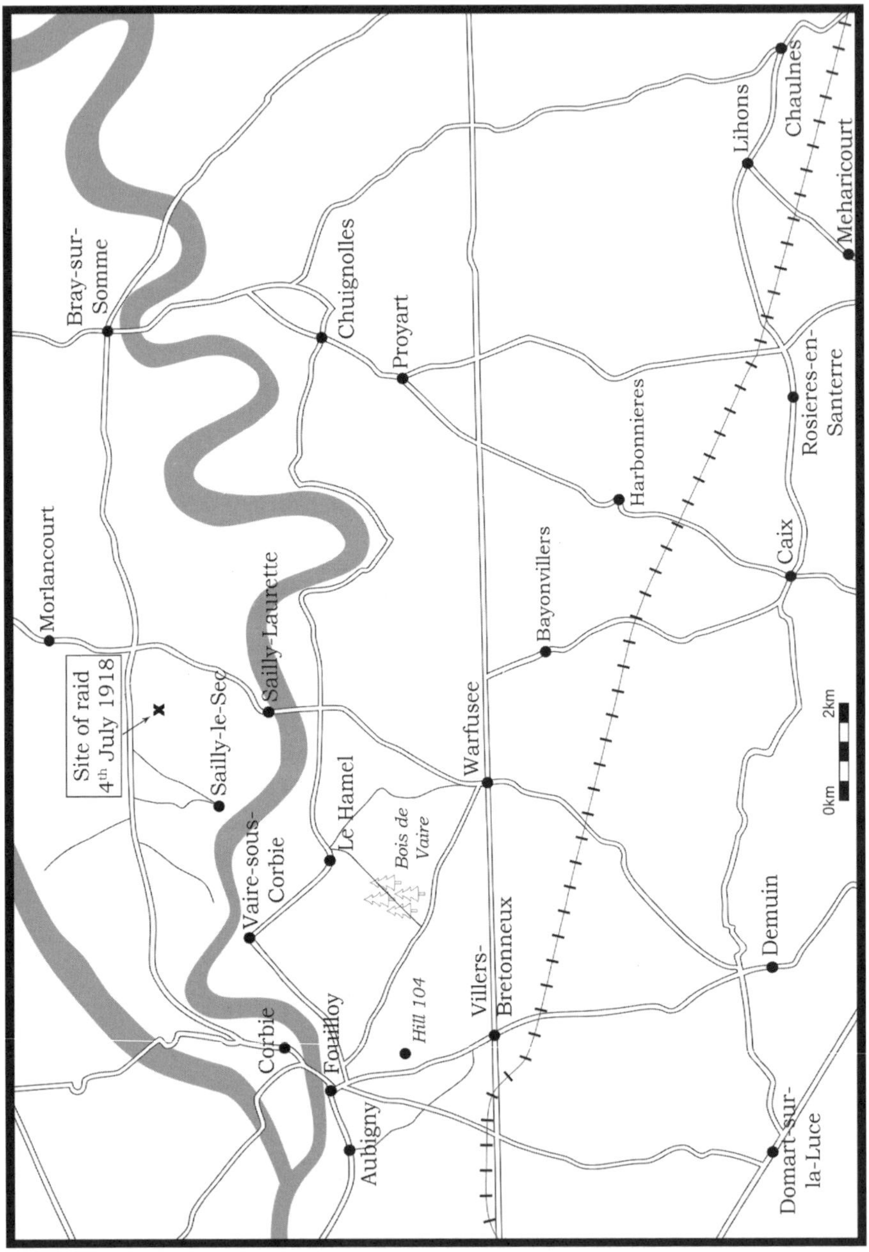

Map 14: Somme area of operations – April to August 1918.

On the morning of 5 April, the unit boarded a fleet of double-decker buses that took them as far as Vecquemont. From there they tramped, via Daours, to Aubigny. Private Street chronicled the journey:

> As we marched through the main street [of Daours] the inhabitants who still held on came out from their homes to greet us. 'Good Luck' and 'God protect you Australians'. [We then marched] across the river, and then through the deep railway cutting, along a good level road with one very slight bend and we were in Aubigny ... shells were falling in the village pretty consistently. Gas shells lobbed now and again, the fumes lurked in corners.

The men rested in the ruins of Aubigny. They could hear savage fighting as German, Australian and British troops wrestled for control of Villers-Bretonneux just over two kilometres away. Private Street wrote of the preparations for the task ahead: 'At dusk [we] loaded up in battle order ... every Lewis gunner carried extra ammo. Bandoliers of ten clips of five bullets were given to the mob ... we could take as much [ammo] as we wanted.'

At 9.00 pm, Lieutenant Colonel Woods received orders for the battalion to relieve the 3rd Dragoon Guards defending Hill 104. Situated a little over a kilometre to the north-east of Villers-Bretonneux, Hill 104 was the highest point along the plateau on which the town lay. The tactical significance of the feature was that it surpassed Villers-Bretonneux as an observation point.[5] At 11.00 pm the battalion, including Private Street, commenced moving:

> We came to the road which crossed us in the direction of Villers-Bretonneux; Fouilloy was to our left. We turned to the right up at the crossroad. From the valley we looked at the black gloomy ridge above us [Hill 104]. This place was going to take some holding.
>
> The night was black ... a halt and we rested ourselves against the bank beside the roadside – twenty yards along the road a fire burned cheerfully. Some British cavalrymen were making tea. Rain was falling and the whole show was trying. Up again and past a big chalk pit ... ahead of us some houses were burning.
>
> Shells crashed and the drizzle continued. On and on we struggled. Hours passed and then we were on top of a broad and fairly flat hill ... cavalrymen lay dead in the mud, cavalry horses strewed the ground ... here and there a rifle or a sword ... After stumbling about in the dark and drizzle for an hour ... we found the mob which we were to relieve.

The battalion discovered that little had been undertaken in the way of defensive works. Corporal George Gill was unimpressed with the arrangements: 'These Tommies appear to be a shiftless lot, there [are] only a couple of possies they seem to have; [they] just hung about with their horses.'

The cavalrymen informed the Australians that enemy troops were in the vicinity and, with dawn only a few hours away, the troops began to dig in before first light exposed them to the German gunners. Limited time permitted only shallow scrapes to be dug and these were not linked in any form of trench system. Morning broke clear and fine. However, instead of being nearby, the closest Germans were discovered to be two kilometres away in the thick woods of Bois de Vaire.

The exhausted men, still wet from the overnight rain, huddled in their pits and tried to sleep. Private Street was:

> ... just dozing off. The sudden scream of a shell. A ghastly crash. A fountain of clods and mud. The shell was the first of many. They crashed on parapet and parados ... mud, clods, crashes, blood, concussion and lacerated entrails lay sprawling in the dirt ... things were in a bad state. Archie Box was dying. [6] Three more men were dead, smashed horribly.[7] Half a dozen wounded.

As the day continued, the shelling eased and the afternoon was passed in leisurely sniping at the distant Germans. With nightfall, reports reached the unit that the Germans were massing for an attack but this turned out to be a false alarm. The rest of the night passed peacefully.

However, on each of the subsequent two nights, the battalion crept its defensive line several hundred metres closer to the Germans. The extreme darkness and periodic rain rendered these advances very difficult and compass bearings were needed to maintain alignment. At the completion of each relocation the men had to dig in, compounding their exhaustion.

The ever-changing location of the front lines confused soldiers on both sides. Several Germans drifted into Australian positions and were captured. On another occasion, a patrol from the 55th became disoriented and stumbled into a German machine-gun post; the Australians reacted first, killing one of the gunners and causing the others to flee. The patrol then beat a rapid retreat to friendly lines.

By now the battalion's exhaustion was apparent to external observers. The 14th Brigade's War Diarist noted on 8 April: '[The 55th Battalion] are to be relieved tomorrow night ... They have had a fairly thin time with the rain and wet and digging each night, and will greatly profit by a few days rest in reserve.'

On the night of 9 April, the battalion was replaced and made its way to the reserve lines. Time away from the front provided the soldiers with opportunities to prowl through Villers-Bretonneux and other abandoned villages and, as Lance Corporal Archie Winter illustrates, 'liberate' anything of value they could find:

> We in common with many others made for the village [Villers-Bretonneux] despite the fact that Jerry was sending over 8" and all types of shells, gas shells included. Of course the wine cellars were soon located and then the wilder element started to shew ... We [soon] all had new cardigan jackets and numerous other warm things to wear. It was quite a common sight to see some of the wild boys in frock coat and top hat, a lady's parasol and silk stockings. We also lived off the fat of the land – such dishes as peacock, rabbit, pheasant were quite everyday affairs.

Private Street also wrote of one item in great demand:

> We had not had a bath for a long time and were getting chatty.[8] Once again the village supplied us. Nearly every man had on women's underwear ... and a few of them, later on, when wounded and in hospital, surprised the nurses by their strange underwear. The new clothes made us feel clean again and that was all that was needed.

On 13 April, the battalion returned to the front line; it was dark and raining. In accordance with normal procedure, listening posts were established in no man's land. At around 11.00 pm a party of six Germans approached a five-man listening post. A shot rang out and the Germans started shouting. In the confusion, and believing that a raid was underway, the Lewis gunners and riflemen in the Australian trenches opened fire. The listening post was lashed with 'friendly' fire that killed Lance Corporal George McMeekin,[9] Private James Donovan[10] and Private William 'Anzac' Harding.[11] Lieutenant Arthur Duprez and the other surviving member of the listening post, Private William Affleck, stumbled back to the Australian lines and alerted their colleagues to the terrible mistake.[12] Shortly after, a party returned to the site of the listening post, finding the bodies of their mates as well as three dead and two wounded Germans.

Another 'friendly' fire incident occurred a week later. An eight-man patrol, under command of Warrant Officer Class 2 Henry Anson, became disoriented and found itself in the middle of enemy work parties. When challenged by an enemy sentry, the Australians threw grenades then retreated with upwards of 50 Germans in pursuit. Thinking only to avoid death or capture, the patrol attempted to make its way back through the Australian wire from an unexpected direction. In so doing they were fired on by several of the battalion's Lewis guns. Private James Burns was killed and two men were wounded. [13] One of the wounded, Private Roy Christie, succumbed to his wounds the following day.[14]

As April progressed, the enemy attitude became more aggressive and all signs pointed to an impending resumption of the German offensive. Hostile artillery fire intensified and in two days the unit lost two officers (Lieutenant Joe Collins[15] and his great friend, Lieutenant 'Big Bill' Clark) and eight soldiers to shelling.[16]

The battalion rotated companies between the front and support lines. On one occasion Lance Corporal Winter and his mates took over a 'possie' the previous occupants were very happy to leave. Winter soon discovered the reason:

> It appears that A Company had a man killed there and had a white cross put up over his grave right in front of our possie. Well Jerry took this as a zero for his guns and consequently we caught a shell. [It] landed right on top of our two stretcher-bearers who were killed on the spot. My pal Fred Carpenter was very badly shaken and I was buried up to the chin.[17] Altogether it was a very bad business as one of the bearers killed was Ned Parryman the pride and joy of the platoon.

Private Street described this tour in the forward areas:

> The wheat fields in front were turning a darker green and the wheat was growing apace ... Several days and nights passed by with the usual work going on. Front line shelling eased up somewhat but reserves and supports received rapidly intensifying shelling. [German] Machine gun fire was maintained. The platoons were feeling in good heart, and soon many of our patrols were roaming in the darkness of the night, up and down No Man's Land, up to the wire Fritz was putting out, and occasionally raiding an outpost.

According to the 14th Brigade's War Diarist, it was not always strict military necessity that influenced the battalion's patrol program: 'There is a great rivalry between the two line battalions [53rd and 55th] and they are out to secure as many prisoners as possible and establish a record. No Man's Land is without doubt completely dominated by our patrols.'[18]

As night fell on 20 April, the unit was relieved by elements of the British Rifle Brigade and the 54th Battalion. The men were weary and needed a protracted break. Lance Corporal Winter described his feelings:

> We were in and out of the line during the greater part of April. Would be brought [out] for 24 hours and then in again. The Colonel [Percy Woods] was very wild at such treatment but the 'brass hats' heeded not his complaints. [One time] the Battalion came out of the line dead beat and the next night we were sent back in again. Colonel Woods kicked up a great row over it but our Brigadier General (whom I may state being a Victorian didn't have much love for New South Welshmen) sent us back in again.

The battalion passed the next few days peacefully. Then, at 5.00 pm on 23 April, the Germans pounded the entire region around Villers-Bretonneux with artillery. The battalion's transport, quartermaster and administrative personnel billeted in Aubigny were subjected to a mustard gas bombardment of such intensity that the ground was running with the oily, toxic liquid. The gas attack inflicted 90 casualties on the battalion, none of them fatal.[19] Nevertheless the burns produced by the substance were so horrific that many of the casualties resulting from contact with gas required repatriation to Australia. One of those exposed to the lethal vapour was the recently promoted Corporal Winter:

> The air was putrid with gas … Jerry sent more gas over and gassed nearly all men in details. The first indication I had [of being gassed] was a dreadful retching and going blind. I was led to a Dressing Station at Daours. After being there some time in agony we moved on. By this time my eyesight had completely gone. After a pretty rough trip I eventually reached No 5 Tommy General Hospital (Rouen).

Everyone was certain that this massive bombardment was a prelude to a heavy infantry attack and the anticipated assault came the next morning (24 April). The Germans seized Villers-Bretonneux but the stout defence of the 14th Brigade, particularly the 54th Battalion, prevented the enemy taking Hill 104. Later that night, the 13th and 15th brigades recaptured Villers-Bretonneux in a brilliant counter-attack, halting the German move on Amiens. Once again, the men of the 55th were fated to play little part in the momentous actions happening around them, their involvement limited to providing unwilling targets for German artillerymen.

On 26 April the unit returned to the forward lines, relieving the battle-weary 54th Battalion. The German artillery continued to pummel the Australians. On one occasion, Corporal Gill was caught in the open laying a telephone cable:

> I had a fairly rough time one night. I helped lay a line to the forward company and coming back we got a fast furious bombardment with whiz bangs lasting about quarter of an hour. Was very pleased when it was over. We soon started off again and had only got about 200 yards when 5.9s [inch guns] started falling around us. I immediately dropped into a shell-hole the OC following none too soon. One shell fell about three yards from us.

A shell claimed the life of the Officer Commanding D Company, Captain John Marshall, the same blast killing Private Charles Partridge[20] and Private George Hughes.[21]

After five days in the forward areas, the troops were relieved. Since the battalion had spent April on the defensive, casualties for the month were high: two officers and 28 soldiers had been killed; a further two officers and 157 men were wounded, including the 90 gas victims. The battalion had 31 officers and 632 soldiers in the field, well down on the establishment strength of over 1000 men.

The unit returned to the front line on the evening of 4 May. To prevent the Germans solidifying their defensive lines a regime of constant patrolling and raiding was instigated. On one occasion, a patrol under the now commissioned Lieutenant 'Dally' Neville rushed an enemy listening post, the Australians killing the three Germans occupying the position. As Neville and his men searched the bodies, another two enemy troops attacked. In the ensuing struggle both Germans were killed. Neville's patrol returned to the Australian lines without suffering any casualties.

Private Street took part in another of Lieutenant Neville's night-time escapades:

> Fritz had a machine gun playing right onto our parapet ... I heard about a raiding party bent on the capture of the machine gun. I walked to the post where the raiding party was to leave from and joined up. There were fourteen of us and we had to take the machine gun and the remainder of the outpost. I could see the machine gun spitting fire as it traversed

> our line, so we set out in a beeline for it. We had gone a considerable distance when a sudden crack was heard. Our bodies disappeared in the wheat as a flare soared into the sky and burst into a brilliant white flame. It was a very powerful flare and lit up No Man's Land over a great area. The flare was falling in our direction; to move was to invite a burst of machine gun fire. We kept a tight hold of ourselves and flattened into the wheat. The flare fell amongst us sizzling and roaring. Fortunately it touched none of us and was soon burnt out. We crept forward until we were within striking distance of the flare king.[22] One of the chaps, who was a roo hunter in pre-war days, crept forward alone and pounced onto the unsuspecting flare king. Our man hit hard with his fist and, as there was a Mills bomb in it, the impact sent the firer of the flares to sleep. On we crept, the machine gun spitting bullets over our heads into our outposts. There was a clear space of twenty yards in front of the Hun outpost. We stopped in the wheat at the edge and waited for the machine gunner to take a spell. We lay low and waited breathlessly in the wheat, our eyes fixed on the outpost. We heard Huns talking, quite unconscious of their danger. At last the gunner left his post and walked along the trench. A low spoken order and we made a silent rush at the post. I was pretty fast in those days and reached the trench first, firing my squirt [hand gun] anywhere. The rest of the mob were in the trench now but we got no fighting. Fritz surrendered most willingly, and the stunt was over in a few seconds. A few Germans had been killed in that rush. We gathered the live ones into a heap, gave them the machine gun to carry and marched home. We collected the flare king; he was just recovering from his stoush up. He looked a most bewildered man and could not grasp what had happened. We reached home and handed over the booty to be taken to the rear.[23]

For the rest of May a week in the firing line was followed by a week in reserves; the stress was unrelenting. Eddie Street captured the hardened professionalism that months of sustained combat had by now engendered in the battalion:

> War became a business to be completed in an as efficient and business-like manner as possible. We steeled ourselves to see the horrible job through, and tried not to think about being knocked ... Heaven or Hell, it still remained to be seen which, was always very close.

Every evening work parties were despatched to the forward areas, carrying stores or to prepare defences. Private John 'Jack' Walker[24] wrote to a female friend of an 'accident' that befell him during such a work party:

> One night while in No Man's Land erecting barbed wire entanglements I tore the rear out of my trousers. The next day Fritz sent over a great quantity of gas which fairly soaked into the ground. Later on I was foolish enough to sit on this ground without any seat in my trouser with the result that the part of my anatomy which touched is now covered in very painful blisters.

On 31 May elements of the 4th Division relieved the battalion. After such a long stretch of protracted fighting, the 55th Battalion, along with the rest of the 5th Division, had been ordered into Corps Reserve to rest. The men, undisturbed apart from occasional training or work details, spent the next three weeks at leisure. They had little to do except enjoy the long summer days. Private Street wrote of the sojourn:

> The mob came out ... and dug-in in the country east of Pont Noyelles and Querrieu ... it was an ideal camp. Tents and home-made humpies filled with good bedding were pitched here and there in the shade of the forest. Sunlight filtered through the branches ... beside the escarpment lay the lake, warm and bright in the sun ... The water [in the Ancre River] was deep and ideal for swimming and punting. Swimming and sunbaking filled our day. No shells fell near enough to disturb our enjoyment.

While they swam, played cricket and went 'fishing' using hand grenades or mortar bombs, the 2nd Division advanced along Morlancourt Ridge, the rising ground between the Somme and Ancre rivers. The ridge was on the opposite side of the Somme River where the 55th had operated over the previous two months. In mid-June, the 8th Brigade (5th Division) took over from the 7th Brigade (2nd Division) on the Morlancourt Ridge. The 14th Brigade was put on notice that they would in turn relieve the 8th Brigade before the end of the month.

On 26 June the battalion's repose ended and the unit moved to a new sector of the front, one kilometre south-east of Morlancourt village. The position was a defensive nightmare; there was no wire and the front line was a series of isolated outposts overlooked by the enemy from high ground south of the Somme River. The weakened battalion was stretched over a two-kilometre frontage.

The enemy was very active. German machine-guns swept the unit's front throughout the night. The constant enemy sniping aggravated the cooped-up soldiers. On one occasion, Private Joe Coughlan volunteered to kill a particularly annoying sharpshooter and, after grabbing a few grenades, he raised his head to locate the flash from the sniper's rifle and was promptly shot between the eyes.[25]

On another occasion, a German rifle grenade struck an outpost, wounding Lance Corporal Henry Reidy[26] and Private Henry Stewart.[27] Stretcher-bearers came for the wounded men and managed to get Reidy back to the dressing station before he died. The bearers carrying Stewart came under sniper fire and had to leave him in the open; he succumbed to his wounds while on the stretcher.

Each night, under a blanket of darkness, the unit sprang into frenetic activity despite the enemy harassment. Those men not on patrols toiled to connect the forward outposts and construct communications trenches. Adding to the sense of urgency were rumours that Lieutenant General John Monash, the recently appointed commander of the Australian Corps to which the 5th Division was attached, was planning a surprise for the Germans. The 55th Battalion was to launch a significant raid as part of the diversionary actions associated with this surprise.

Chapter 12:
July–August 1918 – 'lost a few good sterling lads'

After the failure of their thrust towards Amiens, the Germans were left holding a salient near the Somme River incorporating the village of Le Hamel and its adjacent woods. Uncertain about future enemy intentions, the Fourth Army's General Henry Rawlinson and Lieutenant General John Monash, commander of the Australian Corps, planned a limited operation using the 4th Division to eliminate this salient.

By removing the enemy incursion, Monash hoped to add depth to the defences around Hill 104 and reduce the danger posed by German artillery to the massive Allied counter-offensive being planned for early August. Attached to the Australians for familiarisation training were a number of American troops and, in an adroit political stroke, Monash scheduled the assault to coincide with US Independence Day — 4 July 1918.

The 5th Division was tasked with assisting the 4th Division's attack by launching a series of feints along Morlancourt Ridge designed to persuade the enemy that the major Australian operation was occurring along this feature. The 55th Battalion was to make the 14th Brigade's contribution to the deception operations. The unit was given two tasks: raiding two lines of enemy trenches and undertaking a 'Chinese Attack'.

It was left to the 55th's CO, Lieutenant Colonel Percy Woods, to make the necessary arrangements for the attack. On 1 July, Brigadier General 'Cam' Stewart, commander of the 14th Brigade, approved Woods' concept for the raid, directing that the raiding party was to consist of 200 men and six officers. Woods decided that Captain Ken Wyllie, who had conducted the reconnaissance on which the plan was based, should be in charge of the operation. The raiding party would consist of one officer and 90 men from A Company, two officers and 80 men from B Company and a platoon (two officers and 30 men) from C Company. Zero Hour was set for 3.10 am, the same time as the start of the 4th Division's offensive.

CHAPTER 12

At this time B and D companies, and two platoons of A Company, were in the front line, with C Company and the remaining platoon of A Company in support and reserve. Instructions were issued for C Company to relieve B Company in the front line on the night of 2 July, with two companies of the 56th Battalion moving into the reserve and support positions vacated by A and C companies.[1] On relief, A and B companies, with the designated platoon from C Company, withdrew to the 14th Brigade's reserve area to prepare for the undertaking.

As the objectives of the raid, Lieutenant Colonel Woods selected two enemy trenches that lay parallel to the Australian lines. The first trench, some 200 metres away, was unprotected by wire as the Germans had been unable to move work parties out into no man's land. The second trench, a further 200 metres behind the first, was protected by low wire entanglements around four metres wide.

Captain Wyllie planned the raid to be conducted in two waves. The first was to be led by Lieutenant George Donaldson[2] and comprised 2nd Lieutenant Bill Piddington and 80 men.[3] The second wave was made up of 100 men under the command of Lieutenant Luther Chadwick assisted by Lieutenant Louis Stafford.[4]

The raid plan developed by Captain Wyllie called for the first wave to advance in extended line and 'take the first objective at the point of the bayonet and kill or capture the occupants of the trench'. The second wave was then to pass through them and make their way to the second trench. The leaders were to find gaps in the wire, push their sections through these and into the enemy trench. The sections on the flank were to construct bombing blocks while the remaining men were directed to 'kill or capture the garrison and remove or destroy machine guns and other material'.

The decision on when to withdraw from the second trench was left to the discretion of the two officers in charge of the wave. Captain Wyllie gave strict instructions that: 'Before giving the order to withdraw, the … leaders must satisfy themselves that the task has been completed and that no members of the party, dead or alive, are left in the trench.'

The raiders of the first wave were to remain in the first trench. It was 'only after all dugouts and dead searched, and all guns and material removed or destroyed' that elements of the first wave were permitted to withdraw. Captain Wyllie was emphatic, however, that Lieutenant Donaldson, 2nd Lieutenant Piddington and all of the Lewis gunners and bombers in the first wave were to remain in the first trench until the withdrawing members of the second wave had passed back through them. Only then were they to retire to the Australian lines.

The raid's fire support plan called for six light trench mortars to engage the first trench at Zero, while four batteries of 18-pounder field guns shelled the second trench. Three minutes after Zero, the mortars and 18-pounders were to lift beyond the second trench, suppressing other targets to the rear and flanks and 'boxing-in' the German garrison in this line. Ten per cent of all shells fired by the artillery were to be smoke. Two medium trench mortars were tasked with cutting the wire in front of the second trench. It was hoped that this trench mortar fire, combined with the opening artillery barrage, would blast away most of the wire obstacles confronting the raiders.

In addition to the trench mortars and artillery, several Vickers guns from the 5th Australian Machine Gun Battalion were to fire on targets to the rear and flanks of the operation. As a last measure of support, at Zero Hour all of the Australian garrisons in the front line near the 55th's sector were to fire Lewis guns and rifle grenades on the German front-line defences in front of their respective positions.

The day before the raid was spent preparing for the assault. The weather was fine and sunny. During the afternoon, the officers explained the raid plan, making sure that every participant was conversant with the mission and his part in it. While the raiders were making their final arrangements, work parties from C Company and the 56th Battalion worked through the day and into the night to improve the front line and its approaches to facilitate the movement of such a large raiding force. The Germans observed these work parties and maintained a fusillade of machine-gun fire, wounding two men from C Company.

Under cover of darkness, a party under Lieutenant Norm Wilson went into no man's land to lay the tapes along the 'hopping off' line.[5] By 1.30 am the tapes were in position and pegged down.

By 2.00 am on 4 July all working parties were clear of the front line, returning to the support lines via a different route to that being used by the raiders making their way to the firing line. The RMO, Captain Norm Mackay, also travelled forward and established an advanced dressing station in a front-line dugout.[6]

When all was ready, Captain Wyllie gave the signal and the raiding party 'hopped the bags' and moved silently across no man's land to the tapes. The battalion's after-action report stated: 'Every man knew his position and every man was on it.' The passage to the tape-line went without incident and, by 2.50 am, Wyllie was able to inform Battalion Headquarters that everything was ready. Judging by the 'normal' levels of enemy machine-gun fire and the absence of artillery and *minenwerfer* fire, the assembly had been completed without being observed by the Germans.

At precisely Zero Hour, the supporting artillery and trench mortar fire came thundering down, synchronising with the barrage for the 4th Division attack. Corporal Herbert Harris witnessed the start of the bombardment: 'It was a great sight, the country was one sheet of flame and I would not like to have been on Fritzes side of it.'

Private Bert Bishop was not taking part in the operation but observed the chaotic opening minutes of the raid:

> The air filled with screeching shells and whistling bullets, the swish of mortars and bombs. A horror filled us as we saw what was happening. The barrage, instead of falling on the German lines, was falling anywhere at all. It fell on our own trench, some of it even behind us, while the poor wretches in no-man's-land caught the worst of it.[7]

In the after-action report, Lieutenant Colonel Woods claimed that, within a minute of the commencement of the supporting bombardment, *minenwerfer* fire 'came down close in front of our trenches' where the raiding force was lying unprotected. [8] Most of the men asserted that it was not German fire that lashed them but short shooting by their own medium trench mortars and artillery.[9]

As their covering barrage was not to lift for three minutes, the soldiers had no option but to endure the metal storm. Casualties caused by this fire began to occur all along the tapes. Shell fragments riddled Private Walter Allen; still conscious, he was carried back to the front line.[10] Here, Allen was placed on a stretcher and was on his way rearwards when a shell burst in front of the stretcher party, killing Allen and wounding a stretcher-bearer.

Corporal Ernest Hamilton was blown to pieces when a trench mortar landed on his back.[11] The same explosion wounded Lance Corporal William Hart in the head; he died while stretcher-bearers tended his wounds.[12] Shell fragments killed Private William Harvey as he lay on the tapes, although his body was not found until the following day.

To those caught in this inferno it must have seemed as if time was standing still. Finally the three minutes passed and the supporting fire lifted. The first wave got up and ran towards the first German line, followed by the second. The troops managed to maintain their formation in the mad dash across no man's land, although German machine-gun fire and occasional 'shorts' from the Australian trench mortars and artillery wounded three of their number.

Maintaining direction was problematic. The excessive use of smoke shells and absence of a breeze reduced vision to just a few metres; the air was so acrid that the raiders could only breathe with difficulty. With Lieutenant Donaldson leading the way, the first wave tumbled into the first trench. The raiders encountered two small enemy posts, one of which consisted of a machine-gun crew. Corporal Alec Penny and his section rushed the machine-gun.[13] Penny killed the gunner and a second member of the gun's crew with his bayonet. He then chased another German up the trench, also killing him.

Both enemy posts were quickly subdued, their garrisons killed with the exception of one German who, along with the aforementioned machine-gun, was captured. The trench was searched and found to be clear of Germans; they were manning their front line with very few men.

As planned, the second wave 'leap-frogged' through the first wave and headed towards the next trench. Artillery and trench mortar 'shorts' continued to land in no man's land among the Australians, causing additional casualties. The barbed-wire entanglements were found to be largely intact, possibly because the medium trench mortars and artillery tasked with wire-cutting duties were firing 'short'. Nonetheless, the wire did not prove a difficult obstacle and most of the attackers were able to find a way through and move on to the second trench.

Here they met considerable resistance from a strong German garrison. On the left flank Lieutenant Stafford, standing on top of the parapet, confronted a party of up to 20 Germans. Stafford used his revolver to shoot the nearest German but was in turn fatally shot in the chest by an enemy officer from a range of just two metres. Stafford's batman, Private Clarence Swift, shot and killed this officer.[14] Swift, with two or three men, then jumped into the trench and worked along it, accounting for nine Germans with bayonet and grenade.

On the right, Lieutenant Chadwick's party encountered a machine-gun post. The Germans resisted and a brutal hand-to-hand fight ensued. Chadwick and the men nearest him killed the gun's entire crew.[15] Elsewhere, Private John Hannah was the first member of his section to reach the enemy trench.[16] The nearest Germans to him comprised a post of six men throwing grenades at the raiders. Hannah approached the post unnoticed and flung himself at the grenadiers, killing one of the Germans with his bayonet and shooting another. Surprised by the sudden assault, the remaining Germans fled back down their trench pursued by Hannah, who killed two more with the bayonet, and shot a fifth. The last member of the enemy post was taken prisoner.

The enemy trench system was deep and well constructed. Four dugouts were discovered descending several metres underground; the raiders did considerable damage to these by hurling mortar bombs into them.[17] The raiders also disabled a heavy machine-gun they were unable to remove from its concrete platform.

The fighting in the German trench was confused and thick smoke compounded the bedlam of the swirling combat. As the raiding party fanned out in the trench system, the Germans launched a counter-attack. A fierce bayonet battle ensued.

Private Samuel 'Sam' Jacobs[18] killed three Germans with his bayonet; Private Isacchar Higgs used his bayonet to account for two more.[19] Eventually the raiders halted the enemy counter-attack and commenced 'mopping-up'.

Lieutenant Chadwick, realising that the mission was accomplished and the pause in the fighting was likely to be fleeting, gave the order to withdraw. He lifted the captured machine-gun to his shoulder and walked back across no man's land.

Reduced in numbers, the remaining able-bodied Australians in the second wave had no option but to leave the bodies of their fallen mates (Lieutenant Stafford and Private Charles Sinclair) in the German trenches and direct their energies to carrying the wounded.[20] Private Fred Doherty[21] and Private Tom James distinguished themselves by carrying a wounded comrade several hundred metres to the Australian lines.[22] Indeed, on returning to the relative safety of the friendly trenches, Doherty noticed another wounded man in no man's land and brought him to safety.

The retirement was affected without interference from enemy machine-gun fire. Although it was now daylight the smokescreen concealed the Australian movements and the raiders had destroyed all of the machine-gun posts in the immediate vicinity. However, the smoke could do nothing to mitigate the massive German bombardment that descended on the Australian front line and no man's land about nine minutes after Zero. This shellfire caught the raiders as they returned, resulting in several more casualties.

As the survivors of the second wave passed back through the first trench, Lieutenant Donaldson organised men from the first wave to assist in evacuating the wounded. Once the last members of the second wave had made their way to the Australian lines, Donaldson ordered the retirement of the remaining members of the first wave; during this undertaking, a German shell killed 2nd Lieutenant Piddington.

The retaliatory barrage continued for a further two hours; Corporal George Gill witnessed the enormous strain this placed on the battalion's Signal Section:

> We had a fair amount of trouble with the [signal] lines and the signals office as the relay was blown out, a shell landed on the door way, luckily no-one was hurt. One runner was taken away 'shell shock'. They all had a great shaking up … At 6 a.m. I found them all in a pityable state … The Sgt was in a bad state through the shaking he had got and putting in so much work on the lines.[23]

Private Theodore Foret, a stretcher-bearer, was not a member of the raiding party but was part of the front-line garrison.[24] Much of the German bombardment came down in Foret's stretch of the line and two of his comrades occupying a listening post just 75 metres from the German lines were wounded. In broad daylight, and despite being fired on by enemy riflemen, Foret and another man crawled to their assistance, moving the wounded men to safety.

Concurrently with the raid, Lieutenant William Campbell and five men were tasked to assist the deception operations by launching a 'Chinese Attack'.[25] This was to occur several hundred metres south of where the raid was to take place. Lieutenant Campbell and his party placed 15 papier-mâché figures in no man's land. At Zero Hour, and for the next 35 minutes, these figures were raised and lowered. The operation acted as an excellent diversion, drawing considerable attention from German machine-guns and *minenwerfers*. Some of the dummies were practically destroyed by enemy fire.

After the cessation of the German barrage, those members of the battalion not involved in operations found themselves with a 'dress circle' view of the 4th Division attack taking place in the valley below them on their right. Private Bishop was one witness:

> Fixed up a possie ... which gave us a good view of the hollow where the attack was to be made ... As the show opened it seemed we were up in the gods at a theatre, the battle was on the slope below us ... The barrage lifted to well behind the German trenches. The smoke cleared, and we watched tanks and khaki-clad figures advancing slowly and steadily.[26]

The War Diarist related: 'The tanks engaged were particularly conspicuous and could be seen strongly resembling huge beetles moving in front of the infantry, doing excellent work.'

After the raid, the surviving members of A and B companies made their way to the brigade's support lines. That night, C and D companies joined them after they were relieved in the front line. Over the next three nights, parties were sent out into no man's land to retrieve the unit's wounded and dead.

The raid had been skilfully undertaken and achieved all its objectives. Despite the travails endured by the attackers, the Australians claimed 45 enemy dead, captured three prisoners (all of whom were wounded) and seized two light machine-guns. More importantly, given the tremendous volume of artillery and machine-gun fire put down by the Germans in the sector, it is likely that some of the enemy leadership had fallen for the feint and believed the main Australian attack was occurring along Morlancourt Ridge. The raiders' losses were heavy — 19 killed and 64 wounded. Almost half the assault party were casualties.

Writing after the operation, Lieutenant Colonel Woods was in no doubt that the operation was worth the cost the battalion had paid: 'Considering the nature of the operation and the success obtained in drawing enemy fire, I do not consider the casualties excessive. Every effort was made to reduce shell casualties.' Brigadier General Stewart, at least privately, was in two minds: 'People seem to be quite satisfied with our part of the show yesterday although there is not much glowing in it.'

The tiny enemy garrison encountered in the first trench, and the rapidity of the German artillery and infantry response to the raid, gave Stewart reason to ponder the pre-operation security: 'One prisoner states that he knew the attack was coming as they had been told by a deserter on the night 2/3rd. This was the very night the 53rd lost a man on patrol and the circumstances of finding the Bosch quite prepared are very suspicious.'

Like their brigadier, the troops appeared to have mixed feelings about the raid. Lieutenant Lance Horniman wrote in a letter:

> Have just come out of the line. A short distance for a few days and at last big things are afoot. You have probably read about American Independence Day being celebrated here by a 'stunt'. We have just come back from it ... Lost a few good sterling lads but that is all in the game.[27]

Private Bishop was less upbeat, considering that the operation 'Was all worked out beautifully beforehand, and that's where the beauty stayed.'[28]

Charles Bean partly attributes the rather benign response by German artillery to the 4th Division attack to the feints undertaken by the 55th Battalion and the 15th Brigade. Undeniably, the 55th's raid and 'Chinese Attack' attracted significant German attention. Given the success of the 'Chinese Attack', one cannot help but speculate whether the raid was needed, and whether the more pervasive use of papier-mâché figures, combined with some artillery demonstrations, could have achieved the same outcome as flesh and blood.

The period following the raid was quiet. The weather was fine and hot and this assisted in bringing the Spanish Influenza epidemic sweeping through the battalion under control.[29] With the exception of several large work parties, the unit rested. However, they were unable to entirely relax; the occasional shell landed among the troops and a stray bullet killed Private Ernest Burnaby.[30]

The 55th's time out of the trenches was short-lived. On 10 July the unit was back in the forward areas, accompanied by five men from the US Army attached to gain front-line experience. Private Bishop noted that there was plenty of front-line experience to be had: 'Things very lively. Fritz trying to raid us every night and bombed our post one morning.'[31]

Constant sniping and the occasional barrage reminded every man that he was at war, but these activities resulted in only minimal casualties. Heavy rain and thunderstorms ensured that the Americans became familiar with another characteristic of the Western Front — mud. Lieutenant Horniman wrote of this rotation in the firing line:

> Trench life can grow a trifle monotonous: it's a mix up of three things for the most part, shells and their sundry missiles, rations and sleep. If a shell manages to interfere with the enjoyment of the other two sides of trench life then things are 'no bon'.
>
> During the day we get what sleep we can, occasionally when things are fairly quiet during the day one is able to sleep fairly well.
>
> A few have to go out in No Man's Land on patrol every night as well. It is good sport crawling about in the dark.

On the evening of 17 July, the battalion was relieved and moved to the support lines. Their time here was to be abbreviated — the next night they returned to the front line to replace the 56th Battalion. The short turnaround angered the tired men and the relief was sloppy, extending several hours longer than was usual.

It was to be another mundane stretch in the trenches. Patrolling was very difficult due to the enemy habit of firing machine-guns throughout the night. Bert Bishop wrote of the period:

> Our battalion was at very low strength. Patrols met almost nightly. Bombardments never ceased. At night every man 'stood to' till dawn. It worked out at each man holding twelve to fifteen yards of the western front. On the lower slopes to the left were the ghostly ruins of Albert, nearer was the rubbish heap that had once been Mericourt.[32]

Despite the fact that many troops had not been away from shellfire for over six months, spirits remained high, dampened briefly by the news that Lieutenant 'Dally' Neville had been seriously wounded while on secondment to III Army Corps. Neville's patrol had somehow clashed with a 'friendly' patrol and in the ensuing melee six men had been killed and Neville and others wounded.

On 24 July the 55th was relieved, moving to the reserve lines. A week later, the diggers marched to Bonnay, where trucks carried them to Frémont, well to the rear.[33] The next two days were spent at leisure, the 14th Brigade's War Diarist observing: '[The] rest [was] much appreciated by the men who enjoyed being in towns which afforded the opportunity to purchase comforts beyond those provided by the Canteens.'

The battalion relished the break, for the men knew that their time away from the fighting was likely to be fleeting.

CHAPTER 13

August 1918 – 'an anti-climax'

Once it was apparent that the German Spring Offensive had been halted, Generalissimo Ferdinand Foch, Commander-in-Chief of the Allied armies, considered that it was time to return to the offensive. By now the Allied armies had reorganised and been reinforced after their defeats in March and April. Men and materiel were pouring in from America. Instead of launching a massive assault on a single point, Foch decided to undertake a series of limited attacks against German salients at St Mihiel, Château-Thierry and Amiens. If these attacks proved successful, a general offensive would follow.

For his part, Field Marshal Douglas Haig already had plans in place for an operation near Amiens. Haig had previously directed General Henry Rawlinson, in command of the Fourth Army now astride the Somme, to prepare a blueprint for an attack on the salient. Rawlinson, in consultation with his Canadian, Australian and British corps commanders, had devised a scheme for a combined arms assault on the tank-friendly terrain south of the Somme River.

On receiving the go-ahead from Generalissimo Foch and Field Marshal Haig to conduct the operation, General Rawlinson was allocated more than ten Australian, Canadian, British and French divisions, 600 tanks, and over 2000 artillery pieces. The Australian Corps and Canadian Corps would spearhead the attack when it was launched on 8 August. For the first time on the Western Front, all five Australian divisions were placed in the Australian Corps under the leadership of Lieutenant General John Monash.

For the men of the 55th Battalion, the impending offensive meant that their time away from the front would be brief. At dusk on 2 August the battalion, now numbering only 27 officers and 417 other ranks, was again on the move, this time marching to Longpré les Amiens, a settlement on the outskirts of Amiens. To preserve the operational security of the coming offensive, Lieutenant General Monash had directed that all movement of troops and materiel be conducted at night. Private Bishop described the frustration created by Monash's decision:

> We marched, full pack up back towards the line. It was a very dark night. Leaving the road we struck across fields to the left. The march became anything but orderly. Stop-go took over, velvet blackness enveloped us. It was apparent that our guides had got themselves and our battalion just plainly lost. In full marching order we kept warm, even hot, as we trudged along but during the stops we got cold and miserable and cranky … We plodded on and on. A long halt let us know we were again lost. On again, stop again.[1]

The men spent the next two days catching up on sleep and it was not until nightfall on 4 August that the battalion was ordered to move again — this time to Bussy.[2] The march was unpleasant due to the massive number of vehicles also using the roads. On arrival at their destination, the troops bivouacked in holes and dugouts on the western side of the Hallue River. The weather closed in, and 5 August was punctuated by heavy rain. The unit relaxed, for the strictures against activities during daylight hours remained in place.

The next day the veil of secrecy began to be lifted on the forthcoming operations. The battalion's senior officers received detailed briefings and planning commenced; for reasons of security, the troops were not yet brought into the confidence of their commanders. Still it was no secret that something big was afoot — the unit had been informed to prepare for 'mobile operations' and the nucleus had departed several days earlier.

The morning of 7 August dawned clear and the muddy ground began to dry. The day was spent briefing the soldiers on their roles in the offensive that would commence the following morning. Private Bishop wrote of the briefing he received:

> Our company assembled by the river bank. We comfortably sprawled about in the long grass, eagerly awaiting to learn what was in store for us. Our company [commander] … gave us particulars concerning the overall operation. The attack would commence at dawn the following morning. All five Australian divisions and a smaller force of English and Canadian troops would spearhead the advance. We would have Australian artillery with us, Australian air force above us. And then he said words that electrified us, words which had never before appeared in battle orders.
>
> 'Everything is practically in position now for what we hope will be a huge advance. There is no fixed objective, we go as fast and as far as we can.'[3]

Lieutenant General Monash's plan required the 2nd and 3rd divisions to advance to the 'Green' Line, a distance of three kilometres that included the capture of the village of Warfusée.[4] The 4th and 5th divisions were then to pass through these two divisions and move six kilometres to the 'Red' Line. The exploitation phase of the operation called for the 4th and 5th divisions to advance a further one and a half kilometres, including capturing the village of Harbonnières.

Major General Talbot Hobbs, commanding the 5th Division, selected the 8th and 15th brigades to lead his division's attack. The 14th Brigade was placed in reserve, travelling to the rear of the other two brigades. The 14th was to cross the start line 15 minutes after Zero Hour, set for 4.20 am, with the 55th Battalion on the left, the 56th Battalion on the right, and the remaining two battalions in support. Hobbs expected a German counter-stroke and his orders reflected a desire to keep the potential employment of his division as flexible as possible. Nevertheless, the soldiers were informed that, in all likelihood, at some stage they would pass through their sister brigades to continue pressing home the offensive.

Private Bert Bishop and his mates spent the rest of the day:

> Lounging about, loafing in the warm sunshine, getting our gear in order. Men talked, argued and discussed the morrow's battle. Twilight deepened into darkness. We assembled in battle order to hear our final detailed orders … We were silent and thoughtful.[5]

Each man was going into the coming battle heavily laden: 170 rounds of small arms ammunition, two grenades, two full water bottles, three days' rations, a blanket, waterproof sheet and greatcoat.

At 1.45 am on 8 August the men left their bivouacs and made their way several kilometres to the battalion's assembly area situated on the long, gradual slope that runs from Daours up to Villers-Bretonneux. Once there, they lay down in the dewy grass, waiting for Zero Hour. The War Diarist recorded: 'Everyone keyed up.' Private Bishop recounted the final minutes before Zero Hour:

> Except for the [artillery] salvos to give help to the tanks [the noise of the artillery was intended to drown out the noise of the advancing armour], there was a subdued silence. Troops, tanks, wagons, horses were everywhere. Occasionally some low-voiced Australian cursing and heaving could be heard as a vehicle or wagon was caught in a shell-hole. The air was charged with suppressed low humming and excitement.
>
> A chill breath of air whispered across the fields. Dawn was at hand. Men rose, stretched themselves. 'Prepare to move', the words came, were passed along. No-one spoke, we were too busy thinking.

> In a few moments we would be in a battle which would sway history. We knew it could mean the winning or losing of the war.
>
> From in front, from all around us, countless guns suddenly fired. Sheets and flashes of flame tore the dawn to shreds.
>
> Everything leapt into roaring, screaming, throbbing life.[6]

Private Eddie Street witnessed the opening barrage:

> Suddenly, reverberating through the cool morning air, crashing into the sky and echoing back, came the sound of a discharge from a mighty gun. As the huge shell screamed forward, the whole of the artillery burst forth in a deadly song … In a few minutes, lines of trenches had been taken by the leading [battalions] and they were now walking across German country. Men lit up their waiting cigarettes and went on. My unit followed on through the morning mist. No man could hear himself speak in the inferno of noise made by the guns. They crashed and roared and the daybreak was made hideous. As the dawn appeared the mist was almost dispersed. Fumes from the smoking guns mixed with the mist and settled on the ground, waist high. The mixture quivered and vibrated in response to the crashing guns and German shell-bursts.

The battalion advanced in section-sized columns, the troops awestruck by the scale of men and materiel being thrown at the enemy. By 8.20 am the unit was halted on the high ground one kilometre north-east of Villers-Bretonneux, close to their April battlefields. Here they caught their breath and absorbed the military spectacle underway on the flat ground ahead of them.

At midday the order came to resume the advance and for the battalion to take up a position on the newly captured Green Line, north-east of Warfusée. This move was accomplished without incident by 3.30 pm; with the slow rate of advance by the leading troops of the division, it took three hours to cover the two-kilometre distance.

At 5.00 pm the order to advance was given; this time the battalion was to relocate one kilometre eastwards to a position halfway between the Red and Green lines on the northern edge of Bayonvillers. This task was completed by 6.30 pm. The troops halted and proceeded to dig in. While fighting raged on their flanks and front the men sat, becalmed, and watched the sunset. By dint of circumstance, on a day General Erich Ludendorff later described as 'the Black Day of the German Army', not a single soldier of the 55th had fired a shot in anger.

The next day failed to bring the exasperated diggers any closer to the scene of the action and the battalion remained waiting in the fields near Bayonvillers. To rub salt into the wounds, the unit was by now so far behind the fighting that the YMCA set up a roadside stall nearby, dishing out cocoa and biscuits to all comers. The War Diarist observed: 'Much evidence can be seen of the rapid advance ... in this sector ... Tanks are all over the landscape "skittled" by the anti-tank and field gun batteries of the enemy.'

It was a bright, sunny day with some high clouds. At one stage, a squadron of German Gotha bombers circled over the battalion dropping some 40 bombs and wounding five soldiers, none of them fatally. The War Diarist reported: 'All ranks are eagerly awaiting the expected order to move forward to take their share of "Biffing of the Hun".'

The authorities, however, had other plans for the battalion, none of which involved 'biffing the Hun'. The 5th Division was ordered to concentrate in a rest area just north of Villers-Bretonneux; the 8th and 15th brigades had already moved to conform with this directive. At 2.00 pm the soldiers commenced their march rearwards, the War Diarist noting that they: 'Were very disappointed when orders were received to move back to our bivouacs near Villers-Bretonneux.'

The 14th Brigade War Diarist used similar words to record the feelings of the troops: 'Disappointment at not having actively participated in the operation was rife, but it was realised by most that they would have an early opportunity of meeting the Boche.'

The next day the men rested and received a welcome change of underwear and uniform. During the afternoon, parties of troops took advantage of the fine and very hot weather and made their way for a swim in the Somme River near Fouilloy. The following day, 12 August, passed in a similar fashion. That evening the war intruded on the enforced summertime idyll; Lieutenant Colonel Woods, the 55th's CO, received notification of a proposed attack for 15 August. The attack would involve a combined French-Canadian-Australian operation to capture Chaulnes and re-energise the stalled Allied offensive in the southern Somme sector.

Lieutenant General Monash directed that the Australian element of the operation would be conducted by the 5th Division on the right (advancing next to the Canadians), the 4th Division in the centre and the 2nd Division on the left. The starting line for the assault was the Lihons–Maucourt road, with the concentration area around Caix.

The battalion moved to Caix at noon on 13 August. Vague orders accompanied the passage. No reconnaissance had been made of either the route or the destination. Officers had been despatched at first light for this purpose but the compressed timeframes meant that they were unable to report before the unit got underway.

The march was conducted in choking dust, making it exceedingly unpleasant. The roads were so crowded with tanks, artillery, horse and motor transport that the road surfaces had been ground to a fine, ankle-deep powder. The sheer mass of materiel competing to use the limited road network meant that infantry columns were often blocked and delayed. Officers struggled to keep the battalion in some semblance of military order.

Notwithstanding the tribulations of the march, by late afternoon the unit had taken up residence in the woods south-east of Caix. Here the men readied themselves for a move forward that night in relief of a Canadian brigade occupying the front line. In the late evening, just as they were about to commence the relief, the attack was postponed by 24 hours and everyone told to stay in their current positions.

Next morning another order was received from Headquarters 14th Brigade restating the intention to progress with the operation, probably on 16 August. Once again the 55th would relieve the Canadians that night.

That afternoon the unit received orders for the coming enterprise. The 55th Battalion was to assist the 15th Brigade's attack on Chaulnes by capturing the village's railway station; the 56th Battalion and the Canadian Corps were to advance on the right. For the first time some tanks had been allotted to the brigade and the prospect of working with these armoured behemoths excited the troops. The Germans were not aggressive, and it was believed that the ground would present a greater hindrance than the enemy.

While the terrain over which the unit was to advance was generally flat, it was a patchwork of overgrown shell-holes, loose wire and old trenches, relics of the battles of 1916–17. There was an additional complication. Two sets of railway lines traversed the area, one running in a north-east–south-west direction and the other north–south. Both lines intersected at the Chaulnes station. Private Bert Bishop describes the impact of these railway lines on the plan of attack:

> Our platoon officer [Lieutenant Inglis], whom we all liked was not looking very happy ...[7]
>
> [He] told us the attack would start at dawn, preceded by a brief barrage. When we reached the top of a rise a couple of hundred yards ahead we would come to a railway cutting, two sets of lines running through it [the north–south line]. It was a steep cutting. The Germans held the top

of the opposite bank. We would slide down our side, cross the lines and attack the Germans on top of their side. 'That's about all,' he said.

Silence lasted a few moments. Sergeant Harlock broke it.

'We slide down our side. It's a steep cutting then Sir?'

'Yes. In fact, the Germans couldn't climb our side and we can't climb theirs.'

Silence again.[8]

In the early hours of the morning came word that the operation had been postponed yet again and everyone was to remain *in situ*. The soldiers slept better once news of the delay was passed around. The 14th Brigade's War Diarist's entry of 15 August articulated the frustration and confusion present in the headquarters: 'Attack is on and off again.'

Finally, at noon, information was received from Headquarters 5th Division unequivocally cancelling the proposed operation. New orders were issued directing the brigade to march several kilometres north to the reserve lines of the British 17th Division, just to the west of Proyart.[9] These instructions also specified that on the following night (16 August) the brigade was to occupy the front line east of Proyart and progressively advance the line. To minimise congestion on the roads, the move was not to commence until 9.30 pm.

The unit was briefed on the tactics to be adopted during the approaching time in the forward trenches. Major General Hobbs, under pressure from Lieutenant General Monash, directed the 14th Brigade to 'Maintain an aggressive attitude to persuade the enemy that we are likely to resume the offensive at any moment. Further a line has been drawn some 2000–3000 yards in advance of the line presently held to which it is intended we should gradually force our way during our [six-day] tour in the line.' The 'advance of the line' was to be achieved by opportunistic patrol actions, capturing ground without costly or protracted fighting in an approach referred to as 'peaceful penetration'. The 14th Brigade's War Diarist reflected on the prospects of the forthcoming move to the Proyart area:

> The whole thing to us appears rather as an anti-climax after the anticipation of a genuine battle to which we had worked up … The whole Brigade, having missed the actual fighting on 8th / 9th is ready and willing to 'buy a scrap' where it can be found. The 'peaceful penetration' process in the new area however, offers attractions although it is generally a more expensive process with less satisfactory results.

The 55th commenced the march towards Proyart at the appointed time. Private Eddie Street was happy to leave Caix:

> Three days were passed here and we did not do a thing except to eat, drink, play cards and sleep. Rumours were rife that we had to hop over and take the rest of the Chaulnes sector … All night long and for two nights Fritz had bombed our open possies, shrubs had been whittled down by flying pieces of bomb and many of the big trees were badly scarred. Bombs had crashed down out of the inky blackness and had fallen near us but nobody was killed. We were glad to leave the wood and started off on our northern road with a lighter heart.

Winding their way via the roads and fields east of Harbonnières, the men of the battalion began to arrive in their new location just after midnight. Private Bishop recalled:

> It was a perfect moonlight night, ideal for aerial bombing. German planes seldom appeared in daylight. This night they became very busy. We plodded some miles across open fields to the accompaniment of the noise of German engines above us and the crashing of bombs about us. It was good physical exercise for us, our movements synchronising with the whoosh-whoosh of the falling bombs as we flattened on the ground, up and on again, down again.[10]

The men found accommodation for what remained of the night in an abandoned French trench system.

The next day, 16 August, was spent preparing for the coming sojourn in the line. The Germans were jumpy: enemy gunners fired numerous large calibre shells into the battalion area throughout the day. That night the 55th relieved the 10th West Yorkshire Regiment in the front line, with the 56th Battalion to the right. The unit had two companies in the forward trenches with the remaining two back in support. The shortage of men required the battalion to occupy the front in a series of platoon-sized outposts, each about 150 metres apart. Gaps between outposts were plugged by patrols and listening posts.

The battalion wasted no time getting patrols out into no man's land. The first patrol, under Lieutenant Julian De Meyrick, was underway by 12.10 am; other patrols followed at 1.00 am and 1.35 am.[11]

These routine activities were a curtain raiser for a far bigger scheme. Lieutenant Colonel Woods, always the fighter, took to heart the admonition to 'peacefully penetrate' and arranged for a small enterprise of this nature to be undertaken within hours of the unit's arrival at the front. This operation had the objective of capturing and retaining two advanced enemy machine-gun posts.

The fighting patrol tasked with conducting this mission consisted of 20 men from D Company under command of Captain Harry Wilson. The party set out at 3.00 am, a few hours before dawn. The enemy posts were believed to be some 200 metres away from the Australian front. About halfway across no man's land, the patrol stumbled into a hitherto unknown German position.

Neither party was expecting the encounter. The enemy responded and a short, nasty fight ensued. Using his Lewis gun, Private William McNeich engaged the crew of a light machine-gun, killing all three men.[12] The remaining members of the patrol overwhelmed the enemy, capturing two Germans and killing several more.

But this success had come at a cost. Captain Wilson had been shot in the shoulder during the melee and was forced to retire, leaving Sergeant Victor Anchor in charge with orders to hold the position until further notice.[13] These directions accorded with the 'peaceful penetration' nature of the operation.

The skirmish had alerted the Germans in their main defensive line to the presence of the patrol. Up to eight enemy machine-guns opened up and *minenwerfers* commenced a bombardment. Having stirred the hornet's nest, the remaining patrol members hunkered down and returned fire.

Australian casualties mounted; one of the first to fall was Sergeant Anchor, killed by a burst of machine-gun bullets to the head and abdomen. Corporal John Pringle took over and was able to restore some order to the chaotic situation, directing one of the Australian wounded to escort the prisoners back to friendly lines.[14] Shortly after, Private Patrick 'Paddy' Kerr was wounded in the chest by several machine-gun bullets.[15] Corporal Pringle dressed his injuries and placed him in a shell-hole with another mortally wounded man, Private Walter Caldwell.[16]

If the situation was already grim, it was now about to deteriorate further. The enemy launched a counter-attack to retake the post. Private James Morrison recalled the next few minutes: 'Many of the party were wounded. Then Jerry advanced on us. It seemed as if the whole front line walked out … Jerry got around behind us.'[17] The 14th Brigade's Intelligence Summary for 16/17 August contains an estimate that 200 Germans were involved in the counter-attack.

Realising the danger of being wiped out wholesale, Corporal Pringle gave the order to retire, leaving behind the dead and those among the wounded unlikely to live. A withdrawal in close contact with the enemy is the most difficult of military undertakings but, under Pringle's leadership, a rout was averted. Directing the patrol's Lewis gunners to provide covering fire, Pringle organised the remaining members of the patrol to make their way back to the Australian lines, carrying the lightly wounded. In the confusion, Private Morrison found himself cut off: 'I tried to get back but was stopped. There were Germans all around us. I put up a fight but was knocked out by a bomb and taken prisoner.'

One of the stay-behind Lewis gunners, Private Leslie Hall, noticed a German machine-gun squad trying to outflank the position.[18] He opened fire on the enemy, forcing them to abandon their gun. Private McNeich again came to the fore, passing his Lewis gun to another man and firing the captured machine-gun. Once the main body of the patrol had returned, the Lewis gunners retired, using their guns to provide covering fire for each other with McNeich carrying the captured machine-gun. Soon after, the Germans reoccupied their outpost.

The enemy counter-attack on Captain Wilson's patrol also came close to bumping into another Australian outpost, this one occupied by Private Bishop, manning a Lewis gun, and a few of his mates:

> The Germans to the other side of our post had been spotted first. Then Smith and I glanced about our end. About twelve Germans had bunched about thirty feet to our left front. I got the gun's sights on them. Then we saw another bunch of them, somewhat larger, to our left back, this latter bunch having passed beyond us without seeing us …
>
> We still had the advantage of surprise and I decided we should make use of it. My gun's sights went to the bunch at right front. I hoped to get them all with a half-burst of the full magazine, then swing to the rear bunch and do what we could to them with the other half …
>
> The gun must have been right at the point of firing when a hoarse voice said: 'No, don't fire. Wait till the bastards see us.'
>
> I eased my trigger finger. It was Ernie [Corey] who had butted in. 'But we can get them by surprise. That means everything.'
>
> 'It does. But before you got onto the second lot we'd be showered with bombs and bullets. There's over thirty of them, there's three of us.'
>
> A whistle blew out among the Germans.
>
> When a flare gave light again the whole of the Germans had disappeared. Our own patrol had been almost overrun by the Germans scuttling home.[19]

The night's excitement was not yet over. At 4.00 am, after the firing associated with the enemy counter-attack had subsided, a German wandered into an outpost commanded by Corporal Ron Stewart and was captured.[20] The prisoner informed Stewart that he was out searching for enemy wounded in front of the Australian position. Stewart passed this information to his platoon sergeant, Arthur Kearns, and the two NCOs decided to make their own way into no man's land to try to locate the wounded Germans.[21]

When they had advanced some way from their own lines, Corporal Stewart and Sergeant Kearns observed a large German party, numbering about 25 men. Quickly hatching a plan, they made their way to the rear of the Germans and, with great bluff, called on them to surrender. The enemy were in no mood to give up and made a dash for the safety of a nearby trench. Kearns and Stewart succeeded in cutting off and capturing five of the enemy party and a light machine-gun. The remaining Germans engaged the two Australians with machine-gun and rifle fire. While Stewart used the captured machine-gun to suppress the Germans, Kearns made his way back across no man's land with the five prisoners. After giving Kearns a reasonable head start, Stewart retired, still carrying the German machine-gun. The intrepid Australians and their five prisoners made it back to the Australian lines. For their initiative and courage, both Kearns and Stewart were awarded the Distinguished Conduct Medal.

The battalion's first attempt at 'peaceful penetration' had been a failure and soured the men's appetite for such risky ventures. Only five of Captain Wilson's patrol emerged unscathed: of the casualties Sergeant Anchor and Private Caldwell were known to have been killed and Private Kerr and Private Morrison remained missing. Without the bravery of the Lewis gunners in covering the withdrawal, the entire patrol could have been wiped out.

Two Germans and a machine-gun had been captured and up to 15 Germans killed. The prisoners were identified as belonging to the *21st Division*, a formation assessed by Allied Intelligence as having only 'fair morale'. Considering the thrashing meted out to the patrol, the Australians were fortunate they had not encountered an enemy with excellent morale.

Later that morning observers noted that the Germans had withdrawn from the outpost. Corporal Pringle and a small party of men, under the dubious safety of a white flag, returned to the disputed outpost in an attempt to bring back Private Kerr. They found Private Caldwell's body in the shell-hole where he had been left but found no trace of Kerr. Surmising that the Germans would not have troubled to move a dead man, on his return Pringle reported Kerr as most likely captured.[22]

After the excitement of the first few hours in the front line dissipated, life settled into a familiar pattern; days were spent snatching whatever rest could be found between short, intense German artillery bombardments and incessant sniping. Night was the province of patrols, Private Street noting: 'We gave Fritz no peace in the line. His posts were harried all night long.' Every day brought a steady trickle of casualties: Lance Corporal Lyle 'Gillo' Gilbert,[23] Private Joe Tyler[24] and Driver Joseph Mines were killed by shellfire.[25]

While the troops kept the Germans occupied, Brigadier General 'Cam' Stewart, commanding the 14th Brigade, asked his battalion commanders to submit proposals for advancing the line the several thousand metres demanded by Major General Hobbs. If the concept of 'peaceful penetration' held any allure for the senior leaders in the brigade, the resulting plans did not reflect it.

The bare nature of the terrain in the sector, the aggressive and watchful enemy and the slow ebb in the strength of the battalions convinced the officers that 'the idea of "peaceful penetration" is faulty'. All agreed that a set-piece attack, supported by a massive artillery barrage, was necessary. This proposal was put to Hobbs, who forwarded it in turn to Headquarters Australian Corps.

On 19 August came the news that the attack proposal put forward by the brigade had been approved and the action was to be launched in the next two or three days. Less welcome was the news that it was not the 5th Division but the 1st that was to undertake the operation. The 14th Brigade War Diarist captured the mood following this announcement:

> This will be a disappointment to all concerned and will certainly cause discontent in this brigade which is feeling rather out of it. We would like to have one good battle and then go out for a rest, because the men are tired and do not improve by being misled time and time again and chased about all over the district with a series of promises and disappointments. However, it is the demands of the tactical situation and every effort will be made to make the men understand this point of view.

That night saw a spike in patrol activity. A ten-man patrol under the command of Lieutenant Rupert Ellsmore encountered a similar-sized party of Germans. After an exchange of grenades and rifle fire, the Germans fled towards their lines. A much larger enemy party materialised from behind the smaller one, but eventually this party also took flight. Lieutenant Eric Farmer took out two patrols; one of these was seen and engaged by a German machine-gun, which wounded a man. The final patrol for the night encountered a machine-gun but the two-man crew fled as the patrol approached, allowing the Australians to acquire the gun as a trophy.

On the following night, the moonlight was too bright for patrols to be employed. As an alternative, several listening posts were positioned in front of the Australian line. After midnight, fog rose and this combined with the smoke and dust of exploding shells to reduce visibility. Just before dawn, the fog lifted, revealing that an enemy listening post had been established near one of the Australian positions. The sole occupant of the German outpost was shot and killed. Shortly after, a party of German stretcher-bearers, under the protection of a Red Cross flag, came out and retrieved the dead man.

On the evening of 21 August, the 55th and the 56th battalions were placed under command of the 1st Division, which was commencing the final preparations for its coming attack. The transfer of the two battalions was made to simplify command arrangements. The units were to continue holding the front line on 22 August and then vacate the trenches that evening while the 1st Division troops passed through. After the assaulting troops were formed up on their start line in no man's land, the 55th and 56th battalions were again to reoccupy the front line.

German bombardments increased in intensity on 22 August. Reports from prisoners indicated that the Germans were expecting an attack in the sector around Proyart and were employing the vicious barrages in an attempt to break up the concentration of troops. Private Bishop wrote of an incident during the day:

> Shelling went on without any let-up … A light shell fell on the parapet nearby. It went through two or three feet of earth and fell, without exploding, into the bottom of our trench. It bumped Freddy, who was having a sleep. Freddy woke up, saw the dud shell beside him, sat up, grabbed the hot thing with two hands and tossed it over the back of our trench. 'Bit of a dud,' he said, and went to sleep again.[26]

Not all the men were as fortunate as 'Freddy'. Private James Alcorn was digging a trench when a shell landed in it, fatally wounding him and injuring six men who were nearby.[27] Private Thomas 'Young Tommy' Marsden,[28] Private William McNeich (who had distinguished himself on Captain Wilson's fighting patrol several days earlier) and Private Arthur Jennings[29] were killed when a shell slammed into their outpost, the same shell wounding three others, including Lieutenant Clarence Mahoney.[30] Also killed by shellfire on this day were Private Abraham Keefe,[31] Private Henry Nott[32] and Private Hugh Roberts.[33]

The 1st Division attack was launched on the morning of 23 August and succeeded in driving the Germans back to the valley of the Somme River.[34] Their role in supporting this attack complete, the 55th and 56th battalions reverted to under command of the 14th Brigade. Both of these battalions were subsequently ordered to withdraw from the former front line and link up with the rest of the brigade in an area just west of Proyart. The operations around Proyart had further depleted the battalion's manpower: the unit was now down to 17 officers and 355 other ranks, and these numbers included men in the transport and quartermaster departments.

The next two days were spent resting, the troops' slumber only disturbed by the occasional long-range shell. Rumours continued to circulate of forthcoming 'stunts'. The 14th Brigade War Diarist wrote on 24 August of the brigade's continuing frustration:

> The Divisional Commander visited Brigade HQ during the morning and said it was possible that we should be moved back tonight. No action was taken however as, owing to the number of orders and counter orders we receive as the situation changes, we have become quite blasé to mere suggestions. We are just standing by ready to move in any direction and expecting nothing in particular.

On the evening of 25 August, orders were received warning of an impending move to the front line the next day in preparation for an attack on 27 August. The assault was to be launched by the 54th, 55th and 56th battalions. Little opposition was expected and it was anticipated that the advance would be 6000 metres, although the objective was, to a certain extent, unlimited.

At noon on 26 August these orders were also cancelled, to the intense annoyance of all ranks. Instead of a set-piece attack as originally planned, the 5th Division was ordered to occupy a portion of the line and engage in 'peaceful penetration'. Realising the detrimental impact the changing orders and absence of any real fighting was having on the 14th Brigade, Major General Hobbs offered Brigadier General Stewart the option of the 'peaceful penetration' mission or of remaining in reserve with the promise that the first 'dinkum hop-over' would be spearheaded by the 14th Brigade. Stewart chose the latter alternative and Hobbs ordered the 8th Brigade to the line.

The battalion spent a further three days of inactivity near Proyart. While the men speculated on what they had done, or not done, to be always in reserve, the remainder of the Australian Corps continued to advance, leaving the 14th Brigade languishing further behind.

Finally, on the morning of 29 August, the order to commence moving forward was received. It was a clear, warm day and, by 11.00 am, the battalion was tramping cross-country towards the Somme River. Bert Bishop described the march:

> It was a weird, desolate country, having now changed hands four times since the 1916 offensive. Overgrown old shell-holes occupied by big repulsive frogs, rusty tangled barbed wire, rotting timber of old dugouts,

> worm-eaten sandbags, seared, dead, blighted trees and shattered stumps made it a depressing march. Slimy slugs and creeping things hid in the rank grass. Even birds kept away from it, their song would have been a misfit.[35]

Corporal Herbert Harris was similarly affected by the unpleasant surroundings: 'Passed dead Huns, horses, mules etc. Stench awful.'

Moving via St Martin's Wood and Assevillers, the unit covered a distance of 13 kilometres across the broken landscape before halting for the night in the old trenches that criss-crossed the shattered woods to the east of this latter settlement. The diggers, exhausted after the long march across such taxing terrain, were directed to recommence their advance at dawn.

While the 14th Brigade had been moving forward, the 15th Brigade had leaped-frogged the 8th Brigade with orders to force the Germans back across the Somme River and continue the pursuit. The 14th Brigade was ordered to conform to the movements of the 15th Brigade and provide assistance when needed. Brigadier General Stewart ordered the 53rd and 54th battalions to close up to the rear elements of the 15th Brigade; the 55th and 56th battalions were ordered to support their sister battalions.

The unit got underway at 6.00 am on 30 August, negotiating similar war-torn country to that traversed the previous day. After covering just a kilometre, it became apparent that the 15th Brigade had not moved from its overnight positions. The battalion was also visible to German artillery spotters and shellfire began to build in intensity. As the 53rd Battalion had closed with the rear of the 15th Brigade, Lieutenant Colonel Woods discovered that the battalion had been unable to make any progress. He ordered his men to take shelter in a nearby system of abandoned trenches and await further orders.

The diggers spent the remainder of the day and that night eating, sleeping and making themselves comfortable. For many men, this was to be their last break for, as Private Bishop recalled: 'On the last day of August we hit real war again.'[36] The much anticipated 'dinkum hop-over' had arrived.

Chapter 14:

Péronne – 'no one will be left soon'

From its source at Fonsommes, the Somme River flows in a northerly direction until it makes a sweeping hook and heads off along a westerly path towards its rendezvous with the English Channel. On the eastern corner of this river bend is the town of Péronne. The river, its associated wide belt of marshes, a canal (the Canal de la Somme) and the high ridges bordering the eastern and northern banks of the river protect many of the approaches to the town. [1]

The military advantages offered by the topography were first recognised by the Romans, who established a fortified town on the site and Péronne's defences continued to be strengthened over the centuries. In the late seventeenth century, Marshal Vauban, France's most famous military engineer, incorporated run-down medieval defensive features when he upgraded the town's fortifications. Realising that natural features safeguarded the town to the south and west, he focussed his attention on guarding Péronne's open northern and easterly approaches. Here he had constructed thick ramparts flanked by broad deep moats.

Three centuries later, Vauban's crumbling legacy was unable to prevent the Germans capturing Péronne in 1914. They used the town as a major headquarters and logistics hub during the years that followed. In July 1916, as part of the massive Allied onslaught known as the Battle of the Somme, the French Sixth Army had occupied the entire Flaucourt Plateau just to the west of the Somme River but was unable to recapture the town itself. Over the following months, sustained shellfire reduced Péronne and its surrounding area to a shambles of destroyed buildings and shattered infrastructure.

In March 1917, the Germans razed what was left of the town before falling back to the Hindenburg Line, leaving the ruins for the Allies to occupy. One year later the pendulum of war swung once more in favour of the Germans, and they were able to recapture Péronne. Another six months later the German fortunes had soured, and this time it was the Australian Corps, led by Lieutenant General John Monash, who turned its attention to retaking the town. However, unlike March 1917, there was to be no uncontested handing over of the keys to the Allied forces; this time the Germans meant to make a fight of it.

Mont St Quentin, and the eponymous village on its crown, dominated the area north of Péronne. The Mont itself is an innocuous, gently sloping hill rising just over 100 metres above the surrounding countryside and is the culminating point of a long ridgeline running to the east. Militarily, Mont St Quentin provides an excellent location from which to dominate the northern and western approaches to Péronne, located 1.5 kilometres to the south.

To capture Péronne, an attacker first needed to seize Mont St Quentin — the key to the lock. Acutely aware of this, the Germans had fortified the position, employing dozens of machine-guns firing from entrenchments screened by wide belts of barbed wire. The houses, monastery and other buildings on the Mont had been woven into the defensive fabric.

The German defenders of Péronne, following Vauban's example, had blended the town's natural attributes and existing man-made fortifications into their own defences. The thick earthen ramparts and old masonry walls were almost immune to artillery bombardment and were ideal for use as machine-gun posts.

To link Mont St Quentin with Péronne, the Germans constructed a series of entrenchments protected by thick wire entanglements. Just north of the town, along the road to Mont St Quentin, lay the wood and hamlet of St Denis, a brickworks and a dilapidated sugar factory; all these areas were garrisoned. The hills east of Péronne provided protection for numerous artillery batteries and offered superb observation points. Machine-gunners positioned on Péronne's ramparts and Mont St Quentin had commanding views across the low-lying area between the two bastions and the range of their weapons allowed them to interlock their fire. Altogether, the Péronne–Mont St Quentin complex was one of the strongest defensive positions ever constructed on the Western Front.

The Germans had one of their premier formations, the *2nd Guards Infantry Division*, located around Mont St Quentin. However, the majority of Péronne's defenders were from combat units in a state of organisational chaos following weeks of intense battle and retreat. A captured German officer from the *65th Infantry Regiment (65th IR)*, ordered to hold the north-east part of Péronne, disclosed that he was in command of a scratch force comprising the remnants of the *1st Battalion, 65th IR,* the *65th IR*'s Machine Gun Company and three companies of the *258th Reserve Infantry Regiment.* All told, his detachment numbered approximately 200 men.

At this stage of the war, it was common for German infantry companies to number around 20 to 40 men. On the other hand, the machine-gun companies tended to have their full complement of 12 heavy machine-guns and 50 to 60 men crewing them, an indication of the value placed on these weapons by the German High Command.

Following their victories in early August, the French and Commonwealth forces had pursued their adversary without respite. As their entire Somme front teetered on the brink of collapse, the German High Command decided to withdraw their forces to the Hindenburg Line. The German leaders hoped that, once their battered army was ensconced behind this formidable barrier, time could be gained to refit and possibly negotiate a truce; the approach of winter would also work in their favour. To regain the strategic equilibrium necessary for an orderly withdrawal, the Germans decided to fight a major delaying action at the last strong defensive position before the Hindenburg Line — along the Somme River at Péronne.

The Australian Corps policy was to cross the Somme and harass the retreating enemy, denying them the time they so desperately sought. Early on 29 August, the 2nd Division had driven the Germans from the west bank of the Somme while the 3rd Division had cleared the enemy forces operating north of the river.

The 2nd Division troops were unable to find a suitable river crossing, jeopardising their pursuit of the enemy. In response, Lieutenant General Monash conceived an audacious manoeuvre. By rapidly shifting the bulk of his corps to the north bank of the Somme he hoped to seize Mont St Quentin in a surprise attack, turning the German flank and forcing their withdrawal from positions in Péronne and along the Somme to the south.

During 30 August, the 3rd Division captured Cléry sur Somme and Halle. That evening, the majority of the 2nd Division crossed to the northern side of the Somme using the bridges near Cléry sur Somme and took up positions from which the 5th Brigade could assault Mont St Quentin the next morning.

Brigadier General Elliot's 15th Brigade replaced the 2nd Division along the west bank of the Somme with instructions to cross the river and seize the high ground a kilometre or two south of Péronne. This ground dominated all possible crossing points. Elliot set about discovering a way to overcome the obstacle in front of him.

Next morning two under-strength battalions from the 5th Brigade (2nd Division) began the assault on Mont St Quentin. To the disbelief of their foes, and many of their own leaders, they were able to capture the main German trench system on top of the Mont.

News of the victory was electrifying.

At 8.00 am, Lieutenant General Monash directed the 2nd Division to exploit the opportunities offered by the capture of Mont St Quentin and thrust its 6th Brigade along the ridge towards Lieramont. To protect the right flank of this brigade, Monash ordered the 14th Brigade to cross the Somme at Ommiecourt les Clery[2] with the general intention of striking out from there in a south-easterly direction beyond Péronne. Monash was in a belligerent frame of mind; he told Major General John Gellibrand, commander of the 3rd Division, that 'casualties no longer matter'.[3]

Brigadier General 'Cam' Stewart, commander of the 14th Brigade, held a conference with his COs at 10.00 am, in which he discussed the brigade's role as a flank guard for the 2nd Division. It did not take long to communicate the instructions from this conference to the men and the unit was underway by 11.00 am. The troops were very happy to receive the order to leave the old French trenches near Barleux — the stench from hundreds of unburied corpses rotting in the late summer heat was almost unbearable.

The men had before them a journey of some nine kilometres to reach the assembly area east of Cléry sur Somme. The undulating country they crossed was a tangled mess of decrepit trench systems, shell-holes and rusty wire entanglements. Progress was slow and exhausting — it was impossible to walk more than a few metres without having to detour around some form of obstruction, often barely visible in the calf-high grass. Frequent halts became necessary to restore order to the flagging columns. Recently promoted Corporal George Gill commented that the enemy added to the difficulty of the march: 'We moved off about midday and apparently Jerry saw us for he sent over a good many shells, some landing very close, but fortunately no-one was hit to my knowledge. We had these shells after us most of the way.'

While the unit picked its way across the shattered countryside, German reserves had counter-attacked Mont St Quentin, forcing the 5th Brigade to partially withdraw from the summit.

The 14th Brigade had been ordered to cross the Somme using the bridge at Ommiecourt. Despite the hindrances of the march, it only took around 90 minutes for the advance elements of the brigade to begin arriving in the neighbourhood of the bridge. Here they found the 7th Brigade queuing ahead of them; German artillery had made the bridge impassable and killed or wounded half the engineers tasked with its maintenance.

There was no semblance of order in the milling mob: everyone had a conflicting set of priorities and orders. Men, horses and wagons of all descriptions wandered around aimlessly; the occasional explosion of a German shell added additional noise and danger. Confronted with such confusion, the diggers did what any self-respecting infantryman would do in similar circumstances — dropped their packs, had a feed and settled down to sleep.

At 5.00 pm Brigadier General Stewart diverted his brigade away from Ommiecourt to the footway and pontoon bridge a few kilometres downstream at Buscourt. These bridges crossed both the Canal de la Somme and the Somme River. The 55th Battalion, the last unit in the brigade column, used a small footbridge to cross the canal. However, they were unable to proceed over the river as the bridge was choked by traffic.

Assessing the situation, the decision was made to move the battalion back to Ommiecourt. Following the 54th Battalion, the unit retraced its steps, moving beside the poplar trees lining the southern bank of the river.

During their withdrawal, the Germans had destroyed the bridge that crossed a few metres above the Somme River at Ommiecourt, leaving a tangled mess of steel and stone. Australian engineers, using sandbags and planks, had constructed a makeshift footbridge in spite of the artillery that continued to pound the area. Private Eddie Street described the crossing:

> Some twenty of them [engineers] were filling in shell-holes on the island and repairing the bridge itself. Shells were concentrated on the spot, but the engineers worked valiantly without pause. As we approached the causeway a shell burst and two engineers fell dead beside their work. The others kept on, and as we passed Diggers' marks of appreciation were bandied across. Nearing the bridge itself another shell crashed and three engineers near the burst fell to the ground in various attitudes of death.[4] We passed them as the smoke cleared away, and saw their still forms lying beside the work which they had been striving so hard to complete for our passage ... By careful walking and in single file we managed to get across. A number of shells crashed down around us putting the wind up us, but injuring no one.

Lieutenant General Monash later criticised Brigadier General Stewart for the delay in getting the 14th Brigade across the Somme. The Corps

Commander had intended to throw both the 6th and 14th brigades into battle late on the afternoon of 31 August. However, these delays were predictable given that three brigades (6th, 7th and 14th), with little notice, needed to cross a significant natural obstacle using a few narrow bridges in daylight under enemy fire.

After crossing the bridge, the battalion became caught up in another debacle. In the narrow area between the edge of the Somme marshes and a high chalk cliff, men, vehicles, horses, and equipment from various units jostled as they made their way forward under an ad hoc German bombardment.

Nonetheless, by 7.00 pm the men had reached an assembly area in some captured trenches in the deep gullies around one kilometre south-east of Cléry sur Somme.[5] German shellfire increased at dusk and continued through the night. Corporal Gill described his circumstances:

> We arrived at the far side [of the river] safely and found possies against a cliff with a hill behind. Here we remained all the night. The shell fire was heavy, some bursting about 50 yards ahead, some behind us, others fell into the water, throwing the water high into the air.

At 9.30 pm a conference was held at Headquarters 6th Brigade near Herbecourt attended by Brigadier General Stewart, the three brigade commanders from the 2nd Division, and representatives from the Australian Corps, Headquarters 5th Division and Corps Artillery.

During this meeting the 6th Brigade was tasked with retaking Mont St Quentin the next morning and then exploiting in an easterly direction. In support of this operation, the 14th Brigade was to eliminate enemy resistance in the area between Mont St Quentin and Péronne, in the process capturing the fortress town itself. Once these objectives had been achieved, the 14th Brigade was to hook past Péronne and exploit south for several kilometres, outflanking the German positions holding up the 15th Brigade's advance. There were no tanks available to augment the infantry assault and air cover would be limited.

Little time was available to organise so complex an undertaking. The conference did not end until midnight, only a few hours before Zero which, at Brigadier General Stewart's urging, had been set for 6.00 am. Stewart raced back to his headquarters on the south side of the Ommiecourt bridge, where Lieutenant Colonel Percy Woods, CO of the 55th Battalion, and the other battalion commanders waited. He did not arrive until 3.00 am.

Brigadier General Stewart broke his assault into two phases. Phase 1 involved the 54th Battalion attacking Péronne while the 53rd Battalion on its left was to skirt north of the town, advancing through Anvil Wood, the hamlet of St Denis, and St Denis Wood, ending its advance in a position just to the south of this wood. To assist the 53rd in its difficult task, D Company of the 55th was placed under command of this battalion. The remainder of the 55th Battalion was to constitute the brigade's reserve.

Phase 2 would see the 56th Battalion, hitherto in support of the 53rd, turn south and seize Doingt and the high ground south-west of this village. The 56th would be supported by the 55th (including D Company) which had orders to mop up Flamicourt, a suburb south of Péronne defended by numerous German machine-guns.

The whole attack was so hastily organised that insufficient time was available for the preparation of a creeping barrage. Instead a program of bombardments on known enemy strongpoints would be implemented. Brigadier General Stewart had available to him thirty-six 18-pounders and twelve 4.5-inch howitzers from the 5th Division's artillery; these guns had been light enough to keep up with the rapid infantry advances across damaged country. Few of the heavy artillery pieces necessary to reduce well-fortified positions, such as Péronne's ramparts, were within range.

Shortly before the attack, Lieutenant General Monash advised Major General Hobbs, the 5th Division's commander, that the four field artillery brigades from the other divisions were now available to support the preparatory bombardment. As events transpired, there was insufficient time to use this additional firepower.

By 4.30 am Lieutenant Colonel Woods' orders were starting to trickle down to the company officers and men. Fortuitously, the battalion had spent the night in the assembly area, so that no pre-Zero Hour moves were required except by D Company, which needed to scramble after the 53rd Battalion as this unit was already moving to its jumping-off location in Florina Trench.

According to the 55th's after-action report, 'the troops were in the same high spirits as when the push began on the morning of 8th August, and looked forward with confidence to complete success.' Lieutenant General Monash agreed with the sentiment of the report:

> The men of the 14th Brigade that day had their mettle up to a degree which was outstanding. On the occasion of the great attack of 8 August, and ever since, it had been the cruel fate of this brigade to be the reserve unit of its Division on every occasion when there was serious fighting in hand.[6]

From within the ranks, Private Bert Bishop had a different view of the level of enthusiasm for the coming battle. Bishop relates a conversation with one of his platoon's new recruits who confided in him: 'That other new chap with me took off his helmet, poked his finger down his throat, made himself sick, said he was gassed, and off he went.' Reflecting on the newcomer's admission, Bishop opined:

> I don't think many of us saw a future in the army. We ate, we slept, we fought, we starved. Home and decent living were a dream, that life had never existed. We were becoming automatons, unthinking, uncaring. It was best to be so. Thinking didn't fit, it could cause harm.[7]

The men had breakfast at 5.00 am; it had been an uncomfortable night and they had been forced to huddle together for warmth. The plunging temperature conspired with harassment by German artillery to limit the opportunities for sleep, prompting Corporal Herbert Harris to write: 'What an awful night. God only knows how any of us are alive. The shells were bursting only 10 and 15 yards away at times.' It began to rain and a heavy mist descended, the damp making conditions even less tolerable.

The preparatory artillery barrage commenced at 5.30 am. The German counter-barrage came down minutes later and was particularly severe around Cléry sur Somme, the adjoining roads, bridges and canal.

The remaining three companies of the battalion moved out on the stroke of 6.00 am. With B Company leading, the diggers moved in single file along the Péronne–Clery road, maintaining contact with the rear elements of the 56th Battalion. Progress was slow and the sound of desperate fighting could be heard a kilometre ahead as the 53rd and 54th battalions commenced their advance. Overhead, stray bullets whizzed by.

After marching for half a kilometre, the men crossed the Canal du Nord and, in single file behind the Officer Commanding B Company, Captain Ken Wyllie, entered a storm of shrapnel. The battalion immediately began to suffer casualties: a shell fragment mortally wounded Lance Corporal Arthur Bellchambers, blowing off one of his legs.[8]

The unit passed through the bombardment and kept moving at a snail's pace down the road, at one stage crossing the Albert–Ham railway. For the first time the troops came under long-range machine-gun fire from German gunners in buildings on the southern slopes of Mont St Quentin and the ridgeline sweeping away to the east. This fire came as a shock as most men believed that Mont St Quentin had been captured the previous day. The veterans found it rather amusing to see the recently arrived reinforcements crouching behind small bushes and attempting to squeeze themselves into tiny holes in an effort to dodge the bullets.

Taking advantage of cover afforded by a slight bank running alongside the road, progress continued, with few casualties. Prisoners began to make their way back, the majority in parties of four carrying stretchers. The intermingling of the prisoners and the Australian infantrymen led to a temporary cessation of machine-gun fire as the enemy would not shoot at their own men.

As the head of the column entered a sunken part of the road, word filtered back from the 56th Battalion that the advance of the 53rd had been held up by fighting in Anvil Wood, several hundred metres away.[9] This struggle, combined with the fire from Mont St Quentin, meant that the battalion was unable to continue to move forward.

While waiting for the advance to recommence, the 55th deployed to form a flank covering Mont St Quentin and the intervening ground. B Company was sited along the embankment of the Ham–Albert railway, just north of the road in which the battalion had been sheltering. C Company was positioned to the left of B Company. From here B and C companies covered the open spaces towards Mont St Quentin. Just south of the road, A Company took up residence in some captured trenches.[10] The whole area was criss-crossed by trenches and barbed-wire entanglements.

The battalion had not been in its new locations for long when a large German mine blew up the railway line 500 metres directly in front. The explosion was terrific and the ground trembled with shock. The diggers gazed in wonder at the vast column of thick smoke and dust that rolled upwards before being carried over the enemy's lines by the wind.

The battalion's 'Narrative of the 'Battle of Peronne' describes what the men found as they fanned out across the battlefield:

> The new position presented a gruesome, yet splendid and inspiring sight. Everywhere, in all manner of positions, were to be seen German corpses, left there by the Battalion which had preceded us. These Germans had obviously made a fierce and stubborn defence in an effort to stem our advance … Here, too, were numerous enemy machine guns abandoned at the last moment, by those of their crews fortunate enough to escape wounds or worse. But, although the original crews were gone, it was only a matter of minutes when these handy weapons were wreaking havoc among their late owners. In one short section of trench, one platoon had six enemy machine guns and three Lewis guns in action.

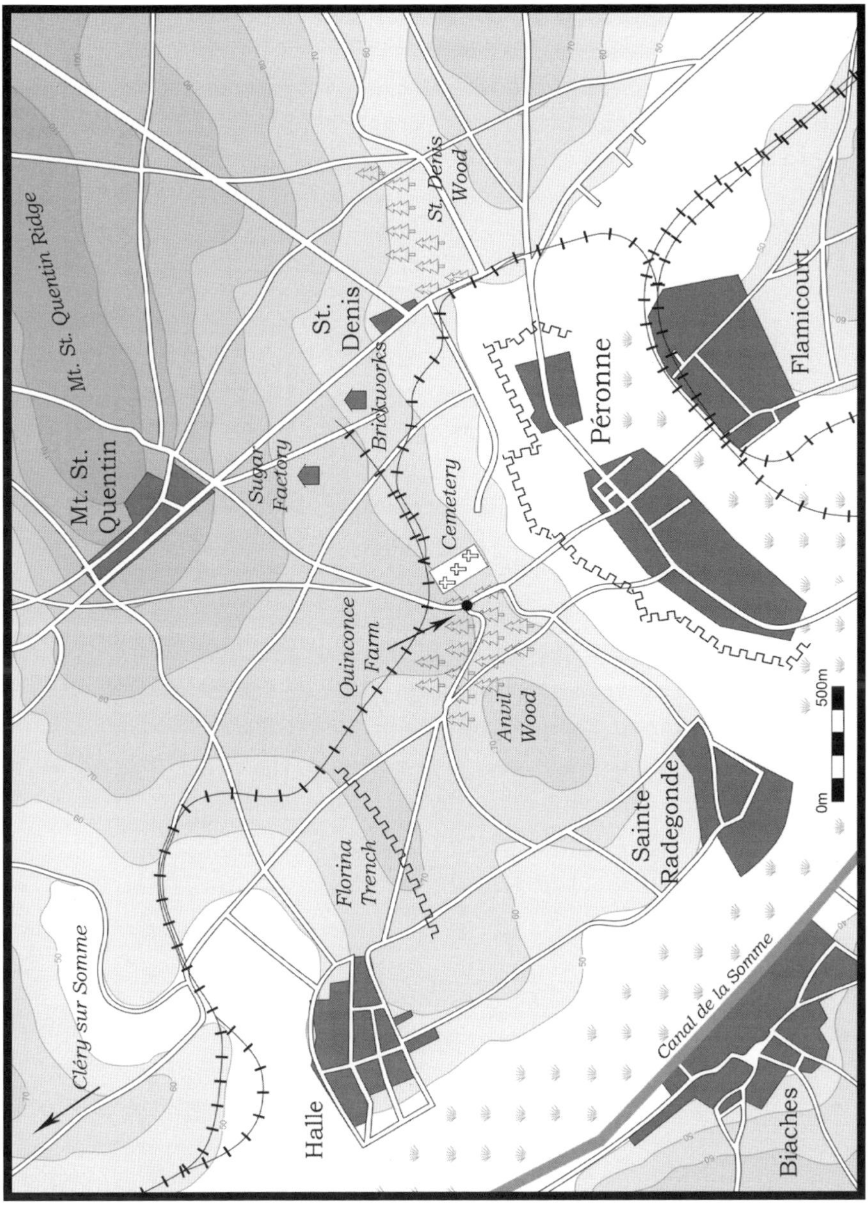

Map 15: Péronne.

It was now 7.30 am and, where they were able, the diggers found a comfortable 'possie' and began exchanging bullets with the enemy machine-gunners holding up the 6th Brigade's advance. Troops from B Company were able to see small parties of the 23rd Battalion (6th Brigade) whose attack on Mont St Quentin had stalled. Realising that the left flank was hanging in the air and that there were still Germans in between the 55th and 23rd Battalions, Captain Sidney Pinkstone manoeuvred his C Company to the left and began clearing towards the 23rd.[11] Private Bert Bishop described the nature of the fighting:

> We saw a road, or rather a track. On our side was a ditch on the other side a trench. German helmets bobbed about in the trench. We crawled out of the wood into the ditch, spreading along it, a man to about every yard. Our officer and sergeant were having a low-voiced powwow. The word was passed to each of our platoon to get ready, we'd rush the trench, no bombs. Watching Mr Inglis' [Bishop's platoon commander] signal, we poised a moment or two. Up went his arm, up we all went and rushed across the road. Keyed-up we were, and we felt punctured as we jumped into the trench. The Germans had gone.
>
> ... For two or three hours it was a game of hide-and-seek. The Germans mostly let us know where they were by showering us with their broomstick bombs. We'd retaliate with our Mills bombs, and as our bombs exploded we'd rush round the corner of the trench. We caught glimpses of them often, but they were running away.[12]

The sun had broken through, but the dust and smoke of battle hid much of what was happening. However, it became apparent that the 53rd Battalion, having lost heavily in clearing the northern section of Anvil Wood, was held up in an adjoining cemetery. Swept by converging machine-gun fire from Péronne, Mont St Quentin and the strongpoints around St Denis, the 53rd was unable to advance. Due to the time delay caused by the fighting in Anvil Wood, the artillery program was almost one hour ahead of the progress by the infantry. This meant that enemy action was in no way constrained by the supporting artillery — the 14th Brigade was on its own.

Despite leaving well before the rest of the 55th, by the time D Company caught up with the 53rd Battalion, this unit was already heavily engaged in Anvil Wood. Lieutenant Colonel William Cheeseman, CO of the 53rd, had intended D Company to follow his battalion's right support company and then form a defensive flank with the 54th Battalion.[13] Whether the

Officer Commanding D Company, Captain Renshaw King, actually received Cheeseman's orders remains unclear, for D Company was drawn into the bitter struggle in Anvil Wood and to the north of it on the 53rd Battalion's left flank.[14]

Captain King was one of the first to become a casualty. He was badly wounded in the face and, during his evacuation, a shell-burst blew off his leg. Lieutenant Stan Hourn took over command of the company despite having been wounded in the hand in the early stages of the attack.[15] Under his leadership, D Company kept pressing the assault, with Hourn wounded a second time, on this occasion in the leg.

Sinclair Hunt, recently commissioned to 2nd lieutenant, was shot through the eye by a machine-gun during the D Company assault. Second Lieutenant Allen 'Sacco' Jackson[16] wrote to Hunt's family describing the circumstances of his mate's death:

> After we had taken the German first system of trenches, we advanced towards the large village of Péronne, and poor Sinclair was leading his platoon, on my right, when he was suddenly attacked by two machine guns. He immediately led his men in a brilliant bayonet charge which resulted in the capture of both guns and 16 prisoners. After that he led another charge against a battery of [machine] guns, but I am very sorry to say he did not go far before he met his end.

Eventually sustained enemy resistance and casualties, particularly among the ranks of D Company's leadership, forced the remaining members of this company to take shelter in some abandoned German trenches which were in a very exposed position well in advance of the other Australians. There was a gap between D Company and the rest of the 55th Battalion several hundred metres to the rear. The men had tenuous links with the 53rd Battalion to their right rear, but the company's left flank was in the air as they were ahead of the 23rd Battalion. Shellfire lashed D Company and losses continued to mount.

Lieutenant Hourn, in severe pain from his two wounds, organised the consolidation of the line, during which he was wounded a third time, remaining with the company until ordered to retire by Lieutenant Colonel Woods. Private Malcolm Robertson[17] was killed when hit in the neck by a shell fragment; Private Arthur Walker was also fatally wounded by shellfire.[18]

While D Company was taking part in the struggle for Anvil Wood, on the right the 54th Battalion had achieved remarkable success, entering

Péronne as the Germans panicked and fled before the Australian advance. By 8.20 am the main section of Péronne town had been seized, leaving only the suburb of Bretagne in the detached north-eastern quarter of the old fortress to be captured; the enemy had left Bretagne unoccupied. But then, inexplicably, the 54th halted its advance. The Germans quickly recovered their composure and began to filter back and re-establish their machine-gun positions.

By mid-morning, the 14th Brigade's advance had run out of momentum and the line solidified. The 54th began consolidating the parts of Péronne it held; the 53rd was too weakened and exhausted to continue the advance. Elements of the 56th moved into the gap between the 53rd and 54th, forced to clear Germans from the southern portions of Anvil Wood as they went. C Company of the 55th Battalion had succeeded in evicting the Germans from the trenches between the 6th and 14th brigades and was in contact with the 23rd Battalion to the north. Curiously, Brigadier General Stewart remained in his headquarters near Ommiecourt and was in no position to assess events and influence the progress of the battle.

During the morning, an incident occurred that became part of the unit's folklore. A private was carrying an incendiary grenade in the pocket of his pants; a bullet hit the grenade, setting the man's clothes on fire and engulfing the unfortunate soldier in a cloud of dense smoke. With the utmost speed, the man divested himself of his garments and, naked except for a pair of boots, he began to make his way to the nearest medical facilities to have his burns treated. Feeling cold and doubtless in a shocked condition, he dressed himself in a discarded German overcoat and helmet. On the way back an Australian officer mistook him for a German prisoner and told him to keep up with his comrades. The 'prisoner' responded in a broad Australian accent, 'Hey, digger, don't be too hard on a bloke.'

By late morning, Brigadier General Stewart was determined to get his brigade moving again. Based on reports from artillery observers that the 6th Brigade had secured Mont St Quentin, Stewart ordered the 53rd Battalion to advance east and seize St Denis and St Denis Wood. The 56th Battalion was to follow the 53rd and, once clear of St Denis Wood, was to connect with the 54th, pass around Péronne and turn south. Zero Hour for the renewed attack was set for 1.00 pm. Once again, there was no time to coordinate artillery fire in support of this assault.

The 53rd Battalion was astonished at Brigadier General Stewart's orders. They could see that the 6th Brigade was not on Mont St Quentin; that brigade's attack had been repulsed and enemy reinforcements were seen streaming into the position. The 53rd Battalion was faced with the prospect of an advance across several hundred metres of grassland with little cover in the face of concentrated fire from the Bretagne ramparts on the right, St Denis in the centre and Mont St Quentin on the right.

Brigadier General Stewart's 1.00 pm deadline came and went with little inclination on the part of the remaining officers and men of the 53rd Battalion to move forward into the enemy machine-guns they knew were awaiting them. To pass the time while waiting for the 53rd Battalion's attack to begin, B and C companies of the 55th continued their long-range duel with their enemies on Mont St Quentin, firing on German reinforcements as they made their way forward.

By 1.30 pm the 6th Brigade had reorganised and was making another attempt to capture Mont St Quentin. On this occasion, after desperate close-quarter fighting, the Australians were successful. By 3.00 pm they had seized their objective and begun to consolidate.

As the volume of enemy fire from Mont St Quentin diminished, B Company left the railway line and took up positions some 200 metres west of the Mont St Quentin–Quinconce road, straightening the Australian line. An old dump and railway yards lined this road and the troops placed themselves where they had clear fields of fire. In this new location, B Company was in touch with the 2nd Division on the left, and D Company with elements of the 53rd on the right. C Company occupied the areas along the railway embankment vacated by B Company.

The capture of Mont St Quentin gave Lieutenant Colonel Cheeseman, under pressure from Brigadier General Stewart, the opportunity to begin preparations for the delayed attack. A Company of the 55th Battalion, which had remained unengaged during the day, was attached to the 53rd to augment this much-diminished unit. At 4.30 pm the company, under command of Captain Fred Cotterell, moved to a position on the left flank of the 53rd, just to the north of the cemetery. The diggers proceeded to dig in while waiting for the order to commence the advance.

From this location, Captain Cotterell would have had an unimpeded view of the ground his company and the 53rd Battalion were to cross. His formidable spirit must have wavered at what he saw.

Before him lay several hundred metres of knee-high grass that sloped down to the moat and ramparts of Péronne, visible to the right. To his front, a kilometre away, was a small rise with the ruined buildings of St Denis visible on the top. To his left he would have seen the other buildings bordering the road running north from St Denis to Mont St Quentin, particularly the two-storey, red-brick ruin that had formerly been a sugar factory. He would also have noted the dirt heaps of the abandoned brickworks adjacent to the sugar factory. To his extreme left was Mont St Quentin, cloaked in smoke and dust.

Closer to where he lay, Captain Cotterell would have observed the embankment denoting the path of the old Albert–Ham railway and a newly constructed railway line next to it heading towards St Denis before looping around to the east of Péronne. A spur line branched off the old track and ran atop another embankment to the sugar factory. In the distance, behind St Denis, wooded hills looked down on the battlefield. With the exception of shell-holes and the railway embankments, there was nothing to protect a man crossing this open space from shrapnel or a high velocity bullet.

At 5.00 pm, the surviving members of the 53rd Battalion, with A Company and a few men of D Company, made a mad dash for their objectives.[19] A deluge of artillery and machine-gun fire descended on them. German gunners in Bretagne enfiladed the attackers with thousands of bullets fired from the ramparts and houses.

Lieutenant Stan Colless was fatally hit in the head by a bullet as he stood atop a railway embankment directing his men towards the brickworks and sugar factory. Private John Lannon was killed as he pressed himself against the railway embankment.[20] Another fatality was Lieutenant Lance Horniman. His brother, Captain Robert Horniman (4th Battalion) provided his family with the details of Lance's death he had been able to glean from Lance's platoon sergeant, Bertie Marshall:[21]

> He [Lance] was leading his platoon forward during an attack in broad daylight, about 5 p.m., and they were held up by machine gun fire from close in front. His platoon took cover behind a bank and himself in front in a shell-hole and with the idea I suppose of locating the gun exactly so as to knock it out with bombs he told his men to keep a good look out while he exposed himself to draw the fire. He stood up clearly to do so and of course the cursed machine gun fired and he fell at once with a bullet through the head. The sergeant got to him at once and bandaged him with his field dressing but death was instantaneous … Poor Lance had made the ultimate sacrifice and laid down his life for his men.

Marshall was awarded the Military Medal for his leadership of the platoon after Horniman's death.

In the midst of the carnage, some local successes were gained. On the left, a mixed party of men from the 53rd and 55th battalions, under command of Lieutenant William Waite (53rd Battalion), reached and captured the sugar factory. However, they were too weakened to advance any further.

In the centre, a few fortunate souls had almost reached St Denis but their lack of numbers precipitated a withdrawal. On the right Lieutenant William Bevan (53rd Battalion), with a composite force of men from A and B companies of the 53rd and a platoon of A Company, had rushed straight for the machine-gun positions on the ramparts. They reached the top of the ramparts but could go no further — by this stage Bevan was down to a dozen able-bodied men. He ordered his remaining troops to return to the cemetery, the jumping-off point for the attack.

In the face of such annihilation, the ill-considered attack petered out. Those officers still engaged in the fighting gave their troops the order to pull back. Everywhere, little groups of men or individuals found whatever protection they could. Driver William Bourke, a member of A Company, crawled into a shell-hole and was fatally wounded when a shell landed in it.[22] To add to the misery of the Australians, as evening fell the Germans saturated Anvil Wood with gas shells.

Throughout the day, as the 14th Brigade struggled for Anvil Wood and Péronne, the German barrage had continued unabated in the rear areas, causing a steady stream of casualties. Private Eddie Street graphically captured one impact of this shelling:

> In the trench where we had been sitting, lay the body of a dead soldier. Further on was another one dead, then some wounded. Cam and I turned over the body where we had been sitting but could not identify it, for there was no head. Then we recognised a torn and bloody Runner's band on a mangled arm. It was Don, our mate.[23] A shell had landed on the back of his neck and the burst had shorn his head from his shoulders as would a blunt knife. We gazed on the poor headless trunk of our cobber, and we were very close to breaking down and howling openly. About a dozen yards down the trench we found Don's head. The face was little marked. Reverently we placed the head in its proper position at the top of the trunk as it lay on the ground …

> We learned that another shell had landed. The AMC Corporal, though thoroughly shaken, bandaged up two severely wounded men.[24] … Shells were still coming down. Cam and I went along to the Corporal to see if we could give him a hand. He completed his work just as we arrived, straightened himself painfully, and then collapsed in the bottom of the trench. 'I'm wounded too,' he said wearily. We found a jagged piece of steel had entered his abdomen low down and had cut its terrible way up through bowel and stomach to near the diaphragm. The fine fellow had calmly bandaged up the other wounded while he himself was suffering the agonies of a mortal wound. We found a stretcher and carried him around to the Doctor but nothing could be done for him. He lived long enough to go back to the Casualty Clearing Station and word came that he smiled as he died.

At 8.30 pm Brigadier General Stewart ordered the retirement of those elements of the 53rd and 55th battalions east of the Anvil Wood cemetery 'so that a thorough artillery preparation of the whole area may be carried out as the necessary prelude to further infantry attacks' already planned for the next day.[25] Darkness also brought a cessation in the shelling. The 55th pushed patrols forward and began reconnoitring the ground to its front. Everywhere men settled down for another uncomfortable night. The wounded from the day's fighting, some walking, others carried on stretchers by prisoners or Australian stretcher-bearers (if they were badly wounded), straggled down the Péronne–Clery road. The historian William Stegemann writes of the compassion shown by some of the Germans carrying wounded Australians:

> Private Bert McDonald received shrapnel wound to the leg.[26] Two German prisoners were detailed to carry him back on a stretcher … On the way the Germans stopped, put down the stretcher, and shared their ration of black bread and wine with Private McDonald.[27]

The fighting had cost the battalion a total of 23 killed and dozens wounded. Little had been gained. Corporal Harris reflected on how the day's events had unfolded:

> Boys getting a bad time. D Company got 30 men left, out of them 5 are unwounded. A Company 6 killed 12 wounded. Poor old Siddons[28] gone, also Lieut. Colless, Lieut. Hourn WIA … How much longer are they going to stand this awful loss and how much longer are they going to keep us in? No one will be left soon.

Lieutenant General Monash now intended to pause for a day while he advanced his heavier guns and gave his exhausted infantry a brief respite. This did not eventuate. Lieutenant General Alexander Godley, commander of III Corps located on the northern flank of the Australian Corps, had received approval to use the British 74th Division in an attack the next day and wanted the Australian Corps to conform to his advance. Reluctantly, Monash agreed to throw his men back into the fight and issued his orders accordingly.

At 9.30 pm Major General Hobbs held a conference at 8th Brigade Headquarters situated at Herbecourt. Brigadier General Stewart, Brigadier General Edwin Tivey (Commander 8th Brigade) and a representative from Corps Artillery attended the conference. Not one of these officers had made a personal reconnaissance of the ground. Brigadier General Elliot of the 15th Brigade was also invited but had somehow become lost on his way to Herbecourt.

Major General Hobbs informed his two brigade commanders that the plan for 2 September required the 2nd Division to continue the exploitation east from Mont St Quentin. This division would form a flank to enable the 5th Division to complete the capture of its original objectives — Péronne and the hills east and south of the town.

The task given to the 14th Brigade was to use its fresher battalions to repeat the operation that had failed so disastrously that afternoon. This time it was the 56th Battalion, reinforced by B and C companies of the 55th, that was to capture St Denis and St Denis Wood. The attack was to occur on a one-company front; that is, the four companies of the 56th and the two of the 55th lined up behind one another. Once St Denis and St Denis Wood had been captured, the two companies from the 55th were to swing around and clear the hills south of Flamicourt. The 54th Battalion, now down to 100 fit men, was tasked with completing the seizure of Bretagne and the remainder of Péronne.

Despite the fact that his battalions were already in place, Brigadier General Stewart again argued that the 14th Brigade would not be ready to attack until 6.00 am and this time was set as Zero Hour. For some reason, there was a critical breakdown in staff work at Headquarters 5th Division and Major General Hobbs did not inform Stewart that the 2nd Division was attacking at 5.30 am.

The 55th and 56th battalions were fated to suffer the same slaughter as had befallen the 53rd that afternoon unless the German machine-guns were suppressed. To achieve this outcome, Major General Hobbs proposed an artillery fire plan similar to the one used that morning, namely the engagement of selected targets. The supporting barrage was to commence at 5.30 am, targeting the brickworks, sugar factory, St Denis Wood and the Bretagne ramparts.

Brigadier General Stewart left the conference and proceeded to his own headquarters, missing the belated arrival of Brigadier General Elliot at the 8th. Elliot convinced Major General Hobbs to halt the 15th Brigade's attempts to cross the Somme and to switch the focus of his brigade's operations to the area around Péronne. With Hobbs' approval, Elliot ordered his 59th and 60th battalions to follow up the 14th Brigade assault. He also generously decided to send the 58th Battalion to Péronne with instructions that Lieutenant Colonel Norman Marshall (CO of the 54th and a former 15th Brigade officer) could use them as he saw fit. To compound the potential for confusion, Brigadier General Stewart was not notified of the changes to the 15th Brigade's mission.

Brigadier General Stewart arrived at his headquarters at 2.00 am and began issuing orders by telephone to his COs. By the time these trickled down, Captain Wyllie commented that 'There was just time to tell everyone what was going to happen then it was off.' According to the 55th Battalion's after-action report, 'The men were now worked up to a great fighting spirit and were all very keen to get to close quarters with the enemy.' Private Bishop recalls there being a somewhat different feel in the ranks as orders were given for the attack:

> [Lieutenant Inglis] seemed reluctant to speak.
>
> 'Come on, Sir,' Corporal Franks said.[29] 'It can't be any worse than we're expecting.'
>
> 'I've never felt so wretched.' Mr Inglis paused. 'I have to give you the plans of our attack.'
>
> 'And the planner is tough on us,' Pojo put in.
>
> 'He certainly is. We go forward at dawn, making a frontal attack on Péronne. The Second Division attacks Mont Quentin. We have no idea what to expect. Orders are just to continue our attack.' He was silent a few moments. 'Now I'll tell you the hardest bit of all … We have no artillery cover or support at all. I think you know what to expect.'
>
> We knew right enough. We knew it would be plain murder for us.[30]

It wasn't just Lieutenant Inglis and C Company who had reservations about the attack plan. Brigadier General Stewart confided in his diary later that night:

> Protested as to the impossibility of the proposal but was overruled. This show is an impossibility until the Péronne ramparts are cleared. The artillery fire we can put down will not keep the enemy m.g. fire down and attacking troops will suffer as the 53rd did today … The Boche has made up his mind he is going to keep their ground and there will be some hard fighting before he is driven out.

The dawn of 2 September was dull and calm. The supporting bombardment was ragged, possibly a result of the orders arriving late. The enemy retaliatory barrage, however, was anything but ragged. From their posts in the surrounding hills, the enemy artillery spotters enjoyed an unimpeded view of the 56th's preparations for its assault; casualties immediately began to occur among the waiting, unprotected men of this battalion.

During the night, the Germans had reoccupied the sugar factory and brickworks, and had strengthened their detachments in St Denis and Bretagne. The Australian troops could see enemy machine-gun crews making their final arrangements to receive the attack. The ramparts gave robust protection to the machine-gunners, and the supporting artillery fire was neither heavy nor sustained enough to dislodge them.

The enemy barrage wreaked havoc on the 56th Battalion; all officers from three of the attacking companies became casualties before Zero Hour. B and C companies of the 55th avoided losses from the shellfire as they were able to remain in the trenches in which they had spent the night, fortuitously situated in the assembly area.

At Zero Hour, the 56th's leading waves crossed the jumping-off line (the Mont St Quentin–Quinconce road) and commenced their advance towards St Denis and the brickworks. The German response was immediate; their artillery increased in ferocity and machine-guns began pouring fire into the khaki-clad line. Private Bishop witnessed the cauldron entered by the 56th Battalion: 'They [the 56th] had gone only a few yards when everything the world held smashed and shrieked and blazed across their front. Smoke and dust reduced visibility.'[31]

B Company was to follow behind the 56th Battalion. According to Captain Wyllie, after the rearmost company of the 56th had passed through the 55th, his company: 'Moved forward in artillery formation under the heaviest enemy barrage we have ever experienced … C Company was just in the rear of B Company and crossing the barrage area in perfect formation.'

Collectively, B and C companies suffered around 12 casualties before they reached the jumping-off line. The Germans were mixing incendiary shells into their barrage, the burning phosphorous condemning those it touched to an excruciating death or painful injury and leaving the able-bodied with tears streaming from their eyes and mucous running from their nose. Private Bishop recalled of the C Company advance:

> Shells of all sizes fell and burst about us … The earth heaved from the big ones, great showers of clods and rubbish fell on us. The smaller ones sharply cracked and threw their venomous pieces of metal about us. The air was shrieking in agony as metal tore through it. The air held a steady, menacing great swish as a myriad machine gun bullets hurtled around. Men were falling everywhere, none but stretcher-bearers allowed to attend them. Those of us unhurt just went steadily forward, our left shoulders down and at an angle as if to protect us from the storm. It was our Polygon Wood barrage coming back to us.[32]

When interviewed after the battle by the historian Charles Bean, Captain Wyllie commented that: 'The spirit of the men was magnificent, they laughed and joked as they pushed forward.' Wyllie omitted to tell Bean that it was gallows humour by men who already regarded themselves as dead. Private Bishop also observed the demeanour of the troops: 'I looked at the men about me. We had never gone into battle like this before. There was no interest, no keenness showing anywhere. We were switched-off automatons, unthinking, uncaring, walking into the valley of death.'[33]

As they advanced across the open field, the battalion continued to lose heavily. Bert Bishop was in the middle of the action:

> Through the murk in front we saw a railway embankment and raced for its protection. I saw our company commander, our platoon commander and Sergeant Harlock there, and one or two others. I joined them. We were all out of breath. As soon as Harlock could manage it he began shouting at the top of his voice 'Twelve Platoon, over here. Come here, Twelve Platoon.' One or two joined us. Then no more.[34]

The surviving members of the 55th crouched along the dubious shelter provided by the railway embankment running to the sugar factory. The rear elements of the 56th were also there, alongside the rigid corpses of the fallen from the previous day. Terrific machine-gun fire from the Bretagne ramparts, St Denis, the nearby brickworks, and the sugar factory swept the embankment and the grasslands.[35] The machine-guns in the top floor of the factory were able to bring constant enfilade fire onto the men pressing themselves into the earth along the embankment. Bullets struck sparks off the rails.

The combination of machine-guns and artillery had inflicted grievous casualties on the 56th Battalion, which almost ceased to exist. Under the leadership of Sergeant Alexander O'Connor (56th Battalion), the remnants of one company were able to push forward over the railway embankment and keep advancing towards St Denis.[36] They reached a point some 400 metres from the hamlet before the sheer volume of fire forced O'Connor's remaining 35 men to halt and commence digging in.

The two companies of the 55th were in better shape than the 56th; none of their officers was yet a casualty and their internal organisation remained intact. In the confusion of the advance, B Company had become slightly disoriented and drifted north, finding itself on the left of the attack. C Company, more aligned with the original axis of advance, moved to a position along the embankment to the right of B Company, where it was better able to support Sergeant O'Connor's party ahead of it.

German artillery fire continued without respite, the occasional gas shell added to the barrage to further hinder the Australians. One blast mortally wounded Private Albert Priddle, fragments from the same shell wounding 16 other men.[37] Another shell disembowelled Private William Flynn,[38] simultaneously wounding Private Patrick Corbett[39] and Private Hedley Webster.[40]

Acting on his own initiative, Private Walter McDonald worked himself into a position where he had a good view of one of the machine-guns holding up the advance.[41] Despite being fired on by German marksmen, he was able to kill the machine-gun crew and render the gun unserviceable.

At 6.45 am, Captain Pinkstone, realising that the attack was faltering, ordered C Company to advance in support of Sergeant O'Connor's detachment, telling his men to 'fix bayonets and get into them'. The open ground in front of the railway embankment was still under intense machine-gun and artillery fire. The 55th's official account provides a version of what transpired:

> The memory of such a charge will long be remembered … There was something distinctly awesome in the way our troops went forward with the billows of machine gun fire and shell barrage smashing against them, and as far as stopping the charge, smashing in vain … they never wavered [but] pushed on until, owing to a terrific concentration of gas, they were forced to fall back to the position from which they hopped off.

Corporal Rupert Campbell of C Company left a different account of his company's push forward:

> We were having a bit of a 'box on' down at Péronne on the 2nd September. The Huns were in a village on a hill about 200 yards away [St Denis] and were indulging in a bit of musketry practice in the shape of gas shells, shrapnel and high explosive. To make the practice interesting and natural, Fritz was firing at live targets. We [author emphasis] were the live targets. It is the luckiest thing in the world that I wasn't a dead [author emphasis] target. Naturally, with so much 'backsheesh' flying around we were lying down flat on the ground fervently wishing we were jellyfish or something. We commenced crawling towards the machine gun manipulators and after I had gone 15 or 20 yards the aforesaid bullet that had my name written on it, arrived. It missed the old nut by a couple of inches and went for a little trip through my shoulder and came out down the middle of my back.

Private Bishop also described the C Company attack, providing insights into the enormous stress the action placed on officers and men:

> He [Pinkstone] went along to our left, disappearing into a tunnel cut through the embankment. Coming back he told us the tunnel ran through the embankment, there were some big shell-holes near its mouth …
>
> We obeyed the order … Just as Harlock was about to enter the tunnel a shell landed nearby. A chunk of it smashed into Harlock's shoulder.
>
> Ernie [Ernie Corey], our stretcher-bearer, had been busy before he came to our embankment. He arrived just as the shell fell. He was knocked over, but got up unhurt. He went to Harlock. I stood beside them. Ernie quickly inspected the damage. One of Harlock's arms was attached to his body by one sinew only. He was bleeding badly. Ernie took a blade razor from his breast pocket, opened it.
>
> 'Here,' he told me, 'hold the arm straight from the shoulder.' He cut through the sinew and got busy trying to stop the bleeding. I had the arm hanging from my hand. I never thought a man's arm was so heavy.[42]
>
> I dropped it and got into the tunnel. Our two officers were in a big shell-hole wondering why no one followed them. One by one we reached the tunnel and made a quick dash for the shell-hole, each man bringing a fusillade of bullets about him. Two big shell-holes joined each other, and with our entrenching tools we made some sort of post.

> Things seemed no happier. The two officers, popping up to see if we could do any better, decided that some kind of ditch or hollow about twenty yards in front would give better shelter. It looked like a long-disused gravel or clay pit.
>
> Our captain [Pinkstone] decided he'd go first and beckon us to come if it was better. Waiting for a quiet moment, he jumped out and ran like a hare. A couple of gunners spotted him and fired, but he disappeared into the hollow. He beckoned to us. Mr Inglis was next, he got there safely. One at a time we ran that gauntlet, safely getting there …
>
> Our company commander was a worried man. All in that shallow shelter were men of Twelve Platoon. Where the rest of the company might be was just a big question mark. By keeping our heads down we were safe enough, but the slightest unthinking movement brought a storm of machine-gun bullets. Our company [commander] kept popping his head up and down trying to think out what best to do. He turned to Mr Inglis.
>
> 'There's a ditch or another gravel pit about twelve yards away. Inglis, do you think you could make a dash for it? Beckon us to go if it's better than this place.'
>
> It was not exactly an order, but the words meant the same thing …
>
> 'Oh, my God,' I thought. We knew several German machine-guns were focussed on our hole. It meant certain death to expose oneself. I looked at our platoon commander [Inglis]. He knew he should not go, only a miracle could get him there.
>
> He took a deep breath, put his right foot in a hole in the bank, and jumped out to race across. He was not really out of our possie when he slid back, a dozen bullets through him.
>
> Captain Pinkstone looked awful.
>
> 'My God, what have I done?' He was sorely distressed. Silently and worriedly he tried to think of some way of getting out of our trap.[43]

Captain Pinkstone had by now realised the impossibility of what he was attempting to do and ordered a withdrawal. He directed six men to stay behind to cover the retirement. The group had several Lewis guns and proceeded to play a dangerous game of cat and mouse with the German machine-gunners — popping up, firing a quick burst, then ducking back in the shell-hole, all the time under artillery and machine-gun fire. Four of the six men in the shell-hole received Military Medals for their bravery: Private Bishop, Private Edward

'Woodsy' Woods (who continued to fight although shot in the buttocks[44]), Private Martin 'Canty' Cant[45] and Private William Christie.[46]

Under the covering fire provided by the Lewis gunners, the C Company survivors dribbled back to the railway bank. Just before he reached the earthen wall, Pinkstone himself was severely wounded, a machine-gun bullet passing through his right buttock. With Pinkstone *hors d' combat*, and most of his men killed or wounded, the attack was extinguished. C Company was no more.

While C Company was smashing itself to pieces in a frontal assault, B Company remained behind the embankment. The company was considerably thinned in numbers. Private Eddie Street had been sent forward with dispatches for Captain Wyllie and described what he saw in the B Company location:

> When I arrived at B Coy post near the railway bank I found it in a dreadful condition. It was the only time I saw dead piled up in a trench. Most of the lads were Diggers and had been killed by MG fire from the ramparts. It was there I saw a wound that haunts me yet. It was not a great gash or dismemberment. The lad lay partly on his side and his tin hat had been pushed back a little on his brow when he fell. A bullet had drilled a neat hole in the centre of his forehead, and a little of his brain oozed out of it. I shuddered to think the same thing might happen to me. A burst of fire and I climbed down among the dead bodies in the trench ... There I handed over my dispatches and returned walking as steadily as I could.

Realising the C Company attack was fading, Captain Wyllie grasped the initiative and directed his company to move further left along the embankment and seize the sugar factory. Under the leadership of Lieutenant Horby Phillips[47] and Lieutenant Rupert Ellsmore, the troops moved inside the factory.[48] In close-quarter fighting they were able to clear the Germans from this position and the adjoining brickworks by 7.45 am.

Once ensconced in the sugar factory, Captain Wyllie was able to make contact with the 2nd Division on the eastern side of the St Denis–Mont St Quentin road; he was also able to speak to Lieutenant Colonel John Scanlan, CO of the 59th Battalion (15th Brigade),whose leading elements had, unexpectedly, begun to appear behind the railway embankment.[49] Private Street recorded the impact of the 15th Brigade's arrival:

> The remarkable 15th Bgde had somehow or other crossed the Somme down near Péronne itself ... The 15th Bgde gave one the impression at this period of the war that they really enjoyed fighting. We were overjoyed when we saw their colour patches joining up with our line, for the unit was nearing exhaustion.

Enemy machine-gun fire was still intense. The 2nd Division, tasked with clearing the ridgeline running off Mont St Quentin, was making slow progress and the troops near the sugar factory and brickworks were still under enemy fire. Meanwhile, the Germans in St Denis, aware of the threat to their flank, focussed the fire of a number of machine-guns on B Company.

From the superb cover and height provided by the factory and the brickworks, troops began to snipe the Germans they could see in the broken buildings lining the road to St Denis. With a clear view towards Péronne, they were also able to identify Sergeant O'Connor's isolated detachment 400 metres to the south.

As their numbers were whittled down by the accurate Australian sniping, the German fire from the vicinity of St Denis diminished and then stopped. During this lull, Captain Wyllie established a perimeter line running alongside the St Denis–Mont St Quentin road just east of the sugar factory. Two Vickers guns were brought up to strengthen this position. Taking advantage of the decrease in the volume of enemy fire from his left and front, Sergeant O'Connor began dribbling men to his left, finally succeeding in gaining touch with Wyllie's troops.

By 10.00 am, the Bretagne ramparts had been cleared of the machine-gunners that had wrought such carnage on the Australian advance. In house-to-house fighting the 54th Battalion, augmented by the 58th Battalion, had captured the prize denied them the previous day. Three companies of the 58th were moving through Péronne to establish a defensive line on the light railway to the east of the town. Brigadier General Stewart's misjudgement in not taking all risks to clear Péronne the previous day had been paid for in the blood of his brigade.

The Australian line now extended from the 54th Battalion in Péronne to the 55th Battalion in the sugar factory and from there to the area held by the 2nd Division. As the intensity of the battle waned, the diggers were able to move about more freely. Private Bert Bishop now found himself having to deal with the after-effects of his traumatic experiences:

> Dazedly I walked along that railway embankment. The German machine-guns in the top windows of that red brick factory [the sugar factory] were silent ... I was the only living creature about that bank of death. My head felt as though it was about to burst. I sat down and pressed my open hands to it. After a few minutes the pain eased somewhat. I looked around. I saw a man moving towards the bank,

> dragging a body. It was Ernie [Corey]. He placed the body in line with those already there. As he turned to go back he saw me.
>
> 'Bert,' he called, 'are you wounded, do you want help?' 'No, Ernie.'
>
> Ernie went back to the body-strewn field. He held aloft a dirty-looking rag, it looked like the tail of his shirt, but no Germans fired at him. He picked up another body, dragged it in.
>
> 'Can't find any more wounded,' he said. 'Look at it, Bert, a whole field of dead men.'[50]

Australian observers began noticing a general German withdrawal towards the woods on the overlooking hills. Mindful of this movement, Captain Wyllie's attention now turned to fulfilling the first objective of the assault — the capture of St Denis and St Denis Wood. In the early afternoon he sent out a platoon under command of Lieutenant Ellsmore to clarify the tactical situation. Crawling across the road and under the protective fire of Australian machine-guns now on the ramparts, the platoon made its way through the buildings along the St Denis–Mont St Quentin road, finding them free of enemy.

Lieutenant Ellsmore and his men pushed on, crossing the St Denis–Aizecourt road. Moving forward another hundred metres, the platoon entered St Denis Wood, a plantation of stunted trees and low undergrowth. Suddenly they bumped into a well-camouflaged German post; neither side was expecting the other. A sharp exchange of fire ensued; an Australian fired first, the bullet hitting the shoulder of the enemy officer in charge of the post. Private Sylvester McMahon used his Lewis gun to silence the two enemy machine-guns in the position.[51] As abruptly as the fight began, it ended. Ellsmore's men captured 11 Germans (including the wounded officer) and the two machine-guns. Three Australians were wounded in the exchange.

Now aware of the Australians' presence, other Germans in and around the wood began to fire on Lieutenant Ellsmore's party. Far from withdrawing, the enemy still occupied the area in some strength. Ellsmore, a long way from any form of support and taking casualties, ordered the withdrawal of his platoon to the brickworks. With their assortment of prisoners, the two machine-guns collected from the post, an additional three machine-guns they found abandoned in St Denis and burdened by their own wounded, the platoon's progress was slow. Private McMahon used his Lewis gun to cover the withdrawal, despite being exposed to heavy machine-gun fire and encumbered by a badly wounded member of his platoon.

As the Australians passed through St Denis, the Germans began dropping gas shells into the hamlet, incapacitating Lieutenant Ellsmore. His platoon sergeant, Forbes Mackenzie, took over command. By 2.00 pm, Mackenzie had brought the platoon, the trophies, and prisoners back to the brickworks.[52]

The effective strength of B Company had been reduced to two officers and 25 soldiers. All of these were exhausted by the strain of intense combat. Lieutenant Colonel Woods directed the remnants of C Company to move to the sugar factory and merge with B Company; he gave this composite company the mission of protecting the right flank of the 14th Brigade and maintaining liaison with the 2nd Division. At the direction of Brigadier General Stewart, A and D companies, which were still with the 53rd Battalion, reverted to the command of Lieutenant Colonel Woods and moved to positions just north of Anvil Wood. Here they were surrounded by acres of corpses from their own and sister battalions.

According to the 14th Brigade's War Diarist, the nature of the day's battle saw the fighting strength of the 55th and 56th battalions so 'considerably reduced and their elements closely interlocked' that Brigadier General Stewart decided to form a single battalion under Lieutenant Colonel Woods. Stewart directed Woods to 'take charge of the situation forward, to consolidate the line held and organise its defence ... no further offensive operations to be undertaken.'

Owing to heavy officer casualties in both battalions, the exact tactical situation was obscure. Woods left his command post sited in the trenches where the battalion had spent the night, and went forward under shellfire to organise the consolidation of the line slightly west of the St Denis–Mont St Quentin road.

The battalion's stretcher-bearers had, by now, been working without respite for over 30 hours. At times under heavy machine-gun and artillery fire, they practised their arduous and dangerous trade. Five bearers were awarded the Military Medal for their courage and endurance: Private Corey, Private William Nicholls,[53] Private James 'Gundy' Marshall,[54] Private James Diversi[55] and Private Bert Schwind.[56] Lieutenant Colonel Woods had recommended Corey for a Distinguished Conduct Medal but he had to settle for a second Bar to his Military Medal.

On one occasion, in the aftermath of the battle, Corey saw two German stretcher-bearers some 70 metres away, preparing to evacuate a wounded Australian on a wheeled stretcher. They beckoned to Corey, who went towards them and called out: 'Can you speak English?' One of the Germans replied: 'We have Australia. You take him. Too heavy.' Corey instructed them to leave the stretcher and, when they walked off, he made his way to it and found Corporal Ronald 'Nugget' Randall, who, standing at 193 centimetres tall and weighing 80 kilograms, was one of the biggest men in the unit.[57] Randall had caught a machine-gun burst in his shoulder and back. Corey picked Randall up in his arms, carried him back 50 metres and placed him in a shell-hole. Meanwhile the Germans came back for their stretcher and waved farewell to Corey.[58] Randall died of his wounds the next day in a casualty clearing station.

The 58th Battalion attempted to move east but was prevented from reaching the Péronne–St Denis road by German machine-gunners in St Denis Wood, Flamicourt and the heights overlooking Péronne. Brigadier General Stewart realised that the Germans had to be expelled from St Denis Wood before the attack could be resumed.

Disregarding the woeful condition of his brigade, late in the afternoon Brigadier General Stewart gave Lieutenant Colonel Woods new orders — his composite battalion was to seize St Denis and St Denis Wood. Lieutenant Colonel Scanlan had earlier advised his brigade commander (Elliot) that a further advance was impossible until Mont St Quentin was cleared and artillery support arranged. Charles Bean notes that Woods supported Scanlan's recommendation.[59] Neither of these two preconditions had yet been satisfied.

Whatever his private misgivings, Lieutenant Colonel Woods, working with the 14th Brigade's other battalion commanders, set about planning an attack to take place after dark.[60] By this stage the 56th Battalion was only able to muster around 100 fit men and the four companies of the 55th were not in much better shape; Woods could commit more than 300 men to the assault. An officer of the 15th Brigade described the condition of the 14th Brigade:

> The white haggard men, tired almost beyond the limits of human endurance, many of them wounded, all suffering from the effects of gas, Lewis gunners too weak to move their guns, the officer in charge tottering from weakness and loss of blood, but affecting a hearty bearing lest the spirits of his men should suffer.[61]

These were the troops being ordered once again into battle.

Given that they had been spared the extreme exertions of the day's fighting, Lieutenant Colonel Woods gave the leading role in the assault to A and D companies of the 55th. Captain Cotterell and his runner, Private Edmund Edwards, set out to conduct a reconnaissance of the ground over which his company was to advance.[62] They never returned: a shell blew off both Cotterell's legs, mortally wounding him; Edwards' mangled body was found nearby, killed by the same blast. Private Street lamented their deaths: 'So passed a kindly, efficient and brave officer, and a game and competent runner.'

Bert Bishop was underwhelmed at the prospect of another attack:

> My own platoon strength was twenty-one when we started, now we were six …
>
> Two men were alive enough to be arguing. One had told the other to 'cheer up, laddie, you'll soon be dead.' …
>
> Other men were talking. They were angry. Things were said that I never expected to hear Aussies say. If they could have got their hands on the heads responsible for our attack the heads would have been torn to pieces. We were all in a very ugly mood.[63]

Dusk was turning to night. Lieutenant Colonel Woods had just finished putting the final arrangements in place when orders were received cancelling the operation. Instead of attacking, the 55th Battalion was to be relieved by the 59th Battalion at midnight and retire.

What had transpired to change Brigadier General Stewart's mind concerning the need for another attack? According to the 15th Brigade report on the operations around Péronne, on hearing of the orders for a continued assault, Brigadier General Elliot contacted Major General Hobbs to point out to him 'the impractability of carrying out this operation in view of the general situation at that particular part of the front.'

Hobbs convened a conference with his brigade commanders, in which it was decided that, as St Denis Wood was still occupied by the Germans, the attack should be cancelled. This was an interesting decision — Stewart had ordered the attack with precisely the intention of clearing the enemy known to be in St Denis Wood. Possibly Hobbs was unaware of Stewart's decision to recommence the offensive until advised by Elliot, and then took action to overrule a poor decision by his subordinate.

The 15th Brigade report on its operations around Péronne contains the following section (one can almost hear Brigadier General Elliot dictating it to a stenographer):

> Had the operation been insisted upon there is no doubt that the troops launched to the attack would have been utterly annihilated. The valley behind St Denis is completely commanded by hills on three sides of it and these were strongly held by the enemy whose machine gun fire was terrific. This again emphasises the grave necessity for personal reconaissances even at great risk for senior commanders before such orders were issued [a 'dig' at both Brigadier General Stewart and Major General Hobbs, as neither officer had visited the front around Péronne]. In this instance the Brigadier General [Elliot] found himself obliged to state that if the operation as laid down was insisted on contrary to his judgement he would have no option but to resign his command the following day.

Given that the 59th Battalion was already in close proximity to the positions of the 55th, the relief commenced promptly at midnight. The 55th was directed to make its way to a rest area on the southern side of the Ommiecourt bridge. As their respective posts were relieved, the men headed rearwards. Private Bishop was nearing the end of a long day:

> We four gunners took our time, we had many rests. It took four hours for us to go four miles.
>
> The battalion Q.M. was calling out platoon numbers. 'Twelve Platoon? Over here, Twelve Platoon.'
>
> There was a big rectangular dixie full of steaming stew. We flopped down beside it. 'Four of you. Where's the rest?'
>
> We squatted around it. We ate. We ate. And still we ate. A transport officer joined the Q.M. The latter turned to him and said in an awed voice: 'My God, Sir, they're not men at all, they're ghosts.'
>
> The Q.M. brought a smaller dixie. It contained tea and rum. Unable to eat more stew we did all we could with the drinking department. Then we just flopped over asleep.[64]

On the morning of 3 September, the battalion reorganised. Casualties had been horrendous: 62 killed and many times that number wounded. Lieutenant Colonel Woods was forced to merge A Company with B Company, and C Company with D Company. Woods was also short of officers to lead his remaining troops, allocating A/B Company to Captain Wyllie and C/D Company to Lieutenant Alfred Hawkins.[65]

The War Diarist reflected on the ordeal the unit had just endured: 'The men themselves say that it was the fiercest fighting that they have encountered and that the quality of the enemy machine gun fire surpassed that of Fromelles itself.' Corporal Harris also contemplated the events of the previous two days:

> Thank heaven we are out of that hell at last. 180 strong. Cpt Cotterell and Edwards had a shell to themselves – blown to bits … We are now in a deep ravine about 2 kilo from our former positions and were greeted by three big shells with gas in them. Got some down my throat again.

Charles Bean wrote of the battle around Mont St Quentin and Péronne:

> This brilliant action, in which, without tanks or creeping barrage … dealt a stunning blow to five German divisions … Where he [German General Ludendorff] had previously intended to hang on [to the line of the Somme around Péronne] he could now however only retreat to the Hindenburg Line.[66]

The 55th's after-action report boasted of the achievement of the Australians in capturing Péronne and Mont St Quentin:

> The memorable work of those two days has been described as the most brilliant thing the Australians have yet done. It may not capture the popular imagination as some of the now historic feats of the Australian Army, such as the landing at Gallipoli, Pozieres and other engagements have done. It lacks the sentimental glamour surrounding the two victories just mentioned, and as the casualties were comparatively small, it does not possess the tragic significance of many other battles. But here the Australians performed a month's work in two brilliant days, and considerably enhanced their reputation for performing the 'impossible'.

David Coombes, in his biography of Major General Hobbs, asserted: 'Never had Australian troops been asked to do more than was asked of these men, their gallantry and determination was magnificent.'[67] In the days following the battle, the Germans executed a general withdrawal to the Hindenburg Line.

The 55th Battalion was battered and chronically under-strength. The men were war-weary and wanted respite from the constant strain of battle, or the thought of future battle. They believed themselves sorely treated by 'the heads' and were increasingly cynical. Corporal Harris wrote of a parade he attended after Péronne:

> General Hobbs said words failed him and nothing he could say would be half good enough, he was filled with admiration for the 55th and better soldiers never marched. Never had such praise as this before but I suppose it means we are going in again.

The sacrifice of the 14th Brigade at Péronne was part of the price Lieutenant General Monash was prepared to pay for future operational success by the Allies. He smelled German blood in the water and was moving in for the kill.

Chapter 15: Bellicourt – 'what a time you must have had'

Following the battles for Péronne and Mont St Quentin, Monash's Australian Corps pursued the retreating Germans with grim determination. Day after day, the corps pressed the German rearguards back towards the Hindenburg Line. By the middle of September 1918, the Hindenburg Outpost Line, a collection of fortified villages and local strongpoints designed to disrupt the momentum of an attacking force, had been pushed in. Australian troops were now confronted with the remarkable might of the Main Hindenburg Line and the subsidiary Le Catelet–Nauroy and Beaurevoir Lines.

Lieutenant General Monash was tasked by General Henry Rawlinson, commander of the Fourth Army to which the Australian Corps was attached, with punching a hole through the Hindenburg Line. This objective, if accomplished, offered the tantalising prospect of fracturing the German Army and shortening, if not even ending, the war. The US Army's II Corps, comprising the 27th and 30th divisions, was placed under Monash's command for the attack.

The section of the Hindenburg defences that the 5th Division was to attack lay a few kilometres east of Hervilly and about 30 kilometres south-east of the Bullecourt segment of the Hindenburg Line where the 55th Battalion had been bruised in the spring of 1917. The fortified village of Bellicourt, situated just behind the Main Hindenburg Line, lay near the centre of the forthcoming Australian operations.

To the south of Bellicourt lay the St Quentin Canal. Constructed by order of Napoleon Bonaparte and completed in 1810, the canal was part of a network allowing barges to travel around France and on to Belgium. A segment of this canal had been built beneath the hilly terrain near Bellicourt: this tunnel began 400 metres south of Bellicourt and emerged near Le Catelet, some four kilometres away.

Napoleon's engineers had dumped the spoil from their tunnel excavations on the surface, forming a mound along the entire length of the tunnel's course. Known as the tunnel embankment, the dirt heap was around 20 metres wide and up to four metres in height. Air shafts had been dug at regular intervals along the embankment to provide ventilation for the bargemen working dozens of metres below. Rising above the surrounding countryside, the embankment provided excellent observation to the east and west.

The canal and tunnel had considerably influenced the defensive architecture of this section of the Main Hindenburg Line. Where the canal was above ground the Germans had constructed the Main Hindenburg Line on its eastern side, using the water and the steep sides of the canal as an obstacle to attacking infantry and tanks. From the perspective of the German planners, the underground part of the tunnel was a wide avenue through their defences — it was also the logical place for tanks to operate. To strengthen this vulnerable spot the Germans constructed layer after layer of defences, making these few square kilometres one of the most heavily fortified areas on earth.

The first layer was the Main Hindenburg Line itself which consisted of two trench lines shielded by up to five rows of barbed wire, each row up to ten or more metres thick. The trenches themselves were typified by the widespread use of deep, reinforced dugouts that sheltered the German garrison. Pillboxes and other concrete-reinforced strongpoints were scattered along the Line to provide protection for machine-guns and anti-tank cannons.

Behind the Main Hindenburg Line, the Germans had sprinkled machine-gun posts along the length of the embankment. Many of these were sited close to entrances to the tunnel itself, allowing the gunners to be reinforced, relieved and resupplied. The protection of these openings was critical because the Germans used the tunnel as quarters for thousands of troops. In turn, these soldiers needed secure egress points.

A wide, gently sloping valley separated the tunnel embankment from the next ridgeline some 700 metres to the east. Along this ridgeline, named Railway Ridge in the northern section where the 55th Battalion was to be engaged, the Germans had constructed a single well-wired trench known as the Le Catelet Line.[1] From this strong defensive position, and the fortified Cabaret Wood Farm on the higher ground just behind it, the German defenders dominated the valley back towards the embankment; any attacker coming from the direction of the Main Hindenburg Line would be forced to cross this killing field. A series of communications trenches, many of which ran over the top of the embankment, linked the Le Catelet and Main Hindenburg Lines.

Aerial photographs had been taken of the area between the Main Hindenburg Line and the Le Catelet Line. From these it was possible to distinguish secondary trenches and saps radiating along spurs and other topographical features. 'Mini fortresses' containing large numbers of machine-guns and the occasional artillery piece could also be discerned. These same photographs showed that barbed-wire entanglements were abundant, although lacking in the depth and formal layout characterised by those fronting the Main Hindenburg Line.

East of the Le Catelet Line was the final defensive layer: the Beaurevoir Line. This was the least mature of all of the Hindenburg Line's positions in terms of the quality of its entrenchments, wire entanglements and strongpoints. Including the Beaurevoir Line, the total depth of the Hindenburg Line's defences was approximately four kilometres.

The first phase of Lieutenant General Monash's assault plan called for his two American divisions to advance 4000 metres behind a creeping barrage with the mission of capturing the Main Hindenburg Line and the Le Catelet Line. Once these objectives were secured, the 3rd Division on the left and 5th Division on the right were to leapfrog through the American troops and seize the Beaurevoir Line and the fortified villages and strongpoints lying between the Le Catelet and Beaurevoir Lines. This second phase would involve an advance of around 5000 metres without the support of a creeping barrage. British divisions were to attack on the right and left flanks. Zero Hour for the operation was set for 5.50 am on 29 September.

In planning the 5th Division's role in the second phase of the assault, Major General Talbot Hobbs intended to pass his division through the 30th American Division with the 8th Brigade on the right and 15th Brigade on the left. The 14th Brigade, including the 55th Battalion, was to make its way to a concentration area near Nauroy, where it was to remain prepared to be committed to the battle where necessary.

The 55th had little rest following the exertions of Péronne. In the days after the battle, the unit joined the pursuit of the retiring Germans, but had no contact with the enemy, apart from being on the receiving end of an occasional aerial bombing or strafing. After a week, the 55th finally moved into recuperation camps around Le Mesnil, just south of Péronne. After two months of intense operations the 14th Brigade, like the rest of the Australian Corps, was undermanned and exhausted, a situation exacerbated by the decline in the numbers of replacements arriving from Australia.

To partially alleviate the manpower shortage, Lieutenant General Monash directed that one battalion from each brigade was to disband and augment the remaining three battalions. The 14th Brigade's commander, Brigadier General Stewart, designated the 54th Battalion as the one to be broken up. In response, the soldiers of the 54th went on a 'strike', refusing orders to merge with their sister battalions. They were not the only battalion to act this way and after several days Monash was forced to back down and postpone the implementation of this policy until after the coming operations against the Hindenburg Line.

The men of the 55th were on the sidelines of the events engulfing the 54th Battalion. Of greater impact on the unit was a directive, championed by the Australian Prime Minister Billy Hughes, that the 1914 enlistees had been granted six months' furlough and were able to return to Australia. In the 55th, this primarily applied to those men who had originally enlisted in the 3rd Battalion and had been transferred to the 55th. Foremost of these was the battalion's CO, Lieutenant Colonel Percy Woods.

On 27 September, the 5th Division prepared for battle. The battalion's nucleus, comprising mainly men identified for furlough, headed to the rear; among them Lieutenant Colonel Woods. The remainder of the unit packed up their equipment and moved eastwards across the 15 or so kilometres that lay between their billets at Le Mesnil[2] and the Hervilly area.

Major Eric Stutchbury replaced Lieutenant Colonel Woods at the helm of the battalion. After his significant role at Doignies, Stutchbury had been promoted major and served as the battalion's second-in-command for most of 1917. In early 1918, he attended a Senior Officers' Course in the UK to prepare him for future command of a battalion. His Syndicate Supervisor, a lieutenant colonel of the Lincolnshire Regiment with the unfortunate surname of Bastard, described Stutchbury as 'A very keen, capable officer, with a good military knowledge. Conscientious and energetic. Good at imparting instruction and gets men to work for him.' Stutchbury returned from his course just in time to take over from Woods.

The unit inherited by Major Stutchbury was very different from the one he had served with at Polygon Wood 12 months previously. Then, it could field over 1100 men, most of whom were combat veterans refreshed after several months of leave and training. Now, the departure of the nucleus, combined with the steady stream of casualties and the shortage of reinforcements, had left the battalion a shadow of its former self. The manpower deficit forced Stutchbury to merge A and B companies into a single company (hereafter A Company), doing the same for C and D companies (hereafter C Company). Captain William Giblett[3] was given command of A Company and C Company was placed under Captain Roy Goldrick. Neither of these combined companies numbered more than 90 men.

In spite of the shortage of troops, the combat power of these two depleted companies was far greater than that of the entire battalion at Polygon Wood. By this stage of the war, each platoon had a minimum of two Lewis guns; a year earlier there was one gun per platoon and at Fromelles there were four to six guns available to the battalion. In the coming engagement each section would have at least one of these automatic weapons.

Like the rest of the 5th Division, it was almost nine months since the 55th had a real break from the fighting and stresses of combat; many men were still rattled by their experiences at Péronne. Private Bert Bishop wrote of his reaction on receiving the order to move forward:

> Our new captain [Goldrick] was just back after wounds. We had known him as a lieutenant, and were quite happy to have him. He did no beating about the bush.
>
> 'We are to smash the Hindenburg Line on the last day of September. The Thirtieth American Division – you have already introduced yourselves to them – is to go in first. We are to leapfrog through them and carry on as far and as fast as we can, there's no objective. I have also to tell you the whole German front is rocking, and Foch says the war will end before Christmas.'
>
> [Goldrick] seemed quite enthusiastic about it. He'd been a long time in hospital with his wounds, and had not become so war-weary as his command. His enthusiasm seemed infectious, anyway.
>
> War over by Christmas? No, don't think about it and what it would mean. Think only of that ugly Hindenburg Line, it had to come first.
>
> Our oncoming battle was discussed in a different way to any other. All previous battles didn't mean much, there were always more battles waiting for us. But this one could be our last, and our thoughts were solemn.[4]

The unit spent 28 September in thick woods to the right of Hervilly, attending to the myriad preparations for going into battle. The preparatory bombardment, delivered by 1000 guns, thundered away to the east. Also located in the forest were some of the tanks slated to assist the Australians in the coming attack. As events transpired, this was the last time the men would be in close proximity to tanks, for there were none available when the time came for the battalion to confront the enemy.

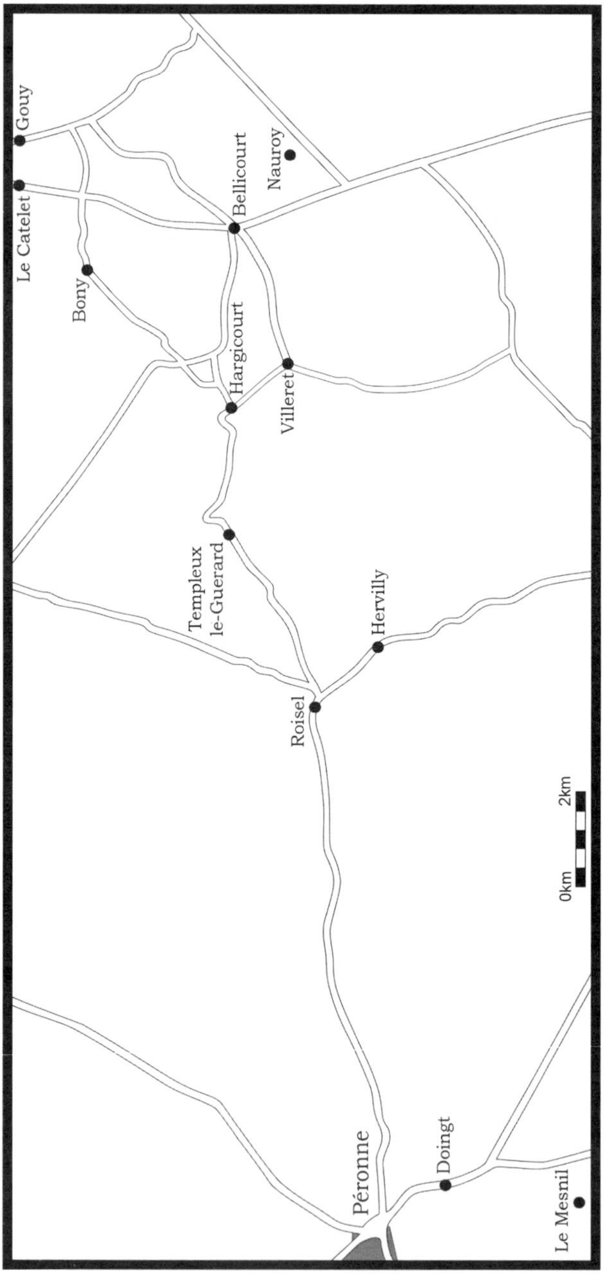

Map 16: Péronne – Bellicourt.

CHAPTER 15

The next day dawned fine and bright. The troops left their bivouac area at the leisurely hour of 8.30 am and, along with the rest of the 14th Brigade, proceeded east along the road linking Hervilly and Bellicourt.

Ahead of them, the 8th and 15th brigades had commenced their advance behind the American 30th Division. But it soon became apparent that events were not proceeding to plan. They encountered unexpected machine-gun fire from German posts that had not been mopped up by the inexperienced American infantry, for whom a daylight frontal assault on the Main Hindenburg Line was their baptism of fire. This tactical blunder allowed the enemy to wait in their dugouts for the creeping barrage and leading waves of the assault to pass by, before emerging to shoot into the front, rear and flanks of the attackers.

Unsettled by this fire, and the heavy casualties it caused, many American units foundered into disorganisation. However, the American infantry that kept fighting succeeded in punching a hole up to 1500 metres deep through the Main Hindenburg Line, clearing through Bellicourt and, in isolated places, pushing on to the west. While not a textbook operation, the American assault laid a solid platform for Australian operations in subsequent days and the ultimate success of the battle.

During the afternoon of 29 September, the 8th and 15th brigades pushed through the scattered American pockets and continued their easterly advance in the face of stiffening enemy resistance. Vicious, disorganised fighting was the order of the day but gradually the Australians passed through the Main Hindenburg Line; by late afternoon elements of both brigades were scrapping with the Germans for control of the southern part of the Le Catelet Line. They were unable to advance further; heavy machine-gun fire from Cabaret Wood Farm to the west and from the dominating German positions to the north extinguished any idea of further progress for the day.

What was happening at the front remained a mystery to the men of the battalion as they plodded towards the sound of the guns. By late morning, the unit had passed through the ruined village of Villeret. A cocktail of dense mist and smoke drifting back from the barrage reduced visibility. It was a tough march as the road was in a bad state of repair and the traffic up and down the route meant the column was continually halting, to the frustration of all concerned.

The Australians began passing evidence of the intense fighting that had taken place that morning. They found American dead everywhere, often in piles. Dozens of leaderless American infantrymen attached themselves to the

55th and other Australian units as they moved along the road. Given the shortage of men in the Australian battalions, these informal 'reinforcements' were warmly welcomed.

At 11.00 am the 55th's slow advance ground to a halt. Artillery columns, the rear elements of the 8th Brigade and the occasional German shell combined to choke the road. Instructing his battalion to seek cover in nearby trenches, Major Stutchbury went to ascertain the nature of the hold-up. By now the weather had clouded over and rain threatened.

In a sunken part of the road, Major Stutchbury met the COs of the 53rd and 54th battalions, as well as the Brigade Major of the 14th Brigade. Stutchbury received confirmation that, for now, the way forward was indeed blocked. As an interim measure until the road was reopened, Stutchbury was directed to move the 55th forward one kilometre and occupy Skin Trench, a recently vacated enemy position. Once there he was to await further instructions.

By 2.00 pm the battalion had began to take over tenancy of Skin Trench. Years of combat had taught the Australians to be cautious. Bert Bishop was wary as he explored the abandoned trench system:

> Moppers-up went down each dugout to make sure they were untenanted before we went on. At dusk we had three of the main trenches behind us, the Germans having fallen back to further trenches. The much-publicised Hindenburg Line astounded us. Masses of barbed wire spread over acres of ground. The dugout steps led down to small towns, it almost seemed. Big rooms, little rooms, passages, corridors – furnished with bunks, tables, chairs – we explored. The Germans certainly meant to stay in that line ... Dugouts were below us, but we had found so many of them mined and booby-trapped that we preferred the discomfort, the drizzle and the mud on top.[5]

The confusion as to what was happening at the front was also evident to those units in the rear. Throughout the afternoon, Brigadier General Stewart received a steady stream of orders, and orders countermanding orders. The result was that his brigade stayed put and waited. The uncertainty filtered down to Private Bishop in Skin Trench:

> It was soon apparent that everything was going haywire.
>
> We talked to scattered Yanks, they all knew nothing. A strange vagueness pervaded the area ... From Bellicourt came galloping several artillery teams, their guns bucking and jumping behind them. German shells fell around them, keeping up with them.

> Several teams pulled up near us. We wanted to know what went on.
>
> A driver told us things were in an awful mess. They had got to Bellicourt, thinking the Germans, chased by Yanks, were at least two miles in front. They found Bellicourt itself to be full of Germans …
>
> German shells were now falling about us. It was chaos everywhere.[6]

For many in the unit, the highlight of their sojourn in Skin Trench was the contact with an American battalion, neatly deployed in half a dozen lines of skirmishers, all with fixed bayonets. This battalion passed through the 55th's positions on an ill-defined mission before disappearing in the mist, smoke and gunfire. Perhaps in the American enthusiasm and amateurism the Australians glimpsed the way they must have been before Fromelles.

By nightfall, the situation at the front had become sufficiently stable for Brigadier General Stewart to issue a coherent set of orders. At 6.30 pm, Major Stutchbury was directed to take the battalion through Bellicourt and reinforce the 57th and 58th battalions in the parts of the Le Catelet Line they held.

It was pitch black and raining hard when, 20 minutes later, the unit began to move. The soldiers made their way across country through a maze of wire, trenches and shell-holes until they reached the Templeux–Bellicourt road. To their right, Bellicourt was being heavily shelled and Major Stutchbury decided to detour north of the village before wheeling towards the intended destination.

By now the battalion was strung out through the middle of the Main Hindenburg Line. Progress was slow as the column picked its way in utter darkness through the labyrinthine landscape. To complicate matters further, they had brought donkeys with them. Twelve of the 'donks' had been allocated to the unit for carrying supplies and water. Through enormous physical effort, accompanied by profuse profanity, these animals were led by their handlers through the terrifying ordeal. On reaching the main trench of the Hindenburg Line, every one of the animals jumped into it; every one except a big black beast that, with the stubbornness for which the species is renowned, refused to leap into the black abyss.

The leading elements of the battalion column had moved about halfway around Bellicourt when Major Stutchbury received a verbal order that the move had been cancelled and that the 55th was to return to Skin Trench. Incredulous, he brought the unit to a standstill while he made his way to a nearby headquarters to verify the message. Private Eddie Street describes the battalion's actions while they waited for Stutchbury's return:

> A halt was called and the troops sat down scattered over a wide area. All around us lay rusty masses of barbed wire. Fritz put over a barrage and he could not fail to hit somebody. Stretcher-bearers went about their tasks with their usual calm. The night was unique for the number of tin hats that were struck by pieces of flying shell. The metallic clinks rang out on the cold early morning air. There was an uneasiness amongst the troops. There was an air that all was not well.

Eventually Major Stutchbury received confirmation of the order. Men and animals did an about face and retraced their steps. When interviewed by historian Charles Bean after the battle, Stutchbury sarcastically remarked: '[the] Battalion moved back in the rain and mud and very dark and gas to Skin Trench – very pleased!'

While the unit was stumbling about the Main Hindenburg Line, Brigadier General Stewart had attended a conference at Headquarters 5th Division. At this meeting Major General Hobbs had stated that the 5th Division could not continue its eastward advance while the 3rd Division and American 27th Division remained engaged in clearing the Main Hindenburg Line. Accordingly, Major General Hobbs directed the 14th Brigade to assist these two divisions by advancing on the right of a renewed 3rd Division push north along the Main Hindenburg Line.

Major General Hobbs' conference finished at midnight. Zero Hour for the attack was set for 6.00 am. For the second time in a month, Brigadier General Stewart found himself directed to undertake a major operation with only a few hours in which to make the necessary preparations.

Before leaving for his consultation with Major General Hobbs, Brigadier General Stewart had arranged to meet with his COs on his return to pass along his own orders. Stewart's meeting was scheduled to be held at 11.30 pm at Headquarters 53rd Battalion situated in Rope Trench, just east of Skin Trench. Major Stutchbury, fresh from his 'Hindenburg Line by Night' experience, did not reach the conference location until 1.00 am. Stutchbury had no reason to be concerned about the lateness of his arrival; Brigadier General Stewart arrived after him, another victim of the dark night.

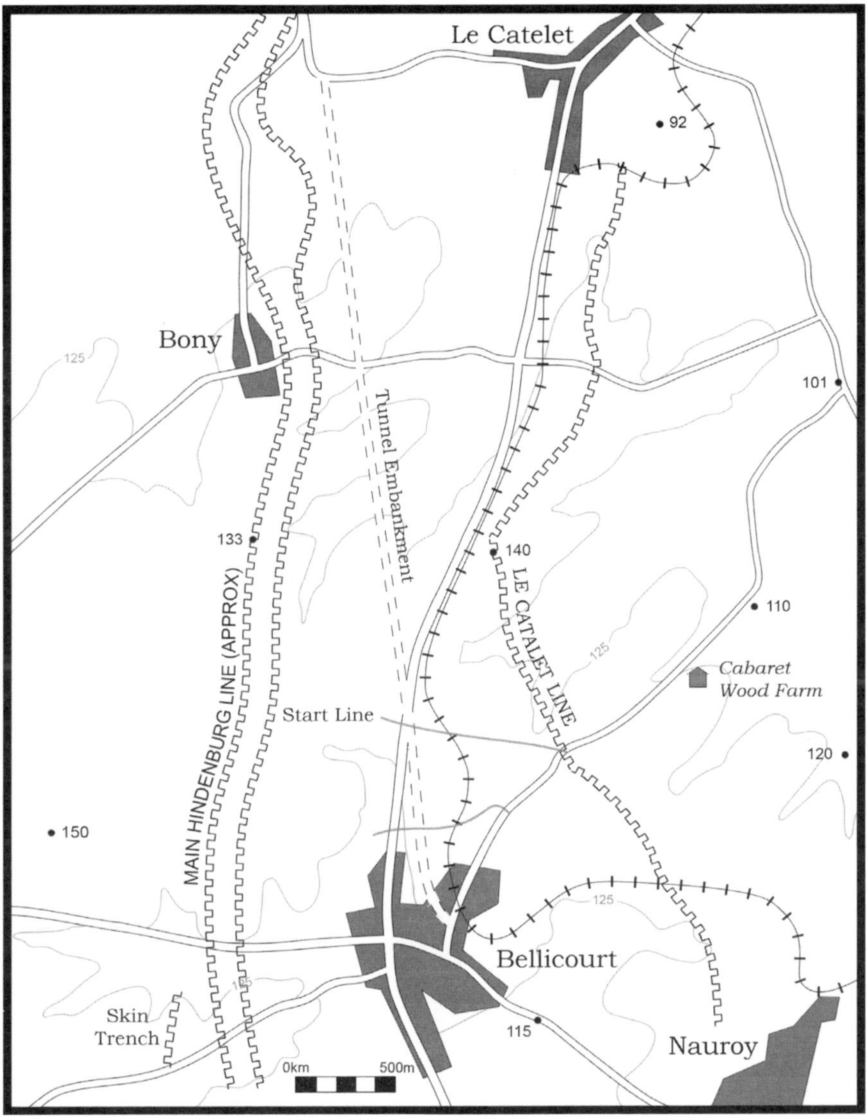

Map 17: Bellicourt – Le Catelet.

Stewart's plan of attack required the 53rd Battalion, supported by the 55th, to advance north in the area between the tunnel embankment and the Le Catelet Line to the right of the 3rd Division's push. The objectives of the attack were to clear the Le Catelet Line and Railway Ridge up to the high ground overlooking Le Catelet village, a total advance of over 2000 metres. Like the 55th, the casualties sustained by the 53rd at Péronne had forced this battalion to reorganise into two under-strength companies.

On the right of the assault, A Company (hereafter A/53 Company) was allotted the task of capturing the Le Catelet Line; C Company (hereafter C/53 Company) on the left was to clear along the tunnel embankment. The 55th's particular tasks were to mop up and occupy the Le Catelet Line once it had been cleared by the 53rd, maintain liaison with the 53rd Battalion to the north and the 58th Battalion to the south, and to send patrols east in the direction of the ruined hamlet of Mont St Martin. Two field artillery brigades, totalling thirty-six 18-pounder guns and twelve 4.5-inch howitzers, were tasked with providing the creeping barrage in support of the 14th Brigade attack; simultaneously, large calibre guns were to bombard the Le Catelet trench.

The start line set for the attack ran from the tunnel embankment (held by the 3rd Division) over the top of a gentle hill just north of Bellicourt to the portion of the Le Catelet Line secured by the 58th Battalion. The 900-metre length of the start line meant that the Australian infantry would be spread thinly on the ground. Ominously, the German-occupied Cabaret Wood Farm overlooked a considerable portion of the start line.

After his battalion commanders left the conference, Stewart articulated his concerns over some aspects of the plan in his diary: 'They [the battalions] are going to have a fair amount of difficulty in getting to their start line by Zero. The country is deplorable and covered with wire. The Hindenburg trenches are 15–20 feet wide in places.'

Major Stutchbury reached the 55th Battalion at 3.15 am. The unit had just three hours in which to plan the attack and complete an approach march of approximately four kilometres over difficult country to reach the start line by Zero. Having learned from their previous experience of traversing the Main Hindenburg Line in the near absolute darkness, Stutchbury and his company commanders used aerial photographs to plot a route chart to the start line that bypassed most of the known obstacles. The Intelligence Officer, Lieutenant Reginald Morris, was tasked with guiding the battalion.[7]

By 4.30 am, the 55th was moving. The column of men was glad to be underway; the rain had settled in and everyone was cold and soaked through. Morris' task was extraordinarily difficult — single-handed, through unfamiliar territory swept by enemy artillery and machine-gun fire, he had to lead the battalion through a confused mass of barbed wire, trenches, and shell-holes in the rain on a dark night. In addition, the lateness of the hour meant that if the unit was to attack on time, it needed to push on with all haste.

As the 55th was in support and therefore positioned behind the 53rd for the attack, Major Stutchbury selected a communications trench several hundred metres behind the designated start line as the assembly area for his battalion. Lieutenant Morris led the battalion around Bellicourt to avoid the heavy shelling which continued to pound the village; machine-gun fire still swept the area. In spite of all impediments, Morris, in a superb feat of navigation, led the battalion to the very trench it was aiming for, albeit several hundred metres too far to the right. Private Bishop takes up the story:

> We had seen the parapet of a trench as we got to within a hundred yards of it. An Australian voice from it yelled out: 'Come on, we won't shoot you'. [Enemy] Machine guns opened up on us. We ran like mad to that trench. A Victorian battalion occupied it [the 58th] …
>
> We should be on the left of this battalion, it seemed, and off we went. We had to progress down in the trench, the top was under heavy fire.
>
> For about three hundred yards we did a horror walk. The trench bottom had no duckboards, or rather duckboards of an unusual kind. Laid at full length, hard up against each other, were dead bodies of the men of the battalion occupying the trench. Mud and blood was their couch to rest on. We had to step on them, and the mud and blood oozed up around them as our weight pressed them deeper into the mire.
>
> We came to a spot where a minor road crossed the trench, the road had not been cut. It was a matter of running the gauntlet on top. The Germans had several guns firing on the crossing. It was take a deep breath, jump and run and drop into the trench the other side.[8]

The battalion was in place by 5.45 am, 15 minutes before the designated Zero Hour. Remarkably, there had not been a single casualty during the approach march. There was, however, one glitch in the unit's move to the start line; the 53rd Battalion had been told to expect guides from the 55th to assist it in making its way to Headquarters 15th Brigade where additional guides could be procured to lead the battalion to the start line.

The 55th's guides were not forthcoming; there had been little time for these requirements to filter down. In addition, when the lead elements of the 53rd caught up with the rear of the 55th's column, no-one in the 55th had any idea where the 15th Brigade Headquarters was situated.[9] The mix-up with the guiding arrangements imposed a significant time delay on the 53rd Battalion.

Soldiers of the 53rd followed the 55th into the communications trench, arriving just on Zero Hour as the supporting barrage began falling in the distance.[10] Now pressed for time, Lieutenant Colonel William Cheeseman, CO of the 53rd, realised that it would be impossible to reach the start line and then deploy into proper formation for the attack. Instead, he lined his battalion up along the parapet of the communications trench and ordered an advance to the start line in artillery formation, with directions to deploy into 'line abreast' formation once this line was passed.

Despite Lieutenant Colonel Cheeseman's best efforts, the 53rd was 15 minutes late crossing the start line, although the delay proved of no consequence. The promised creeping barrage failed to materialise; Major Stutchbury estimated that only two or three guns were firing. As at Péronne, the scattered shelling served only to alert, not restrain, the enemy. Few, if any, of the dozens of German machine-gun positions scattered throughout the attack area were destroyed or suppressed by the bombardment.

The utter failure of the artillery barrage meant that as soon as the 53rd Battalion crested the small hill across which the start line ran the men came under intense machine-gun fire from Cabaret Wood Farm, the Le Catelet Line, Bony and the enemy posts along the tunnel embankment. The converging machine-gun fire, coupled with shelling by enemy field guns engaging over open sights, inflicted a considerable number of casualties. The attack pressed on in the face of this stubborn resistance.

Only once the 53rd crossed the start line did the 55th leave the communications trench and commence its advance. The morning was dull and it was still raining. A Company was in front, with C Company just behind it. The troops advanced some 200 to 300 metres in artillery formation and had just crossed the start line when, like the 53rd, they came under heavy machine-gun fire from the flanks and front. The unit changed into 'line abreast' formation. The ground before them was open, covered with grass, self-sown crops and rank weeds; there were few places to find cover apart from shell-holes and behind the piles of spoil, ordure and rubbish that dotted the landscape.

Here the battalion split. A Company moved to its right to trail A/53 Company along the Le Catelet Line. C Company, which was tasked with following C/53 Company, edged to the left where the tunnel embankment provided some protection from enemy bullets and shells.

C Company continued its advance until it reached another trench, finding the 58th Battalion in residence and some of the 53rd catching their breath. Casualties were already starting to occur: Private George Bush[11] and Private Albert Bedson were both killed by machine-gun fire.[12] Shrapnel fatally hit another man, Private Edward Holden.[13]

After pausing for a few moments, the men of C/53 Company clambered over the trench's parapet and resumed their push. There was a valley just north of the trench and C Company provided covering fire for their comrades as they moved into this shallow depression; enemy fire was particularly heavy from Cabaret Wood Farm.

Lieutenant Luther Chadwick was leading the right-hand platoon of C Company. After entering the trench, Chadwick travelled along it to his left until he met Captain Goldrick, his company commander. Goldrick informed Chadwick of his intention to move the company out of the trench to a railway cutting north of where it was currently sited and position it for an attack east towards the Le Catelet Line.

C Company was starting to leave the relative safety of the trench for this cutting when Lieutenant Reginald 'Diggs' Hill of the 53rd appeared.[14] Hill and six others had been waging a private war. Finding themselves isolated next to the embankment, the seven men had moved along it, in the process killing over 35 Germans and capturing five machine-guns.

Lieutenant Hill was probably in the mood to continue on to Berlin but sanity prevailed when his party identified a German strongpoint on top of the tunnel embankment, some 20 metres beyond where a road cut through the spoil heap. He prudently ordered the withdrawal of his troops to the trench, where he bumped into Captain Goldrick and Lieutenant Chadwick, making them aware of the 'very hot water' lying 200 metres ahead of them. After passing on this information, Hill took the remainder of C/53 Company to join the rest of his battalion in the Le Catelet Line.

Realising that fire from this strongpoint would enfilade his planned easterly attack, Captain Goldrick ordered C Company back into the trench and then set about dealing with the situation in front of him. Keeping one platoon engaged in providing supporting fire towards Le Catelet trench, he ordered Lieutenant Chadwick's platoon to make its way left along the trench, cross over to the western side of the embankment and ascertain whether a covered approach to the strongpoint existed on that side of the mound. As they topped the embankment, Chadwick's men met an isolated party of Americans — when asked for information the Americans knew nothing of the situation. After crossing the embankment, Chadwick's party was subjected to sustained machine-gun fire from Bony and the Main Hindenburg Line.

After placing his men in defensive positions along the trench, Lieutenant Chadwick and his batman began crawling along the side of the embankment to reconnoitre the strongpoint. They approached the road cutting without being seen, finding a spot where they could observe the enemy; they could discern the machine-guns but not the gunners. Having seen enough, the two men made their way back to the rest of the platoon, where Chadwick set about gathering a Lewis gun and four additional men with the intention of returning and attacking the German position.

Just as Lieutenant Chadwick arrived back, 2nd Lieutenant 'Sacco' Jackson, Sergeant Eugene Sullivan[15] and Captain Goldrick's batman came up and asked where the strongpoint was. Chadwick pointed it out to them, and the three men decided to attack the position. Retracing Chadwick's route along the embankment, the party reached the bottom of the bank opposite the machine-gun post unseen by the Germans. From here they began inching their way towards a spot where they could attack the strongpoint with grenades. Unknown to the patrol, another German machine-gun post was in a section of trench 20 metres to their left. The enemy gunner spotted and engaged the crawling men, bullets thudding into Jackson and Sullivan and killing them both instantly.[16]

Captain Goldrick's batman managed to slip away and move back to Chadwick's post. Another patrol was despatched comprising the batman and four others, this time armed with a Lewis gun. This patrol got as far as the road cutting through the embankment when the same machine-gun opened up on them, halting all forward movement. The Lewis gun was brought into action and, under the covering fire it provided, one of the patrol was able to make his way to 2nd Lieutenant Jackson and Sergeant Sullivan, confirming that they were dead.

It proved impossible for the patrol to take out the German machine-gun; other machine-guns further to the left were also firing on the group, making any move to a flank impossible. The patrol members withdrew to the trench where Lieutenant Chadwick was waiting for them. From there, they made their way back over the embankment to brief Captain Goldrick on the situation.

While Lieutenant Chadwick's platoon on the left was occupied with the task of attempting to silence the strongpoint, on the extreme right of the company, Private William McAlister noticed that the A Company advance was held up by enemy machine-guns.[17] Moving his Lewis gun team into a better position, he was able to fire on the German gunners, killing several — the remainder, pinned down by McAlister's fire, were subsequently killed by A Company.

With the failure of Lieutenant Chadwick's endeavours, the momentum of the C Company attack had, for the moment, dissipated. It was now after 9.00 am and the company had not advanced for over two hours. Consequently, a wide gap had opened between A Company and C Company. The 53rd was engaged in bombing its way up the Le Catelet Line and A Company was clearing the low ground to the left of that trench system. Captain Giblett, the overall leader of the 55th's attack, instructed Captain Goldrick to catch up.

Captain Goldrick issued his orders: C Company was to advance along either side of the embankment. There was no objective — just keep pushing. Private Bishop wrote of the events that followed:

> Our company was to go out onto the open ground in front … Positioned correctly, we were a long line of Aussies stretched across a big field of grass. All was ready.
>
> [Goldrick] placed himself out in front. The signal came. 'C Company, advance.'
>
> Our long line of Aussies moved steadily forward down the grassy slope. Everything went well and quietly for about two hundred yards. A rifle fired, machine-guns opened up, bullets hissed about us. Throwing ourselves to earth we found we were opposed by a long line of Germans, they also lying in the grass. For a few minutes we shot at each other. [Goldrick] wasn't satisfied with this type of warfare. He suddenly jumped up, waved his revolver in the air and yelled:
>
> 'Come on, C Company, charge the bastards!'[18]
>
> We sprang up, the whole line sprang up, and straight at the Germans we went. Someone cooeed, it was taken up along the line. For the first time in my life I learned what it meant to 'see red'. We all went mad. The war had taken so much out of our lives we were sick of it. And if killing

> those Germans meant ending the war, we'd kill them, every bloody one of them. Men were making all varieties of horrible screams and shouts. Sheer madness held us.
>
> The Germans waited not on the order of their going, they just went. Abandoning machine-guns and rifles, even equipment, they ran down the slope. They outpaced us, but we kept after them.
>
> Suddenly they disappeared into a trench we knew nothing about. The Germans garrisoning the trench opened up on us with heavy machine-gun and rifle fire. Some light artillery, standing in the open behind the trench, joined in. It was our turn to disappear.[19]

As related by Private Bishop, C Company reached about as far as the cutting in the embankment where Sergeant Sullivan and 2nd Lieutenant Jackson had been killed without taking much fire. As they crossed the cutting, which was at the top of a slight rise, and began to charge into the valley on the other side, they came under very heavy machine-gun fire from multiple points along the tunnel embankment (including the strongpoint mentioned previously), from Bony, the Main Hindenburg Line and from Railway Ridge. The Germans also brought *minenwerfers* and field artillery to bear on Captain Goldrick's men.

Lance Corporal George Campbell was in the van of the wild charge.[20] Seemingly impervious to the machine-gun fire, Campbell led his section in pursuit of the Germans, killing without remorse whenever the opportunity presented itself. The Germans panicked and fled before Campbell's onslaught; the Australians were able to capture the troublesome strongpoint to the north of the cutting and continued clearing along both sides of the embankment.

Casualties mounted. On the right of the embankment, Captain Goldrick fell, badly wounded by machine-gun bullets that fractured his hip and right leg; nearby another officer, Lieutenant Sydnie Cohen was also wounded, shot in the head.[21] Realising that all the officers to the right of the embankment had been killed or wounded, Sergeant Ken Clark took over and continued to lead the troops forward in the face of heavy machine-gun and rifle fire.[22] He personally attacked several German machine-gun nests, killing the crews. Bayonet work was the order of the day.

As the scattered survivors of the charge entered the still-occupied German trench (the one described by Private Bishop), Private James 'Gundy' Marshall found himself with just one other Australian in 250 metres of trench; everyone else in his section was either dead or wounded.[23] Marshall and his companion set to and began bombing and bayoneting their way along the length of the trench, killing or wounding a number of enemy infantry. Lewis gunners dealt with those Germans who chose to flee from the trench in the face of the reckless onslaught by the two men.

Moving along the right-hand side of the embankment, Lance Corporal Arthur Auckland,[24] accompanied by Private Henry Klein, observed a machine-gun post to their front enfilading the other members of C Company and slowing the advance. [25] Although they were under sustained fire, both men used grenades to force the enemy gunners to retire, abandoning two light machine-guns.

Lieutenant Chadwick, on the left of the embankment, was so short of men that he was forced to bypass active enemy positions. Eventually Chadwick's platoon, confronted with diminishing numbers and no pause in the volume of German fire, reached a high water mark where further advance was impossible.

Lieutenant Chadwick gave the order to retire several hundred metres to the trench from which the attack had been launched. Just as Chadwick made it back into the trench, a shell landed nearby. Private Eddie Street described the officer's fate:

> Lt Chadwick was hit and fell still and silent on the foot of the trench, apparently dead. Bombers and riflemen passed up and down the trench. There was no time for niceties and the body was trampled into the mud. Some half hour later a large digger strode down the trench and trod on the mud-covered face in the bottom of the trench. The corpse of Lt. Chadwick came to life, and from the muddy mouth came voluble remonstrance. Lt Chadwick thus survived his wound under the boots of his platoon.

Chadwick, although alive, was badly concussed and played no further part in the battle.

On the right, the survivors took cover in whatever shelter was available and waited. Sergeant Clark finally realised that the tactical situation was hopeless and ordered the withdrawal to the starting trench. Private Bishop had thrown himself in a nearby shell-hole as the charge began to peter out:

> Quickly following me came another chap, he was big and heavy, and he almost knocked my head off. It was Mick, our company sergeant major.[26] We squirmed as low as possible into the bottom of the hole. A whizz-bang landed on the side of our shelter. Mick slumped into the side of the

> hole, as if he were going to sleep. He was. A few inches before my eyes I saw a hole in the back of his head, just below his tin hat. Blood filled the hole, hesitated, spurted out …
>
> I peeped up to try to see what went on. A dead silence settled over it all. Here and there I saw a man jump out of a hole, race to another one. No firing was coming from the Germans. A bigger shell-hole was on my left, I decided to give it a go. I raced to it. Not a shot was fired at me. Two of our men were in it, both dead. I seemed to be the only living creature …
>
> That morning's wild mad rush at the Germans had taken the last vestige of strength from me, and my thoughts were ugly. I pushed myself out of the shell-hole. Surely some of my company had survived, I'd have to find them. No-one shot at me as I walked so very slowly back.[27]

The shaken survivors gathered; over 50% of C Company had become casualties. Lieutenant John Pye was the only uninjured officer; the men were unstrung and exhausted.[28] For the second time in a month, C Company had ceased to exist. It was 10.00 am.

Back with A Company, Captain Giblett and his men came under enemy fire as they crossed the start line but made their way to the same trench that was to figure so prominently in the fortunes of C Company later in the morning. Here they paused; there was heavy fighting on their right as the 53rd Battalion fought its way up the Le Catelet trench.

By 8.30 am the 53rd had made sufficient progress for A Company to resume its advance across the open ground lying to the left of the Le Catelet Line. Methodically, the troops advanced, adjusting their pace to conform to the slow progress of the 53rd's troops in the adjacent trench system. Sergeant Tom O'Rourke, an armourer, was hit in the arm and sought shelter in a trench — he later died. Private Street wrote of O'Rourke's death: 'In the recess I saw a man kneeling in prayer, hands clasped and in them a Mill's Bomb. The man was quite dead. He was a decent, friendly man above military age.'[29]

The fighting experienced by A Company was intense, the enemy resisting with unexpected ferocity. The structure of A Company broke down and everywhere small groups of men gathered around a leader, often a man without rank but a known 'fighter', to attack a strongpoint or trench. Once achieved, the soldiers would move off to the next objective. One of these outstanding leaders was Corporal Ed Woodhart, who led his men in a vigorous assault on one of the first German trenches encountered in the advance, seizing the trench and killing the occupants.[30]

A Company's officers moved among the men, rallying the scattered parties. Lieutenant Francis Armstrong was in the forefront of the Australian advance. When the attacking troops were held up by heavy machine-gun fire from a 'mini fortress', Armstrong stood up and, with a shout of 'Come on, lads', made a rush for the post, accompanied by the rest of the company. The defenders were not expecting something as audacious as a full-blooded frontal assault and were overwhelmed, most dying on the point of a bayonet.[31]

The Australians kept pressing forward, clearing their way with bayonet, grenade and Lewis guns. No quarter was given; this was not a day for taking captives, for no man could be spared to guard them and the diggers were in no mood to take prisoners. Gone were men who were once farmers, labourers, office clerks or teachers — this day they were simply executioners. They advanced slowly and methodically, killing everything in their path. Captain Giblett seemed to be immune to enemy fire and was everywhere on the battlefield, urging his men forward.[32]

After four hours, A Company had advanced almost one kilometre and Captain Giblett was able to report the capture of a 77mm field gun to Major Stutchbury in Battalion Headquarters. The Germans kept several of these artillery pieces in action until the last possible moment and they were captured after desperate fighting. So far A Company had been fortunate, suffering only three fatalities.

Shortly after 10.00 am a runner arrived from C Company informing Captain Giblett of Captain Goldrick's wounding and that the company was down to half-strength. Giblett ordered C Company to reorganise where it was and await further orders.

Gradually Captain Giblett's men noted a change in the tempo of the battle. The German resistance became increasingly determined. Just after the field gun was captured, the already fierce barrage from enemy machine-guns, artillery and *minenwerfers* intensified, causing a number of casualties. One of those killed was Sergeant Harry Barker — a bullet ignited a smoke bomb he was carrying and the phosphorous in the grenade burst into flame as it came in contact with the air, incinerating Barker in an agonising death watched in horror by his nearby mates.[33] His body was so badly burnt and contaminated with residue phosphorus that after the battle it was necessary to bury his remains where he fell.

This increase in enemy fire was a precursor to a massive German counter-attack that broke on A Company from the direction of the Main Hindenburg Line. Under enormous pressure from German bombers, and with his left flank 'up in the air' due to the inability of C Company to advance, Captain Giblett had little choice but to order his men to seek sanctuary in the Le Catelet Line.

A Company's retirement was even more chaotic than its advance. Desperate to extricate themselves, the Australians became involved in running skirmishes. The relentless enemy pressure forced the men backwards, until they tumbled into the protection offered by the captured sections of the Le Catelet Line. Once there, the company took up a position to the rear of the 53rd, still attempting to push up the line.

The German tsunami smashed into the Le Catelet Line at various places; the Australians were spread too thinly and were too disorganised to keep the enemy at bay. If it were possible, the fighting became even more frenzied. Germans suddenly appeared around corners or emerged from sections of trench believed to be clear of enemy troops. Bayonets, knives, grenades, spades and pistols were the only weapons of any use in such a melee. Private Hector MacCallum, a Lewis gunner, was killed by the concussion of a grenade explosion, his body sustaining no noticeable wounds.[34]

At one stage of the enemy counter-attack, volunteers were sought to leave the comparative safety of the trench and assist the wounded lying in the open. Private William Donohue put up his hand.[35] On jumping out of the trench, machine-gun bullets smashed both of his legs. He was dragged back into the trench, bandaged and left on a stretcher for collection by stretcher-bearers when the opportunity arose. The Germans subsequently snatched the line at this point, forcing the Australians to abandon the still-conscious Donohue. A determined push shortly afterwards led to the section of the trench being recaptured. It was too late for Donohue — he had bled to death in the intervening period.

The Germans started bombing down the Le Catelet Line from the north. Another large enemy party established a position on an open bank to the south-west, threatening the rear of the Australian troops. Under attack from front and back, the Australian situation was critical; the 53rd and 55th battalions were at risk of being cut off and destroyed piecemeal.

It was at this tipping point in the battle that a hitherto unremarkable soldier came to the fore. Appreciating the dire nature of the situation, Private John 'Jack' Ryan led the men near him in an attack against the Germans on the bank.[36] Ryan's party rushed the flank of the German position, fusillades of enemy bullets reducing the size of the assault force to Ryan and just three others.

Undeterred by the apparently hopeless situation and the fact that the enemy forces were several times more numerous, the Australians flung themselves on the Germans, bayoneting three of them and sowing confusion. Ryan, alone, then moved along the bank, flinging bombs at the remaining Germans, in the process driving them back across no man's land, where they were easy pickings for the Lewis gunners.

Eventually Ryan's good fortune abandoned him; he was shot through the shoulder and took no further part in the battle. His actions extinguished the German threat to the Australian rear and enabled the recapture of the lost parts of the Le Catelet Line. He was awarded the Victoria Cross for his reckless bravery.

Private Ryan was not the only member of the 55th who was at times waging a personal war with the Germans. In one section of the Le Catelet Line, Corporal Ron Stewart found himself to be the only unwounded member of his Lewis gun team.[37] Undaunted by the machine-gun fire sweeping the parapet, Stewart clambered out of the trench and, positioning the Lewis gun out in front of the trench, began firing on the advancing Germans, killing or wounding many of them. Another enemy party, however, made its way towards the Australian positions along a communications trench, shielded from Stewart's fire. Realising the threat, Stewart grabbed a handful of grenades and joined a group bombing its way towards the Germans. Deciding that prudence trumped valour, the enemy withdrew, leaving several of their dead in the trench behind them.

Corporal Woodhart was again prominent. Realising that all the officers and senior NCOs in his vicinity had become casualties, he took command of those soldiers around him, directing the fire of the Lewis gunners and riflemen onto the Germans. The volume of fire from his section of the trench was so devastating that many of the assaulting Germans were killed, and not one of them reached the trench.

Sergeant Tom Buckingham became involved in a bomb fight with a large enemy force in the Le Catelet Line.[38] After an hour's engagement, during which Buckingham continuously hurled grenades, the Germans began to retire back down the trench. This was the signal for Buckingham to leap out over the parapet and pursue the Germans with more grenades as they pulled back.

Gradually the German counter-attack lost momentum and then faded altogether. Seeking to press home any tactical advantage arising from the German failure, Lieutenant Colonel Cheeseman ordered the recommencement of his battalion's advance along the Le Catelet Line. The Germans held on and the Australians were brought to a halt without further capture of ground.

It was now approaching midday and the infantry fighting reduced in intensity, punctuated by occasional German bombing forays that were easily repulsed. The Australians built a 'bombing block' to separate the section of the Le Catelet Line they held from that retained by the Germans.

At this time the Australian locations were as follows: the 53rd Battalion plus A Company occupied the Le Catelet Line and C Company remained licking its wounds in the trench into which it had retired after its abortive bayonet charge. On the left of the tunnel embankment the 3rd Division had progressed some 200 metres along the Main Hindenburg Line before again being stalled by German machine-gun fire.

Brigadier General Stewart took stock of the situation. To obtain maximum results from his dwindling resources, Stewart instructed Lieutenant Colonel Cheeseman to take overall charge of the two battalions. Cheeseman retained the 55th's company-level command arrangements and left Stutchbury in charge of his battalion's administrative and logistical arrangements.

Stewart was aware of the perilous situation, writing in his diary:

> Casualties have been heavy. 53rd strength is now 180 men and 55th only 120 so that the combined battalions do not make up two decent companies. 55th Battalion have got 1 x 4.2 inch gun and one 77mm and about 25 machine guns. The number of machine guns shows the rotten nature of the fighting.

Realising that the positioning of C Company offered him no tactical advantage, Lieutenant Colonel Cheeseman directed this company to make its way to the Le Catelet Line and join the rest of their battalion under command of Captain Giblett. This move was promptly undertaken.

During the early afternoon, the Le Catelet Line was again subjected to intense German artillery and *minenwerfer* fire; enemy machine-guns continued firing from the north and west. Private James Granville was killed in this shelling.[39] Lance Corporal Edwin Neild had taken charge of a party of signallers and was able to establish a telephone line between Battalion Headquarters and Captain Giblett's forward headquarters — a distance of over 900 metres.

Neild and his companions had to run the line across an area under heavy shellfire, and at times the group found themselves under direct machine-gun and rifle fire. The telephone connection allowed Giblett and Major Stutchbury to control the battle more effectively, particularly in terms of providing supplies and organising artillery support.[40] Before the telephone link was established, all communication was by runners. Moving between the firing line and Battalion Headquarters, three soldiers distinguished themselves in this role, in the process earning the Military Medal: Private Albert Cadd,[41] Private Lionel Freebody[42] and Private Street.

Brigadier General Stewart, realising that there was no longer any possibility of attaining his brigade's objectives, ordered Lieutenant Colonel Cheeseman to link up with the 3rd Division on his left flank. Elements of the 53rd Battalion, moving along a communications trench running west from the Le Catelet Line, tried several times to achieve this during the afternoon. Each attempt ended in failure owing to heavy German fire, particularly from a 'miniature fort'. This particular 'fort' was reported by a prisoner to contain at least 30 machine-guns.

While the Australians were unsuccessful in reaching all their objectives, their achievement in capturing a one-kilometre stretch of the Le Catelet Line posed a serious tactical conundrum for the Germans. With each metre of trench held by the Australians, the enemy defence of the Main Hindenburg Line became more tenuous. The main German supply and communication lines ran from east to west and the seizure of significant lengths of the Le Catelet Line was the equivalent of drawing a knife across the throat of the enemy garrison in the Main Hindenburg Line.

The Germans could not let this threat go unchallenged. At 3.00 pm, in pouring rain, the *1st Guard Grenadier Regiment (Kaiser Alexander)* of the *2nd Guards Infantry Division* launched a counter-attack down the Le Catelet Line and from the vicinity of the tunnel embankment. This Regiment was one of the elite units of the German Army, equivalent in status to the British Guards Regiments.

In a day characterised by bloody close-quarter battles, the German attack was astonishing for its savage intensity. The 53rd Battalion was forced to cede the bombing block and withdraw 200 metres down the Le Catelet trench.

The diggers had never witnessed anything resembling this counter-attack, and for many it was a dream come true. They were in a strong defensive position confronted by lines of Germans advancing across open ground. It was an opportunity tailor-made for Lewis gunners to wreak havoc.

Private Stan Yelds realised that from within the Le Catelet trench he would be unable to position his Lewis gun to bring effective fire to bear on the Germans.[43] He clambered over the parapet and found a suitable spot in front of the trench. From here, he unleashed a stream of continuous fire on the attackers, often at close range. In the words of the citation for the Military Medal subsequently awarded to him, Yelds 'literally mowed down their advancing waves'.

Private John McInnes[44] and Private Andrew Lewis[45] moved their Lewis gun to a position where they were able to enfilade a strong party of Germans who were seen endeavouring to outflank a small group of Australian infantry. Discovering that the gun could not be used effectively from ground level, McInnes placed the gun's barrel on his shoulder, in which position it was fired by Lewis, both men standing upright in the face of German machine-gun fire. They continued operating the gun in this unconventional manner until the enemy party was broken up, leaving scores of dead and dying scattered on the ground. The German infantry could not sustain such slaughter. They withdrew, leaving behind an estimated 200 dead and many times that number wounded.

At 4.30 pm, Major Stutchbury received a message from Captain Giblett that the Germans were again massing, possibly in preparation for another counter-attack. Giblett told Stutchbury he was confident that his Lewis guns would deal with anything the enemy could throw at him. To resupply Giblett before the looming attack, a 25-man working party was obtained from the American 119th Infantry Regiment and despatched to take ammunition and water to the front line. The precautions turned out to be unnecessary — an artillery barrage descended on the Germans, smashing the troop concentration.

This demise of the prospective enemy attack failed to bring the curtain down on the day's hostilities. Private Bishop wrote of the fighting continuing along the saps and trenches around the Le Catelet Line:

> It was a strange warfare … The German trenches just seemed a huge maze. Often we'd be stalking them and, around a bend, they had disappeared. At one spot my platoon found itself cautiously creeping along one trench and discovering that on each side of us were trenches full of Germans. The Germans on our right had a machine-gun, and every time they glimpsed a tin hat they let a burst off …
>
> We went a hunting on our own. We threw Mills bombs down the dugouts, we took pot shots whenever we saw a German helmet. At one stage we found ourselves confronting half a dozen Germans. The first four 'Merci Kamerad'-ed [i.e. surrendered], the rest ran. This business continued until dusk.[46]

As usual, the stretcher-bearers were in the thick of the action. Ignoring all threats to their personal safety, the stretcher-bearers moved about no man's land, ministering to the wounded, including several Americans found lying on the battlefield. Four bearers were awarded Military Medals: Private Arthur Olive,[47] Private Henry Lemon,[48] Private William Muir and Private Stan Murray.[49]

One stretcher-bearer again stood out among his peers: Ernie Corey. Now a corporal, he had been placed in charge of the battalion's stretcher party. This responsibility did not stop him getting into the thick of the action and rescuing wounded from almost impossible situations. However, after years of near misses, Corey's luck finally ran out. At around 11.00 am he went to bring in Captain Goldrick, whom he had earlier bandaged.[50] A nearby shell-burst sent shrapnel ripping into his right groin and thigh, the force of the explosion also tearing away his first aid bag. Seriously wounded, Corey crawled five or six metres to his bag and put a tourniquet over his femoral artery to stop the bleeding. He then proceeded to crawl rearwards, covering about 300 metres before being found by Lieutenant Chadwick and Sergeant Ken Clark, who carried him to a dressing station.[51]

The actions of this remarkable man were again deemed worthy of official recognition and he received a Bar to the Military Medal, his third. He is the only man to be awarded three Bars to the Military Medal. Writing after the war, Lieutenant Colonel Woods described Corey as

> a splendid soldier whose temper remained unruffled even in most adverse circumstances … Corey was of powerful physique and, invariably while out stretcher-bearing, he wore white shorts, carried his stretcher perpendicularly, but seldom made use of it, preferring to pick up his patients under one of his strong arms, and walk back with him, still holding the stretcher perpendicularly with the other. He had an undaunted spirit, and worked almost up to the enemy wire, rescuing wounded – foe as well as friend.

The logistics elements of the battalion also had a busy day. Corporal Herbert Hodges was given the unenviable task of looking after the battalion's pack mules and donkeys.[52] He had already endured an arduous time the previous evening trying to move his equine charges through the Main Hindenburg Line. Once the battalion was involved in battle, Hodges' job became even tougher. Despite the rain and ever-present shellfire, Hodges and his animals made constant trips to the firing line, carrying ammunition, hot meals and water.

At 8.25 pm, Brigadier General Stewart directed Lieutenant Colonel Cheeseman to leave a garrison in the Le Catelet Line and, using this as an anchor point, form a flank to the tunnel embankment. Permanent contact was to be established with the 3rd Division on the left. Given the limited numbers available to him, Cheeseman elected to establish a series of platoon-sized outposts rather than form a continuous defensive line.

Around midnight small patrols started filtering out of the Le Catelet Line and making their way towards the tunnel embankment. It was another pitch-black night, raining and with the occasional storm passing through. Four platoons followed the patrols, establishing positions some 100 metres apart. Occasional firefights erupted with isolated parties of Germans and a few prisoners were captured. In general, the creation of the outpost line occurred without incident, as by now both sides had fought themselves to a state of near exhaustion.

The morning of 31 September had opened with Germans, Americans and Australians intermingled on an ill-defined battlefield, a situation precluding set-piece attacks supported by massive artillery bombardments. The day's operations by the 53rd and 55th battalions, along with those of the other Australian and American forces on their flanks, had stabilised the front sufficiently to allow the Allied planners to contemplate a return to more conventional warfare the following day.

As his focus for operations on 1 October, Major General Hobbs chose the area to the right of Railway Ridge. An attack was to be launched by the 8th Brigade on the right, the 54th and 56th battalions in the centre, and the 15th Brigade on the left. The 53rd and 55th battalions were ordered to consolidate the ground already taken. As part of its attack, the 15th Brigade was to capture Cabaret Wood Farm, the source of much of the machine-gun fire that had poured into the 53rd and 55th.

The next morning dawned overcast but the rain had ceased, a small mercy for the sodden troops. The men found themselves in premium seats to observe the operations of the 15th Brigade. The attack was an overwhelming success, the Germans melting away in the face of the advance. By 7.30 am, the 15th Brigade had cleared Cabaret Wood Farm and was exploiting to the north.

The 5th Division's attack drew the majority of German attention away from the Le Catelet Line and the platoon outposts. Machine-gun fire had practically ceased and the morning passed without incident but for the shelling of one outpost by a single German field gun. This artillery piece was to cause the battalion's last fatality of the war; a shell landed in close proximity to 2nd Lieutenant William Dillow, severing both his legs.[53]

As the morning progressed there was a substantial reduction in longer range machine-gun fire. Patrols were pushed out and discovered that the Germans had withdrawn one kilometre to defensive positions just south of Le Catelet. Acting on this information, Lieutenant Colonel Cheeseman ordered the 53rd and 55th to advance around 700 metres along the top of Railway Ridge to positions from which they could overlook this village. The 53rd Battalion's after-action report described the nature of the advance:

> Platoons were reorganised in the Le Catelet Line, supplies of ammunition and bombs replenished, Lewis guns cleaned, and all preparations made.[54]The Battalion left the trench one section at a time and took up its position in artillery formation at right angles to the trench. When liaison was established with the [flanking battalions] the whole line moved forward. The movement afforded a fine sight from the point of the spectator, and as a study of manoeuvre was nothing if not daring, when it is realised that the battalion was moving across open country in broad daylight, practically into the teeth of the enemy. Except for some stray shelling which was not responsible for any casualties the advance was unmolested. And 1.30 p.m. found the line formed, posts organised, in fact an outpost line that would have taken a lot of breaking.

The men of the 53rd Battalion occupied the left-hand segment of the outpost line, overlooking Le Catelet. In this position, they linked with the 3rd Division on their left. To their right, the remnants of the 55th occupied the high point of Railway Ridge, the Le Catelet Line (now vacated by the Germans) and the ground sloping down Railway Ridge to the south-east for a distance of several hundred metres.

C Company remained in the Le Catelet Line. Private Bert Bishop described the condition of the trench and his mates occupying it:

> It was a wide, comfortable trench and we flopped about it. Our company now numbered seventeen men ... I looked around at the men of my company, the whole seventeen of them. A year ago, when Haig inspected us before Polygon Wood, we numbered two hundred and fifty, fit, strong, proud of our splendid battalion. And now we were seventeen, and we were neither fit nor strong ...
>
> Eyes were sunken, lacklustre, dull. Cheeks had fallen in, their grey colour enhanced by the stubby black and brown hair about them. Shoulders seemed shrunken. Dark blue patches and pouches showed beneath the eyes of every

> man. It was heartbreaking to look at them. No one moved or talked. It needed effort to do so.[55]

The afternoon and night passed without incident, providing everyone, except those detailed for patrols or sentry duty, with a chance to rest. The Germans also seemed weary, limiting their aggression to lobbing a few gas shells.

The next day, 2 October, was fine and clear. The tranquillity continued, broken only by desultory shelling of the Australian positions. The 6th Royal Inniskilling Fusiliers relieved the unit that evening. Private Bishop recorded the handover discussions:

> Our relief arrived on time. Inniskilling Fusiliers took over our sector.
>
> An officer spoke to George [Sergeant George Peters, acting company commander].[56] 'I'm looking for C Company, where are they?'
>
> 'You're looking at C Company, Sir.'
>
> The officer was shocked. 'A handful of men? You're a company?'
>
> George said that was so.
>
> 'Oh, my God,' the officer said. 'What a time you must have had.'[57]

After relief, the men made their way back through Bellicourt to Villeret, settling into shelters just north of this latter village by midnight. The troops slept and rested until midday the next day, relishing the sunny and clear weather. After lunch, the battalion, along with many other units of the division, began marching to their former camp at Le Mesnil. Ellis describes the state of the troops as they tramped back along the road:

> Troops more fatigued had rarely been seen and yet, by sheer determination, they overcame the weakness of the body and marched back in excellent order to their new positions. But their strained, pallid faces revealed what they had passed through, and numerous transport units along the roads respectfully and in silence pulled their vehicles to one side that the war-worn men might not have an extra step to march. It was the mute and eloquent testimony of brave men to heroes.[58]

At Le Mesnil Camp, the soldiers had three days of eating, sleeping and cleaning up. The weather remained fine and sunny. The 55th's War Diary records: 'The men were very cheerful despite the hard times they had gone through during the recent tour of the line.' Lieutenant Colonel Percy Woods rejoined the battalion from the nucleus and resumed command.

The 'hard times' had come at a heavy cost to the battalion: 19 killed and 61 wounded. While far fewer numerically than the unit had suffered in many previous engagements, proportionally it was almost half the fighting strength of the battalion. In quantitative terms, the battalion had captured 18 prisoners, one 77mm gun, one 4.2-inch howitzer and 29 machine-guns. The low number of prisoners was testimony to the brutal intensity of the battle.

The unit was spent. Private Bishop assessed that there were only 80 fighting men left in the battalion. After Bellicourt, the 55th Battalion was no longer fit for battle and would need a protracted time to rebuild.

Chapter 16: Peace – 'La guerre est finie!'

After the intense action in the Hindenburg Line, the surviving members of the battalion were given the prolonged break they craved. Private Bert Bishop wrote of the journey to the rear areas:

> On the sixth day of October we entrained at Péronne and headed away from the war. Next day we detrained at Oisemont, marched through Martainneville to a tiny village named Vismes-au-Val, where we were to have our long-awaited and much overdue rest.
>
> The countryside was beautiful. Little woods and open fields spread all around us. It was paradise to us and within a few days we began coming back to thinking human beings. Peace negotiations were going on, and each day we eagerly read English papers and read the communiqués posted up outside our battalion orderly room.[1]

A policy was promulgated granting all original Anzacs six months' furlough, during which they were able to return to Australia. The 55th's long-serving CO, Lieutenant Colonel Percy Woods, was eligible and marched out of the unit in early October; he left without much fanfare, his departure rating only a passing mention in the War Diary.[2]

The recently promoted Lieutenant Colonel Selwyn Holland was appointed to succeed Woods, but he was almost immediately sent away to a school for COs. In his absence, Captain Ken Wyllie and Major Eric Stutchbury took turns as temporary CO.

Writing from Vismes-au-Val on 11 October, Private Bishop was in a reflective mood:

> I don't believe it's possible to beat the record the five Australian divisions have made this year, take any troops, any nation, any war.

> We've just about reached our limit, though, they have worked us right to the death. My battalion came out of the line on Oct 3rd eighty-odd strong, just a battered wreck of what was as fine a battalion as could be found anywhere. However, we've given Fritz a hell of a drubbing and turned him right upside down and it seems if he wants peace right now. All the Australian divisions are out now, and it looks as if our long promised rest has come at last. I hope the war is over before we have to go near the line again. I'm sure I never want to see any more of it. During the last two months we've seen more war than all the rest of my time put together. We've found out what it really is, what it is to be fighting days and nights on end, nothing to eat, no sleep, heavy casualties, till one wonders how he manages to retain his sanity, wonders why men were brought into the world for this, wonders if he is a human being or the most heathenish animal that ever lived. However, there seems to be good prospects of it ending before long and I hope to heaven it does.[3]

Private Eddie Street also heard the murmurings of peace: 'A faint rumour began to circulate that the war was about to end. We did not put much faith in it, partly because if it didn't end, the disappointment would be awful.'

On 11 November, rumours became reality. The war was over. Bert Bishop recorded this historic event:

> A fellow put his head into our barn. He yelled to us: 'the Armistice has been signed.'
>
> I don't think anyone believed him.
>
> I went to the battalion orderly room. On the noticeboard by the door was pinned a broad octavo piece of paper. On it was typewritten in capital letters: 'Hostilities will cease at 11a.m. today'. The date at the top was the eleventh.
>
> I read it over and over. I walked off. I came back to see if it was still there. It was. I should have felt happy, but I did not. It was over, but so much had gone from life, the dreadfulness of it all seemed to overwhelm everything else. I thought of Ray and Billy, of Harold, my cousins. And I thought of all the grand fellows I had known, the friends I had made, who would not be coming home with me. I went off by myself, I walked for miles. However could I adjust myself to the new world which was opening for me?[4]

Private Eddie Street was also nonplussed at the news:

> On the afternoon of the 11th November four of us strolled down to a little railway station. The Station-Master was on tenterhooks, running into his office every few minutes. He began talking to us, and then his telephone bell rang. He raced back along the platform into his office, was there but a moment then he came racing back to us waving his arms and shouting excitedly. When he quietened we were able to understand him 'La guerre est finie!' he quavered, and then blew his little horn in ecstasy, and off we went to tell all of whom we could see. Our reactions? We went to [a nearby estaminet] had two coffees and rums, and then coffee straight, and a generous feed of eggs and chipped potatoes and so to our straw [to sleep]. Peace! It seemed incredible, but we wanted to believe it, and we knew it was true, we felt it.

Corporal Herbert Harris was almost melancholy at news of the war's ending, writing:

> Hostilities ceased 11 a.m. today. Not as much excitement as at a tame football match. Many amongst the boys all miserable. Flags flying from a few houses but tame as a mothers' meeting. Am sitting in our barn playing patients [sic]. We can't realise yet that it is all over. Wonder what I am going back to. More war I suppose.

The War Dairy for this first Armistice Day records the outbreak of peace almost as an afterthought:

> Training was continued within units, but the great event of the day was the receipt of news that the armistice had been signed by Germany and that hostilities ceased at 11a.m. today. The news was received by the troops without any actual display of enthusiasm, but there is no doubt that it is welcome news.

The next fortnight the battalion spent training, placing more emphasis on the ceremonial rather than combat skills. There was talk of the unit becoming part of an occupying army but in the end this proposal was shelved; with the AIF's reputation for wildness and indiscipline when out of the firing line, cooler heads realised that stationing Australians in Germany would be a recipe for disaster.

By December, the battalion was again on the move, this time to parts of France and Belgium formerly occupied by the Germans. Private Street recalled:

> Across the old battlefields where so many of our gallant cobbers lay ... We detrained and marched through places made famous in the first year of the war, weary miles we tramped, relieved only by the welcome given

> by the people of released villages. Here and there we were made very welcome, the whole population of towns and villages turning out.

The unit spent time in Busigny, Sivry and Le Sart. Finally, the men arrived at the town of Rance in the Walloon area of Belgium. Here they spent Christmas and the early months of 1919.

With the coming of the New Year, the battalion began to 'thin out'. Week by week small groups of men departed for Australia, leaving behind the mates who had sustained them, and the battalion that was their home. By 8 March, the numbers had shrunk to an extent that it no longer made any sense to retain the 55th Battalion as a separate identity; it was directed to combine with the 53rd Battalion. A month later the merged 53rd/55th Battalion joined with the 56th Battalion[5] to create the 14th Infantry Brigade Battalion. Two weeks later, this was also disbanded. In the end there was no welcome home parade, no fine speeches or bittersweet celebrations of victory and loss; the 55th Battalion simply disappeared.

Yet in the oft-troubled minds of the survivors, the unit never disappeared. In the 1920s, as another Armistice Day approached, John McCallum, no longer a lance sergeant but a brilliant student of history at Sydney University, wrote:

> The Old Battalion lived in memory and in memory those who had died lived younger, better, stronger, braver than those [left alive]. And it is of those whom we left dead at Fromelles or Pozières, at Flers or Passchendaele that we think now ... We knew them young, alert, strong and truer than steel – and perhaps, for some of us, it would be better to be with them. At least 'They shall not grow old'.

Roll of Honour

This roll includes the names of the officers and men who died of all causes while serving with the 55th Battalion outside Australia, including those attached to the unit. While every effort has been made to ensure that the Roll of Honour is as accurate as possible, errors may exist.

2146	Abrahams, Albert
1902	Agassiz, Theo
3001	Ainsworth, Joseph
2990	Akhurst, Cyril Wilton
1611	Alchin, Oliver Henry Gordon
3004	Alcock, Thomas Frederick
3346	Alcorn, James Carson
3226	Allen, Ernest Thomas
2124	Allen, Samuel Leslie
1612	Allen, Walter Otto Herbert
4728	Amour, Percy William
2557	Anchor, Victor Bernard
2983	Anderson, Esmond Lewis
3753	Anderson, George
2123	Arnett, Alexander Samuel
3755	Arnold, David Edward
2555	Arnold, John
3002	Ayers, Charles Thomas
25655	Babbage, John Frederick
3475	Bagge, Robert Nelson
1710	Bailey, Henry Robert
2128	Baker, Alexander
3007	Baker, Vinton Battam
3247	Baker, Vivian Ernald
1614	Bald, Thomas John
1057	Baldwin, William Herbert
1618	Ball, Michael James
5341	Banwell, Harry
2125	Baragry, Edmund Joseph
2129	Baragry, William Francis
2582	Barber, Alexander Herbert
3007	Barber, Richard
3014	Barker, Harold Francis
2781	Barker, Patrick
3216	Barnard, John Osmond
3031	Barrett, Rossiter Alfred
5339	Barry, Patrick
1520	Barton, David
3004	Bates, Albert
3954	Baugh, Francis Coll
3759	Beaven, Francis Wilson
5331	Bedson, Albert Leopold
25573	Belford, Harold Joseph
4745	Bell, John William
3760	Bell, Sydney
2372	Bellchambers, Arthur Amos
1879	Berryman, Gilbert
3376	Bingham, Albert Charles
1626	Bingley, Charles Stanley
1622	Birch, Chamberlain Edward
3211	Bird, Frank Hilton
3761	Bishop, Raymond Charles
2375	Black, Norman Stanley
5332	Black, Peter Anthony
3657	Blackburn, Horace Samuel
4747	Blunt, George Alfred
3009	Bolt, Herbert Thomas
1709	Bolton, Leon Harry
1086	Bourke, William
1621	Box, Archie Lorne
3109	Boyce, Samuel John
2574	Boyle, Edward Francis
2907	Boyle, Jerry
2786	Bradford, Apsley
3119	Breen, William
4765	Bridges, William Francis

3239 Broadhurst, Jack
3013 Broadhurst, Leonard
3212 Brown, Francis Cecil
4758 Browne, Alfred
2896 Bruce, John Kensell
2266 Bryant, William Henry
5333 Buckenham, Robert Victor
2133 Buckley, John Joseph
2618 Burch, Arthur
2625 Burgess, Sidney Edwin
5330 Burnaby, Ernest Arthur
3015 Burns, James William
2380 Burns, Robert Howell
1625 Burt, Albert George
2870 Burton, Arthur Herbert
3770 Burton, Charles Joseph
3105 Bush, George William Ham
5334 Butler, Thomas Patrick
2262 Byrne, Thomas Francis Joseph
2617 Byrnes, Cecil John
5344 Caldwell, James Dell
1286 Caldwell, Walter
2647 Cambey, Michael Matthew
3028 Cameron, Roy
2135 Campbell, Harold Sidney
3221 Cane, Arthur Samuel
2974 Cantwell, Michael Henry
1917 Carmichael, Alfred John Everleigh
2638 Casey, Ernest George
Chapman, Percy Wellesley
2272 Christie, Roy William
3033 Churchill, James Edward
2621 Clark, Frederick Goodwin
3381 Clark, Harry Septimus
Clark, William Frederick
1896 Clarke, Edgar George
2643 Clarke, Granville George Thomas Coden
4773 Clifton, Wenford John
3040 Cochran, John Gemmell
3378 Cohen, Sydney Israel
1889 Coleman, Leslie Raswell Elton
Colless, Stanley
2637 Collier, Lawrence Peter
Collins, Arthur Joseph
2139 Collins, Walter Gerald
2603 Commelin, Anton Adolf Constant
2134 Conlon, William Edwin
2886 Connelly, Alexander
1165 Conroy, Thomas Frederick
2142 Cook, William Henry
5060 Cooney, Thomas Albert
2629 Cooper, Samuel
2499 Cornick, Robert
2619 Cosgrave, William Patrick
Cotterell, Fredrick James
2170 Coughlan, Joseph Francis
5206 Cowan, Charles William
1911 Cracknell, Reginald William
2339 Crane, Hector Philip
3383 Crane, Stanley
3262 Crawford, Clarence Roy
3045 Crebert, Walter Henry
2388 Crocker, James Arthur
2617 Crockett, William James
3043 Crooks, Thomas
2383 Croucher, Frank Cyril
3784 Cullen, Andrew
5349 Cullen, Thomas Henry
1908 Dalton, Michael George
4787 Dark, Norman
2626 Davies, John Templeton
2523 Davis, Varney Eric Stanford
5207 Day, Charles William
4794 De Jongh, Philip
2893 Delany, William Wallace
3743 Dennis, Gordon Harris
5488 Detmers, Oscar Waldermar Frank
3047 Dewar, Robert Arthur
5362 Dickman, John
Dillow, William
59728 Dirksen, William

4677 Dixon, Benjamin
1637 Dixon, Bertrand James
2633 Dobson, William
2397 Donoghue, Thomas
5357 Donohue, William
2632 Donovan, James
3827 Donovan, Stephen
3246 Doughan, William Roy
915 Downer, William Charles
2600 Duffy, Frank
3538 Earl, Belford Wellington
2145 Eccleston, Napoleon Sylvester
3056 Edwards, Edmund John
4802 Egan, Edwin
1542 Egan, James Michael
2400 Ellis, Raymond Stanley
3777 Ellis, Verner Murray
2897 Elphick, Stephen Henry
3228 Elrick, James Robertson
5366 Erskine, Reginald Orme
5368 Eyre, Reginald Henry
3059 Ezzy, Charles
3060 Fahey, Patrick William
1291 Fairweather, Andrew Carnegie Booth
1921 Farrell, George Charles
1920 Farrell, Timothy Oxley
1904 Fenwick, Percival Roy
3061 Ferguson, Peter
3957 Ferris, Edward
2407 Field, David Ignatius
4939 Field, Harold Ernest
2404 Finch, John
2147 Fisk, Glynn Moulton
2150 Flanagan, Herbert Ellis
17184 Flanigan, Percy
3310 Fletcher, Frederick
1653 Flynn, William Agustine
2900 Ford, Alfred William
2646 Foret, Theo Ambrose
2902 Fox, Walter
2403 Franklin, George Edward
3803 Franks, Edgar Russell
2148 Freebody, Edwin James Thomas
3051 Freeman, Norman Lister
3520 Fryklund, Charles
Gardner, Archibald Ramsay
225 Garrett, William
4811 Geason, Percy
3808 Gell, George Lenton
5373 George, Frederick William Arthur
Gibbins, Norman
2157 Gilbert, Lyle Ernest
2661 Ginns, Albert Ernest
2158 Girvan, Roderick Roland Charles
2410 Gleeson, Thomas James
3517 Glynn, Harry
3408 Goodman, Richard Garfield
3146 Goodwin, James
2413 Granger, Edgar David
2907 Granville, James Thomas
5094 Gray, William Fordyce
1908 Greathead, Frederick
1907 Greathead, Henry Percy
2412 Grene, James
2909 Gribble, Percival Patrick
2659 Griffiths, Ernest Sydney
4953 Guillaume, Louis
3803 Gunn, Henry
3411 Hackman, William
3156 Haggar, Charles William
1667 Hall, Leslie John
1918 Hall, Reginald Eric
4820 Halpin, Lawrence
3605 Hamblin, William
2914 Hamer, Edward Arthur
1665 Hamilton, Ernest
3240 Hancock, Eric Floyd
5212 Harding, William
3507 Hare, David Edward
2431 Harpur, Patrick Bertrand
3819 Harris, James John

2674	Hart, William Barker
5380	Harvey, Leslie Ambrose
2684	Harvey, William Raymond
1920	Hauber, Henry Ernest
2163	Hayden, John Arthur
3315	Haydon, Herbert Ernest
53588	Henrickson, Anton Wicktor
3243	Hester, William
231	Higgins, Samuel McCulloch
5110	Hill, William Richard
25530	Hinson, Rupert Sykes
2609	Hocking, Francis Almund Wearne
1660	Hogan, Edward Thomas
2848	Holden, Edward
2152	Holliday, Sydney Charles
2424	Hollingsworth, Clyde
2919	Holloway, Leslie James
	Horniman, Lancelot Vicary
3090	Horsfield, Frederick
1542	Hughes, George
4826	Hughes, James Augustus
25546	Hunt, Austin George
1969	Hunt, Frank
5112	Hunt, Richard George
	Hunt, Sinclair Edward
	Hunter, David de Venny
5386	Hutchings, Leslie John
2419	Hyde, William James
3539	Ibbott, Albert Edward
	Inglis, Archibald John
2985	Ipkendanz, Eric Pirival
1658	Ireland, George
5478	Isedale, Harold Richard
	Jackson, Allen Raymond
1346	Jackson, William Henry
3078	Jaques, Arthur Henry
1924	Jennings, Arthur Frederick
2435	Jennings, Leslie Raymond
4831	Jennings, Walter James
1928	Johnson, Edgar Bootes
1927	Jones, Harold William Robert
2433	Jones, Hugh Henry
4710	Jones, Roger Arthur
4830	Jones, Waldo Emerson
3253	Joy, James
3101	Kanaan, Edmund Michael
1682	Kasel, Joseph
2682	Kavanagh, Ambrose
3154	Kay, Sydney Deakin
1941	Kearney, Edward David
3164	Keefe, Abraham
2695	Keeling, George Henry
2022	Kellond, John Henry
2684	Kelly, Philip
3143	Kelly, Robert William
5120	Kennedy, John Herbert Ross
1953	Keown, Sidney John
	Kerr, James
5121	Kerr, Patrick James
3163	Keyte, Robert
2701	King, George
3167	Kinsela, James Frederick
1665	Knight, Albert Lilley
2160	Kumsay, Cecil George Laurence
1696	Kurtz, Harold Thomas
2712	Lakeman, Leslie Joseph
5392	Lannon, John
3258	Lawrence, Francis John
4840	Leister, Leslie
4775	Leneard, Edward William
4228	Lennard, Henry
4229	Lennard, Samuel Robert
25562	Levy, Coleman
5896	Lewis, Francis Thomas
2937	Lewis, Raymond Henry
1686	Lidden, William
3170	Loader, James
2175	Longhurst, Vincent James Martin
3169	Lovejoy, Charles George
5397	Luby, Arthur Alfred Harold
2055	Lulham, Herbert Otto
2705	Lund, William James

3349	Luxford, Elexus	5220	Mills, Andrew
3964	Lyons, Frederick Norman	2450	Mines, Joseph
2861	Lyons, George	3378	Mitchell, Albert Edward
3187	MacCallum, Hector Archie	4848	Mitchell, Patrick
3186	MacDonald, John	2739	Moore, Eric Osborne
2691	Mackenzie, William Guilar	2457	Moore, Robert Thompson
270	Maher, Linus Edward	1945	Moore, William Martin
2644	Marsden, Thomas Samuel	3599	Morgan, Pritchard Llewellyn
	Marshall, John James	1949	Moriarty, Merion Morton
2718	Masterton, Alexander	3691	Morrison, Joseph
2873	Mayer, Henry	2454	Morrow, George Frederick
2462	McAlister, Hugh Wallace	3182	Mulvihill, Lawrence Rupert
2699	McCarroll, Frederick		Munro, Stuart
3447	McCormack, Ernest Albert	3875	Murphy, Edward Peter
4855	McCulloch, Thomas	2716	Murphy, Harold Richard
2200	McDonald, Leslie	2537	Murray, Harold
2469	McGill, James Laurence	2712	Naylor, Joseph
3873	McGuarr, Richard James	2477	Neave, William Arthur
838	McIlwain, Andrew Eyles	4860	Nelson, Alexander Edward
1189	McIlwain, Joseph Oscar	5421	Nelson, John
3672	McIntyre, Dugald	3107	Nelson, John McDonald
2497	McKenna, Edward Francis	1709	Newman, Edric
3099	McKeown, William Arthur	2204	Nichols, Henry
1943	McLaughlin, William Gerard	3193	Noakes, Abraham Theodore
3273	McLeod, Hector John	2475	Noble, Joshua Oswald Earl
1708	McMeekin, George	2423	Noldart, Albert Stanley
2740	McMillan, William	1956	Nott, Henry William
1704	McNeich, William Henry	1954	Nowland, Frederick Ernest
2470	McPherson, Leonard Donald	4865	O'Brien, Thomas Francis
2848	McRae, George William	4866	O'Neill, John Joseph
5405	Meale, Stanley Victor	2551	O'Neill, Patrick William
	Mendelsohn, Berrol Lazar	2205	O'Reilly, Donald Henry
2525	Menser, Leslie Maurice	2482	O'Reilly, Edward Henry
2864	Merritt, Hurtle Augusta	3452	O'Rourke, Thomas
823	Metcalfe, Arthur Charles	2770	Oakes, Arthur
2693	Mewburn, Earl Roland	845	Oliver, Wilfred Roland
3264	Meyer, Ernest Sydney Rudolph		Palmer, Herbert Leopold
3138	Meyer, Frederick Herman George	871	Pardey, Leslie James
5406	Miles, Leonard	2212	Parker, Hugh Archibald Davidson
3880	Millard, Charles Westbrook	870	Parryman, John Edward
1736	Milligan, Francis Frank	2969	Partridge, Charles

2758	Partridge, William Matthews
5428	Passmore, Arthur
4874	Paul, Albert Raymond
2748	Payne, Edgar Ashton
3204	Pegram, Albert George
2214	Pegram, Arthur Joseph
2668	Penfold, Neil
2217	Penney, William Benson Percy
2751	Petitt, Albert Edward
3458	Phillips, Harry Rudolph
5482	Pickersgill, Reginald
	Piddington, William Thomas
268	Poole, Edwin William
1204	Poole, Frederick Alexander
2711	Pratt, Gilbert Harding
2973	Priddle, Albert James
2209	Primmer, Ernest Montie
2841	Pringle, George McDonald
3455	Pritchard, Nathaniel
2215	Pryce, William Louis
2218	Quigley, Leslie Joseph
2721	Quinlivan, John Patrick
2720	Quinlivan, William Joseph
2225	Randall, Ronald James
2222	Rankin, Hugh Alexander
1718	Rankin, Leo Vincent
3905	Raymon, Robert William
4885	Reay, John Gilbert
25533	Reeves, William John
2765	Reidy, Henry Stephen
5225	Renouf, Albert
3907	Rex, Edgar Hartzenberg Bain
1966	Reynolds, Frank Bishop
2493	Reynolds, William Morton
5168	Rixon, John Osbourne Montague
2763	Roberts, Hugh
1168	Robertson, Alexander Christopher
2226	Robertson, Harry
3619	Robertson, Malcolm Clyne
1427	Robinson, Arthur William
1967	Roddan, Alexander Reginald
2987	Rogers, Walter
2989	Rootes, Henry Leonard
1087	Rose, William Henry Christian
4961	Roth, Harold Wallace
3913	Rue, Albert Edward
3211	Ruse, Samuel Boley
4877	Rutherford, Walter Roy
2221	Ryan, Cyril James
3006	Sargent, William Edward John
2528	Schafer, Frederick James
5432	Scott, John Russell
2924	Seaman, George Alexander
5446	Seater, William Albert
1735	Seckold, Percival Frederick
5434	Sewell, Joseph Augustine
1729	Sharkey, Edward Francis
179	Shephard, Leslie Clarencc
3922	Sheridan, John Francis
1632	Siddons, Bruce Lionel
3469	Simmonds, David John
2982	Simmonds, Edward Ernest
4865	Simons, James Lovel
5182	Sinclair, Charles Robert
2831	Sinclair, George Watt
2784	Smith, Alexander Robert
	Smith, Arthur Henry
3135	Smith, Colin Lindsay
2740	Smith, Darcy Benedict
5447	Smith, Harold Edward
180	Smith, Leslie John Shaw
2236	Smith, Reginald Norman
2241	Smith, Reuben
2741	Spindler, Albert Ernest
3924	St Smith, Herbert Newey
	Stafford, Louis Norman
5441	Stapleton, John
1291	Starr, William Gordon
3171	Steavens, John Edward
2772	Stephen, William John
1740	Stevens, Gus
2228	Stewart, Henry Wallace

2732 Sticker, Paul Oscar
1663 Stirling, Richard
2503 Stormont, Fulke Greville Le Poer
3467 Stott, James
2836 Stringer, Arthur Frederick
1177 Stubbins, Walter
2731 Sullivan, Eugene
1737 Sullivan, George
3013 Sullivan, John Thomas
184 Sykes, Henry Alexander
1745 Syran, Karl
2797 Taylor, Frederick Andrew
2510 Taylor, Jason
3190 Thatcher, Sidney Rupert
2825 Thompson, Alfred
1978 Toone, William John
2245 Tozer, Alfred Sydney Albert
2246 Turner, Simon Henry
3943 Twining, James George
2774 Tyler, Joseph James
2694 Urquhart, William
2469 Vass, David
Vogan, Arthur Henry
1838 Vogt, Peter Christian
2269 Walker, Arthur Washington
3952 Walsh, Patrick Joseph
3196 Walsh, Thomas
3654 Watson, Clarence Sinclair
3242 Wearne, Richard
2257 Webb, John
2997 Webster, Harold George
3650 Weir, John Francis
1981 Wensor, Edwin John
2515 West, Edward Henry
2378 Westlake, Harold Bass
3226 Whipp, William Arthur Stanley
1258 Whitehouse, William
2256 Whiteman, Vernon John
3308 Whitnall, Alexander Charles
2514 Whittaker, Thomas
2753 Williams, Benjamin Victor
2529 Williams, Frederick John
1759 Williams, Lloyd E C
3235 Willis, Ubert Victor
2802 Wilson, David
2775 Wilson, George Henry
4925 Wilson, Henry O'Brien
3473 Wilson, Oliver William
3309 Wilson, Thomas William
2252 Winner, Arthur James
3243 Wolfe, Thomas Edward
2039 Wood, George
5103 Wood, Harold Hewson
4914 Woodward, Edgar Hubert
Woolley, Charles Russell
3517 Yeark, Alfred John
2812 Young, James

Summary of Casualties

	Officers	Other Ranks	Total
Killed in Action	23	459	482
Died of other Causes (wounds, accidents, disease)	_	49	49
Total Deaths	23	508	531
Wounded in Action*	71	1804	1875
Prisoner of War #	3	102	105

*Wounded in Action describes the number of individuals requiring hospitalisation as a result of combat operations. Of the number shown, many subsequently died of wounds and are also counted among 'Died of other Causes'. Where an individual has suffered multiple woundings, these are counted as a single entry.

#Of the number shown, three other ranks died of wounds after capture, and two of illness while captive. These are also counted among 'Died of other Causes'.

Honours and Awards

The honours shown are those awarded for services with the 55th Battalion.

Victoria Cross

1717 Ryan, John

For conspicuous gallantry and devotion to duty, and for saving a very dangerous situation under particularly gallant circumstances during an attack against the Hindenburg Defences on 30th September, 1918. In the initial assault on the enemy's positions this soldier went forward with great dash and determination, and was one of the first men of his Company to reach the trench which was their objective. Seeing him rush in with his bayonet with such exceptional skill and daring, his comrades were inspired and followed his example. Although the enemy shell and machine gun fire were extremely heavy, the enemy trench garrison was soon overcome. In the assault the attacking troops were weakened by casualties, and, as they were too few to cover the whole front of the attack, a considerable gap was left between Private Ryan's battalion's left and the unit on the flank. The enemy counter-attacked soon after the objective was reached, and a few succeeded in infiltrating through the gap and taking up a position of cover in ear of our men, where they commenced bombing operations. The section of trench occupied by Private Ryan and his comrades was now under fire from front and rear, and for a time it seemed that the enemy was certain to force his way through. The situation was critical and necessitated prompt action by someone in authority. Private Ryan found there were no officers or N.C.O.'s near, they had become casualties in the assault. Appreciating the situation at once, he organised the few men nearest him, and led them out to attack the enemy with bomb and bayonet. Some of his party fell victim to the enemy's bombs, and he finally dashed into the enemy position of cover with only three men. The enemy were three times their number, but by skilful bayonet work they succeeded in killing the first three Germans on the enemy's flank. Moving along the embankment, Private Ryan alone rushed the remainder of the enemy with bombs. It was while thus engaged he fell wounded, but his dashing bombing assault drove the enemy clear of our positions. Those who were not killed or wounded by his bombs fell victims to our Lewis gunners as they retired across 'No Man's Land'.

Distinguished Service Order (DSO)

McConaghy, David McFie
Slater, Henry Ernest

Bar to Distinguished Service Order

Woods, Percy William

Member of the Order of the British Empire (MBE)

Murray, James Edward

Military Cross (MC)

Armstrong, Francis Edwin
Chadwick, Luther
Chapman, Percy Wellesly
Clark, William Frederick

Colless, Stanley
Cotterell, Frederick James
Denoon, William
Donaldson, George Shirley
Duprez, Arthur Alexander
Ellsmore, Rupert
Goldrick, Roy Arthur
Gow, Carl Beeston
Hourn, Midford Stanley
Mackay, Norman John
Miller, Harald
Morgan, William Black Stewart
Morris, Reginald Sydney
Neville, Dalton Thomas Walker
Pinkstone, Sidney Albert
Woods, Percy William
Wyllie, Hugh Alexander
Wyllie, Kenneth Robert

Bar to Military Cross

Giblett, William Norman
Stutchbury, Eric William

Distinguished Conduct Medal (DCM)

1701	Anson, Henry
1517	Bourke, Edward Alexander
3026	Buckingham, Thomas Andrew
1887	Clark, Kenneth Arthur
	Colless, Stanley
3134	Dowling, Ronald Guy
2816	Dunne, John Herbert
2491	Hannah, John Leonard
2609	Hocking, Francis Alumnd W.
1676	Kearns, Arthur Andrew
743	Lee, Leslie William
3519	Mackay, Alexander Leslie
3093	Mackenzie, Forbes Burnside
1690	Marshall, James Percival
153	Matthews, Albert Reginald
3098	McPhee, John Alexander
4851	Mortlock, Harold
	Neville, Dalton Thomas Walker
2415	Penny, Alec Gordon
3604	Peters, George Luke
2535	Stewart, Ronald Esmond
3204	Winter, Archibald Thomas

Military Medal (MM)

2260	Arnett, Charles Victor
3347	Auckland, Arthur
	Biden, Eric Broomfield
3762	Bishop, Walter Herbert
2913	Buckeridge, John Wilfred
2133	Buckley, John Joseph
1635	Cadd, Albert Ernest
3030	Campbell, George McIntyre
2138	Cant, Martin Hiliary Clement
	Chadwick, Luther
3736	Christie, William Neil
1630	Curran, Leslie William
2821	Dawe, Ernest Robert
5477	Dawes, John Francis
5067	Diversi, James Leopold
2633	Dobson, William
5076	Farrell, Coulson Montague
2646	Foret, Theodore Ambrose
2151	Freebody, Lionel Edward
3510	Freeman, Francis John F.
4806	Fuller, Claude Elliott
2001	Gillett, Arthur James
243	Groutsch, Francis James
8422	Hall, Arthur Thomas
1667	Hall, Leslie John
3078	Hardy, Arthur Henry
3129	Higgs, Isacchar Pearson
3822	Hodges, Herbert
132	Howell, Frederick Walter
1673	Jackson, Leslie Alfred
2167	Jacobs, Arthur Samuel
3254	James, Thomas
2684	Kelly, Philip
4834	Kerin, Michael
1681	Klein, Henry Charles
1940	Lemon, Harry Marcus
1168	Lewis, Andrew
3860	Marshall, Bertie
1690	Marshall, James Percival
1695	McAlister, Frederick
5418	McCabe, Christopher
651	McCluskey, Peter
1188	McDonald, Walter
3188	McInnes, John Alexander

2698 McMahon, Sylvester Luke
1704 McNeich, William Henry
2605 Muir, William Beresford
2846 Murray, Stanley
2849 Neild, Edwin
2704 Nicholls, William Gourley
2745 O'Brien, James Patrick
3144 O'Dea, Denis John
2181 Olive, Arthur Horace
3150 Perkins, James Alfred
3604 Peters, George Luke
2854 Pringle, John Leslie
3900 Puckeridge, Joseph
3285 Rosborough, Archibald
3121 Ryall, William Patrick
1738 Schwind, Bert
2743 Sharpe, Thomas John
2012 Smith, Abraham Gorham
2535 Stewart, Ronald Esmond
1570 Street, Edmund Harrington
5452 Turner, Albert Harold
2019 West, Frank
2917 Wolrige, Edgar Claude
194 Wood, Louis Cecil
2546 Woodhart, Edward Grey
1756 Woods, Edward George
2258 Yelds, Stanley Darwin

Three Bars to Military Medal

2143 Corey, Ernest Albert

Meritorious Service Medal

167 Petts, William
4800 Jenkins, William Harold
3027 Collens, George Percy

Foreign Awards

Belgian Croix de Guerre

3860 Marshall, Bertie
2184 Monck, Peter Eugane
Slater, Henry Ernest

French Chevalier of the Legion of Honour

Holland, Austin Claude S.

United States Distinguished Service Cross

2487 Parkes, Thomas

Mentioned in Despatches

The number of times mentioned is shown in brackets after each name.

2575 Bennett, Austin Marchant
1517 Bourke, Edward Alexander
Cowey, Robert Orlando
5362 Dickman, John
Gow, Carl Beeston
2414 Granger, Hubert Stanley
5388 Hollis, Roy Edward
4800 Jenkins, William Harold
3437 McColl, James
McConaghy, David McFie
Neville, Dalton Thomas Walker
4957 O'Sullivan, Vincent Dominic
Panton, Alexander Waugh
Peake, Robert Algarnon (2)
3200 Pearce, Oliver Austin
Pinkstone, Norman Ewart
3464 Saxby, Charles
Slater, Henry Ernest
Smith, Alfred Edward
Wilson, Norman Lewis
Woods, Percy William (3)

Summary of Honours and Awards

VC 1
DSO 2
MBE 1
MC 22
DCM 22
MM 71
MSM 3
Foreign Decorations 5

Bars
DSO 1
MC 2

3 Bars
MM 1
MID 19
MID (twice) 1
MID three times) 1

Nominal Roll

This roll includes the names of the officers and men who served with the 55th Battalion in Egypt, France and Belgium. It includes personnel attached to the unit. The roll has been compiled from official and private sources and does not include men of the 53rd Battalion who amalgamated with the 55th in 1919. While every effort has been made to ensure that the Nominal Roll is as accurate as possible, errors may exist. Reinforcements joining the unit after the end of hostilities are not included. The names of known Aboriginal servicemen are denoted #.

3104	Aberline, Arthur Ernest
2146	Abrahams, Albert
1054	Ackroyd, William
4729	Acton, George William
3765	Adams, Alfred
2856	Adams, Charles Frederick
4727	Adams, William John
59842	Adkins, Leonard Killingworth
2858	Adlam, Linford Charles
3102	Affleck, William John
6	Agassiz, Cecil Travnor
1902	Agassiz, Theo
3592	Agland, Adolphus Doran
3007	Ainsworth, Ernest Alfred
3001	Ainsworth, Joseph
3506	Ainsworth, William Harrison
2613	Airey, William Henry
1614	Aitchison, John Dixson
2990	Akhurst, Cyril Wilton
2369	Alchin, Clyde John
4354	Alchin, Nevil Victor
1611	Alchin, Oliver Henry G
3004	Alcock, Thomas Frederick
3346	Alcorn, James Carson
3348	Alder, Thomas William
3353	Alexander, James Allan
1615	Alford, Francis George
2612	Alford, Frederick William A.
3798	Allan, David William
3752	Allan, Duncan Gillies
1713	Allen, Edward James
3226	Allen, Ernest Thomas
	Allen, Harold Raphael
2553	Allen, Herbert Daniel
	Allen, Keith Wickham
3591	Allen, Leo
2367	Allen, Percy James
2552	Allen, Reginald Edward
2124	Allen, Samuel Leslie
3002	Allen, Samuel William
1612	Allen, Walter Otto Herbert
3101	Allport, Henry Albert
3731	Allsbury, Joseph
1867	Allsop, William
2958	Allt, Percy
4053	Alston, John Linden
2611	Ambrose, James Francis
4735	Amey, James Alfred
4728	Amour, Percy William
2554	Amourin, Alfred Gusto
1869	Amourous, Leopold
2557	Anchor, Victor Bernard
59698	Anderson, Archibald
2983	Anderson, Esmond Lewis
4732	Anderson, George Peter
3753	Anderson, George
4733	Anderson, James
3352	Anderson, Ormond Stanley
2370	Anderson, William Frederick T.
2558	Andrews, Leonard

NOMINAL ROLE

803	Angove, Ernest
2261	Angus, Frank
3354	Angus, Robert
5236	Anker, Samuel Peter
1701	Anson, Henry
2122	Apps, Alfred Arthur
2857	Archbold, George Walter
2121	Archer, Halloran
3355	Archinal, William Henry
1868	Armitt, Herbert Cecil
3349	Armstrong, Edward William
	Armstrong, Francis Edwin
2366	Armstrong, Harold George
5474	Armstrong, John Bird
1052	Armstrong, John William
2368	Armstrong, Noel Raeburn
1613	Arndt, August James
2123	Arnett, Alexander Samuel
2260	Arnett, Charles Victor
5031	Arnold, Allan John
3755	Arnold, David Edward
2555	Arnold, John
2371	Arnst, Robert
2368	Arthur, Victor Alexander
1509	Ash, John McMillan
5032	Ash, Sydney John
3210	Ashcroft, Herbert Alfonse
126	Ashdown, Stanley Alfred
4356	Ashworth, Arthur
3103	Ashley, Raymond Leslie
2782	Ashton, Arthur William
4355	Askew, George Michael
	Askham, Albert Charles
1866	Atkins, John Phillip
3756	Atkinson, George Edward
5326	Atkinson, Kenneth Alaton
3347	Auckland, Arthur
1866	Audubon, Leonard Benjamin
3230	Austin, Merton James D. A.
3002	Ayers, Charles Thomas
25526	Ayres, Frederick Stanley
59699	Azoulay, Leo Joseph
25655	Babbage, John Frederick
	Back, John
2859	Backhouse, Leslie James
3475	Bagge, Robert Nelson
1710	Bailey, Henry Robert
169	Bain, John
2128	Baker, Alexander
2567	Baker, Alfred
2127	Baker, Charles Robert
3363	Baker, Edward Warrington
4741	Baker, Harry Knox
59703	Baker, John
2560	Baker, John Halero
53525	Baker, Stanley Vincent
1623	Baker, Thomas Norman
3007	Baker, Vinton Battam
3247	Baker, Vivian Ernald
1614	Bald, Thomas John
3020	Baldwin, Harry Bertram
59704	Baldwin, Reginald John
1057	Baldwin, William Herbert
3118	Baley, Henry
1618	Ball, Michael James
53683	Ball, Percy Randolph
2578	Bamford, George
2918	Banbury, Gavin Irving
3366	Band-Scott, David
3316	Banks, Albert William
3793	Banks, John William
25539	Bannerman, David Christopher
1880	Bannerman, David Sutherland
3362	Bannet, Nathaniel
5341	Banwell, Harry
2332	Baptist, John Eric
2125	Baragry, Edmund Joseph
2129	Baragry, William Francis
2582	Barber, Alexander Herbert
3007	Barber, Richard
2619	Barham, Leo
3766	Barker, Edgar William
3014	Barker, Harold Francis
1914	Barker, Henry
4742	Barker, James Lochee
2781	Barker, Patrick (alias P. Hogan)

3594 Barlogio, John James
Barlow, George Rolla
3216 Barnard, John Osmond
5327 Barnes, Ernest Walter
2562 Barnes, James Joseph
5473 Barnier, Harold Robert
1615 Barratt, James
59713 Barrett, John Patrick
3031 Barrett, Rossiter Alfred
3375 Barrett, Thomas William
3757 Barrie, Andrew Robert
2374 Barrington, Henry Edward V.
3122 Barrow, Robert
59705 Barry, George
2382 Barry, John
5339 Barry, Patrick
5033 Bartle, William Henry
2381 Bartley, Lawrence Joseph
1520 Barton, David
1878 Barton, Eric Arthur
Barton, John Hampton
1880 Bassingthwaite, Roy Shiel
1874 Batcup, Reginald Leslie
2007 Bateman, Lawrie Marsh
3018 Bateman, William Joseph
3004 Bates, Albert
2379 Bates, Thomas John
4743 Bates, William David
1070 Batten, Leonard
3358 Batten, Robert
3954 Baugh, Francis Coll
3318 Bawdon, William Tatteley
3008 Bayley, Ernest Arthur
1741 Beale, Robert
3759 Beaven, Francis Wilson
1624 Beaver, John
2565 Beaver, Lawrence Ernest
3700 Beckman, Ernest George
2861 Bedford, William Edward
5331 Bedson, Albert Leopold
5343 Beech, Frederick William
3015 Beeche, Joseph Alfred Murray
3633 Beesley, George
1913 Beesley, William Gladstone
2532 Beileiter, Arthur Ernest
25573 Belford, Harold Joseph
1304 Bell, Eric Walter
3697 Bell, John Henry
4745 Bell, John William
3760 Bell, Sydney
5035 Bell, William Joseph
2372 Bellchambers, Arthur Amos
4746 Bellew, Alan George
2795 Benjamin, David Henry
2575 Bennett, Austin Martin
4751 Bennett, John Thomas
2547 Bennett, Joseph
3006 Bennett, Lionel William
5036 Bennett, William George
4752 Bennett, William Ponsford
3021 Bentley, Phillip Percival
3365 Bernard, Bert
3460 Berry, James
1990 Berry, Lester
1879 Berryman, Gilbert
2908 Bevan, Arthur Frank
3024 Bevan, Edgar David
Biden, Eric Broomfield
3376 Bingham, Albert Charles
3121 Bingham, Evelyn Francis
1626 Bingley, Charles Stanley
2130 Bingley, Harold David
2875 Binks, Ernest Henry Thornly
1622 Birch, Chamberlain Edward
3211 Bird, Frank Hilton
Birks, Eric Napier
59709 Birt, Arthur Samuel
3761 Bishop, Raymond Charles
5328 Bishop, Richard Henry
3762 Bishop, Walter Herbert
3238 Black, James
2375 Black, Norman Stanley
5332 Black, Peter Anthony
2863 Black, Thomas
3657 Blackburn, Horace Samuel
3258 Blackett, Hilton Sidney

NOMINAL ROLE

2864 Blackman, Samuel
Blackman, Walter John
2620 Blake, Charles William
3361 Blake, Richard John
3367 Blatchford, Leslie
3518 Blatchford, Norman Clyde
3114 Blewitt, Bertie Edgar
2271 Block, Herbert
4744 Blomfield, Frederick Gordon
3368 Blood, Frederick Thomas
4747 Blunt, George Alfred
3116 Blyth, Robert
2275 Blyton, Ernest John
2033 Boag, John
4736 Boardman, John
2623 Bofinger, Jack Stanley
1100 Boland, Frank Corrie
3490 Bolder, Alfred Vincent
3489 Bolder, Horace Dudley
3009 Bolt, Herbert Thomas
3474 Bolton, Clois Philip
1709 Bolton, Leon Harry
2274 Bonaba, Victor Charles
4001 Bonner, John
53886 Bonnett, Archibald
25527 Booth, Harold Arthur
3478 Booth, William
2615 Borman, Frank Henry
2337 Boston, Edwin
3020 Boswell, Reginald William
2378 Boulter, Harold Holmsdale
1517 Bourke, Edward Alexander
3360 Bourke, Frank
2276 Bourke, Thomas
1086 Bourke, William
1887 Bourne, Alfred
2373 Bourne, Arthur John
4737 Bouveret, Alfred John
1877 Bowden, Eric Richard
Bowe, John Stewart
2797 Bowen, Frederick
59841 Bowler, Percy James
53445 Bowles, Eric Clarence R.
25680 Bowles, Leonard Frank
1661 Bowtell, Alfred John
1715 Bowman, David McQueen
1621 Box, Archie Lorne
3470 Boyan, Arthur
3109 Boyce, Samuel John
3369 Boyd, David
4757 Boyd, David
2520 Boyd, James Samuel
2574 Boyle, Edward Francis
2907 Boyle, Jerry
5336 Boyle, Michael Silvester
2865 Boyton, Gordon Staley
2786 Bradford, Apsley
2011 Bradford, Frederick William
1871 Bradley, Henry
1620 Bradney, Wentworth Edward
3593 Brady, Thomas (alias A. W. Dodds)
4749 Bradshaw, Frederick James
5491 Brady, Walter Phillip
5432 Bragg, William Charles
3111 Braggs, Phillip
5043 Braitling, Albert Victoria
2866 Branson, Douglas Gerald
3492 Brasdon, Frederick
3473 Brassard, Arthur Louis
25652 Bray, William Malcombe
4659 Braybrook, Charles William
3119 Breen, William
3112 Brenell, John#
2560 Brennan, James Clyde
4741 Brennan, William Arnold
5475 Brennan, William Henry
Brent, Lindsay Peregrine
59708 Brereton, Edward Ernest
4750 Brett, Joseph George
2493 Bridgham Edward James (aka Thompson E.J.)
4765 Bridges, William Francis
4949 Briggs, Claude
63 Briggs, Henry Francis
1873 Bright, John James
2796 Brightfield, Henry James

2519	Brissenden, Lynton Edward
3011	Broadhurst, Bert
3012	Broadhurst, Frederick Cornelius
3239	Broadhurst, Jack
3013	Broadhurst, Leonard
	Brock, Kenneth Henry
3357	Brodie, George Oliver
4759	Brodie, William John Thomas
3231	Bronger, Claude
3110	Brooke, Walter
1619	Brooker, Charles
3117	Brooker, Clyde James
2905	Brooker, Eugene
2006	Brooker, Henry Stafford
3010	Brookes, Edwin Nicholson
2376	Brooking, John Edwin
5329	Brooks, Albert Joseph
2867	Brotherton, Charles Edward
1869	Brotherton, James
2621	Brown, Aubrey Thomas Ray
5044	Brown, Charles
2868	Brown, Charles Frederick
3013	Brown, Charles Joseph
3016	Brown, Eric Herbert
17681	Bruce, Eric Robert
3212	Brown, Francis Cecil
3108	Brown, Frank
3647	Brown, Frederick
5045	Brown, Harry
4754	Brown, Henry
3766	Brown, John
3017	Brown, Lachlan
2616	Brown, Louis Nivin
1872	Brown, Morley Thomas
5337	Brown, Roy
1623	Brown, Sydney
1689	Brown, Walter Ernest
3115	Brown, William Charles
5046	Brown, William John
4758	Browne, Alfred
1627	Bruce, Cecil
17681	Bruce, Eric Robert
1876	Bruce, Frederick
2896	Bruce, John Kensell
1870	Brumby, Joseph Thomas
	Brumwell, Donald Stanley
3767	Bryan, Richard Alfred
53571	Bryant, Leslie James
1878	Bryant, Samuel George
2266	Bryant, William Henry
394	Bryce, Charles Edward
2132	Bryce, William John
3234	Bryson, Joseph
2909	Buchan, George
5333	Buckenham, Robert Victor
2913	Buckeridge, John Wilfred
5338	Buckingham, Arthur Edwin John
3026	Buckingham, Thomas Andrew
2133	Buckley, John Joseph
4761	Buckley, Joseph
1625	Buckley, Patrick
1065	Buckman, Harry Albury
4763	Bull, Lewis Marsden
3573	Bullman, Thomas King
2725	Bulmer, Herbert Henry J.
2869	Bunton, William Wallace
2618	Burch, Arthur
3768	Burden, Frank
4762	Burford, John Henry
2625	Burgess, Sidney Edwin
1616	Burn, Charles Francis
5330	Burnaby, Ernest Arthur
2914	Burnett, Gordon John
2538	Burnett, James
	Burns, James
3015	Burns, James William
2380	Burns, Robert Howell
5202	Burrows, Frederick Charles
1625	Burt, Albert George
2870	Burton, Arthur Herbert
3770	Burton, Charles Joseph
3359	Bush, Charles
3105	Bush, George William H.
3113	Bush, Michael Raymond
3107	Bush, Roland Elijah
2626	Bushell, Martin Gregory D.

3630	Butcher, Charles Henry	3596	Campbell, John Allan S.
3373	Butcher, Henry Luke	2872	Campbell, John Cameron
2559	Butler, Daniel Arthur	3127	Campbell, John Gordon
3771	Butler, Edward Malcolm	3955	Campbell, John James
2871	Butler, Samuel George	3494	Campbell, Robert
1875	Butler, Sydney	5350	Campbell, Robert James
5334	Butler, Thomas Patrick	3253	Campbell, Robert Neil
3106	Butler, William Ennisford	3270	Campbell, Rupert Joseph
2571	Buttsworth, Ernest Albert	707	Campbell, Valentine Darcy H.
2895	Byerlee, Cyril Alfred		Campbell, William Ernest
5476	Byers, Robert Howard	2873	Campbell, William James
3237	Bynon, James	2622	Campbell, William Ross
3356	Byrne, Clarence	3221	Cane, Arthur Samuel
2536	Byrne, Edward Robert	3379	Canning, George
53680	Byrne, George Bernard	2616	Cannon, Charles Thomas
	Byrne, John Samuel	4769	Cannon, Frederick John
2131	Byrne, Raymond John S.	2116	Cansdell, Dudley Smart
1617	Byrne, Richard	2799	Cant, Cecil Henry
2262	Byrne, Thomas Francis J.	2138	Cant, Martin Hilary Clement
2617	Byrnes, Cecil John	2974	Cantwell, Michael Henry
3015	Byrnes, Robert Reginald	3741	Capon, Charles Beresford A.
2614	Cabban, Henry Percy	3508	Capper, John Richard James
1635	Cadd, Albert Ernest	3802	Carey, John
5048	Cain, Ernest	1288	Carlisle, Thomas
3387	Cairns, Thomas	3597	Carlson, Charles Albin
5344	Caldwell, James Dell	1917	Carmichael, Alfred J. E
1286	Caldwell, Walter	5353	Carnellor, Nicholas
3007	Callinan, Denis	1628	Carney, William
4766	Callinan, Eugene Patrick	3029	Carpenter, Frederick Lloyd
2647	Cambey, Michael Matthew (alias Burke)	1629	Carpenter, Harold Charles
2367	Cameron, Alexander Rankin	1883	Carpenter, Henry
1991	Cameron, Angus	2393	Carpenter, Keith Thomas
1882	Cameron, Donald McGregor	3035	Carpenter, Kenneth James
1636	Cameron, Hugh Allen	2248	Carr, Charles
3315	Cameron, John	3252	Carr, Harold
3504	Cameron, John William	59724	Carr, John
3028	Cameron, Roy	3042	Carroll, John Stinson
3795	Cammeron, Malcolm Alexander	2938	Carruthers, Robert William
2136	Campbell, Albert Edwin	2874	Carter, Albert Jerome
3008	Campbell, George	3386	Carter, Joseph
3030	Campbell, George McIntyre	3694	Cartwright, Arthur Edward
2135	Campbell, Harold Sidney	4771	Cartwright, George Henry
2390	Campbell, James Charles	2638	Casey, Ernest George

3027	Casey, Michael John
109	Casey, William
2875	Cash, John Richard
3396	Casimir, Edward
1886	Caspers, Joseph Henry
1715	Cassidy, Stephen Joseph
2387	Cassilles, Henry
2876	Castles, Timothy Leo
5354	Catterall, Frank Alfred
2877	Causer, Thomas George
25638	Cavill, Walter
2608	Cawsey, Thomas Henry
25628	Cayill, Walter
1073	Cazanave, Joseph
3219	Chadwick, Luther
3382	Chalker, Charles
2630	Chalker, Charles Edward
2631	Chalker, Frederick William
3130	Chalker, John
59844	Chamberlain, Ralph James
3799	Chamberlin, Hector Llewellyn
3031	Chambers, Alfred John
3032	Chambers, Frederick Joseph
3125	Chambers, Walter
5346	Champley, Raymond
3491	Chaplin, Ernest George
3485	Chaplin, Frederick William J.
	Chapman, Percy Wellesley
3489	Chapman, Thomas Leslie
3264	Charters, John
2878	Cheeseman, Frank
2760	Cheeseman, Joseph Dennis J.
3391	Cherry, Luke Tarver
3392	Chesterton, Sidney Herbert
3493	Chilcott, Clifton
3775	Christie, John Clarence
2272	Christie, Roy William
2632	Christie, William
3736	Christie, William Neil
4772	Church, Walter Clifford
2013	Churcher, Charles Hubert
3033	Churchill, James Edward
1893	Clancy, Harold George
2048	Corsie, Phillip Clarence
2621	Clark, Frederick Goodwin
3381	Clark, Harry Septimus
56055	Clark, James Harold
1887	Clark, Kenneth Arthur
	Clark, William Frederick
1721	Clarke, Arthur James
1888	Clarke, Arthur James
2879	Clarke, Arthur Thomas
1904	Clarke, Deighton
2880	Clarke, Cecil Richard
1918	Clarke, Cecil Sydney
1896	Clarke, Edgar George
3693	Clarke, Frederick
2141	Clarke, Frederick John
2643	Clarke, Granville George T. C.
2642	Clarke, Herbert James
4768	Clarke, Herbert Joseph M.
2881	Clarke, James Edward
1340	Clarke, John Edward
2627	Clarke, Norman Henry
2137	Clarke, Richard William Alexander
3659	Clarke, Victor Nelson
3129	Clarke, William John
3377	Clarke, William John
2394	Clayton, Herbert Frederick F.
2882	Cleary, James
	Cleland, Leslie William H.
1892	Clifford, George
433	Clifton Alfred James
4775	Clifton, Edward William
4773	Clifton, Wenford John
3394	Clinch, Sydney John
3029	Clinton, Harold David James
2883	Clough, Alexander Claude
58543	Cluett, Cecil Redvers
4767	Clulow, Reuben Edward
2893	Coady, William Joseph
3043	Coates, John Bulman
2601	Coates, John William
2636	Cobb, Bertie Whitby
3126	Cobban, Richard Gordon
3040	Cochran, John Gemmell

2539 Cochrane, Henry
Cochrane, John Meston
2800 Cochrane, Weston
3378 Cohen, Sydney Israel
Cohen, Sydnie Lionel
2884 Cole, Edward
2525 Cole, Harold Patrick F.
1919 Cole, Reginald Norman
919 Cole, William Walter
3729 Coleman, George Harold
4774 Coleman, Henry George
1889 Coleman, Leslie Raswell E.
3027 Collens, George Percy
Colless, Stanley
3380 Collett, Arthur William
2384 Colley, Henry
2389 Colley, Wilfred Joseph
2637 Collier, Lawrence Peter
2644 Collier, William Henry
728 Collins, Arthur
2641 Collins, Arthur Ernest
Collins, Arthur Joseph
3388 Collins, Edward
1633 Collins, Oliver Harold Cecil
2139 Collins, Walter Gerald
4779 Collis, John Cornelius
5057 Collison, Reginald Hercules
3026 Colvin, William Cargill
2885 Comans, William Francis
2784 Comfort, Frank Herbert
2603 Commelin, Anton Adolf C.
637 Commens, Christopher Addie
1885 Compton, Robert
3385 Condran, Michael Hugh
3777 Conlon, Reginald G (aka Clinton R.G.)
1631 Conlon, Ralph Joseph
2134 Conlon, William Edwin
2886 Connelly, Alexander
4776 Connelly, Leo
2892 Connelly, Patrick
59720 Connor, Rov Gordon
1165 Conroy, Thomas Frederick
3037 Cook, Arthur Fletcher Croft
3034 Cook, Harry William
2887 Cook, John
4780 Cook, John William
5347 Cook, Joseph
3695 Cook, William
4781 Cook, William Edward (aka Engel W.)
2142 Cook, William Henry
3131 Cooley, George William
4785 Cooney, John Thomas
5060 Cooney, Thomas Albert
4088 Cooper, Ashley James
2628 Cooper, Frank Lane
2645 Cooper, Harry Porterfield
2629 Cooper, Samuel
5061 Cootes, William Samuel#
2521 Cope, Edward
3042 Cope, Thomas Herbert
2635 Copson, Edward John
2391 Corbett, Patrick Joseph
2143 Corey, Ernest Albert
2640 Cormack, William James
3123 Cornford, Arthur Foster
2499 Cornick, Robert
1993 Cornish, Thomas
53528 Corrigan, Patrick John J.
2619 Cosgrave, William Patrick
3486 Costello, Herbert Frank
3389 Cotter, James Frederick
1716 Cotterell, Ernest
Cotterell, Fredrick James
3049 Cottier, Edmund Roy
3050 Cottier, Walter Charles
1891 Couch, Alfred Gordon
2010 Couch, John James
2170 Coughlan, Joseph Francis
2633 Coulson, Thomas Leslie
5355 Coupland, John Robert
3648 Courbarron, Frederick Hamilton
2597 Covil, Arthur Leonard
5206 Cowan, Charles William
Cowey, Robert Orlando
1761 Cowley, Francis Patrick
1631 Cowper, Percy

2902	Cox, Henry Sydney	4783	Currie, Robert
2385	Cox, Horace	4786	Curtis, Harold John
5348	Cox, James Thomas	1890	Curtis, Henry
3493	Cox, Jesse	3220	Curtis, John
2263	Coyle, John Joseph	2888	Cusack, Denis Aubrey
3652	Crabbe, Reginald Archie	1634	Cusack, Thomas
1911	Cracknell, Reginald William	1642	Dade, Roy James
1727	Cragg, Charles John	3499	Daisley, Henry William C. S.
3128	Craig, Charles Volney	3402	Dalby, Joseph
2886	Craig, Robert	1908	Dalton, Michael George
2887	Cramp, Cecil	3040	Daniel, Clive Percival
3675	Crane, Arthur	2090	Daniher, Joseph
2339	Crane, Hector Philip	1897	Dare, Francis Ellison
3383	Crane, Stanley	4787	Dark, Norman
3780	Crapp, Albert	2890	Davey, William Henry
3781	Crapp, Arthur Hilton	2582	Davidson, David Andrew
3262	Crawford, Clarence Roy	1894	Davidson, John Ingleton
3045	Crebert, Walter Henry	3399	Davies, Charles William
3586	Creed, Samuel Albert	3787	Davies, Frederick Charles
2634	Creighton, William Joachim	2626	Davies, John Templeton
3721	Creith, Bert William	2340	Davies, William
1881	Cresswell, William	3224	Davies, William Alfred
2140	Crisp, Percy Lawrence	3398	Davis, Arol Bentley
3783	Crisp, Septimus Stanley	3724	Davis, Barney
2612	Criss, Joseph	4377	Davis, Cecil Eric
2388	Crocker, James Arthur	2296	Davis, Cecil Rex
2617	Crockett, William James	53707	Davis, Charles Henry N.
2890	Crook, Arthur Bertram	2777	Davis, Lionel
1632	Crook, Frank McArthur	2341	Davis, Robert
3043	Crooks, Thomas	3397	Davis, Sydney
3034	Cross, John	3132	Davis, Thomas
2383	Croucher, Frank Cyril	2523	Davis, Varney Eric Stanford
3384	Crowe, James Robert	1895	Davis, William
3784	Cullen, Andrew	3499	Daw, Charles
5349	Cullen, Thomas Henry		Dawe, Ernest Robert
	Cummins, George	3501	Dawe, Harold James
1203	Cummins, Orton	5477	Dawes, John Francis
3124	Cunningham, Clement Francis	2398	Dawson, Alexander
3263	Cunningham, Ernest	2635	Dawson, Herbert
1018	Cunningham, John	3400	Dawson, Hubert Clyde
951	Curran, Edward Michael	2891	Dawson, Kenneth Ewen
1630	Curran, Leslie William	3791	Day, Charles Edward
1901	Currie, George Audley	5207	Day, Charles William

3133 Day, John Patrick
2764 Day, Leo George
2892 Day, William
4794 De Jongh, Philip
De Meyrick, Julian Frank
2972 Deacon, George Herbert
3505 Deal, Alfred
3489 Dean, Charles
5056 Dean, George Henry
4792 Deans, James Frederick
2395 Dear, Harry
2396 Deeney, John
2762 Degenhardt, John Alfred W.
2893 Delany, William Wallace
4975 Dempsey, John Alais Hepple
1117 Dempsey, Joseph
3514 Denangle, Francis Charles
3009 Denman, Victor
3743 Dennis, Gordon Harris
3135 Denny, Herbert John
Denoon, William
3599 Dent, John William Hugh
5360 Derrey, Frederick Henry [aka D'Errey F.H.]
5488 Detmers, Oscar Waldermar
1638 Devenport, John
3136 Devereaux, Martin
3047 Dewar, Robert Arthur
3017 Dibden, Frederick Robert J.
5361 Dickens, Joseph
3683 Dickenson, William
5362 Dickman, John
1096 Dickson, Ian Clarence
Dillow, William
3138 Dilnutt, Francis George
4397 Dind, Foster George
59728 Dirksen, William
5067 Diversi, James Leopold
4677 Dixon, Benjamin
1637 Dixon, Bertrand James
4797 Dixon, Herbert
3690 Dobbie, Roy Robert
2633 Dobson, William
4770 Dodd, Downie
2477 Dodd, George
1898 Dodd, James Aaron
3351 Doherty, Frederick William
2267 Doherty, Jeremiah
1639 Doherty, John
3717 Doig, Clyde
1922 Doig, James
2962 Dominick, Frederick Thomas V.
2358 Donald, James
2193 Donald, James Harold
3282 Donaldson, Alfred
Donaldson, George Shirley
1538 Donnelly, Edward
5068 Donnelly, Thomas Francis
2397 Donoghue, Thomas
5357 Donohue, William
2632 Donovan, James
3827 Donovan, Stephen
4798 Doolan, Patrick James
2911 Doonan, Thomas Michael
5489 Dormer, Harry
2144 Douch, Roy
3246 Doughan, William Roy
3279 Dougherty, Albert
3041 Douglas, Frederick James E.
3037 Douglas, Frederick John
909 Douglas, Granville Peter
Douglas, Harold Graham
3250 Dove, Edward Alfred
3139 Dover, Thomas
5364 Dow, Frank Robertson
3134 Dowling, Ronald Guy
3280 Down, John
915 Downer, William Charles
2763 Downes, Patrick
3794 Downey, Michael
3479 Downey, Robert Walter
3488 Downie, John Robb
2894 Downing, Allan John
1637 Downing, Charles Edward
2631 Downs, John Archibald
1906 Dowton, Cecil Gordon

9717	Doyle, Frank James Mayne R.
3805	Doyle, Harold
3050	Doyle, John
3776	Doyle, Norman Allister
5359	Drennan, Joseph Morris
2399	Drew, Haldane Victor Reece
3039	Drew, Sydney William
3795	Drew, Valentine John
5358	Drew, William Donald
2480	Drewe, Francis Clifford
53715	Druhan, Walter James
4788	Drum, Francis
3038	Drummond, William
2815	Druyve, Francis Malcolm T.
53710	Duck, Hiram Harold C.
4789	Duckett, William Charles
2600	Duffy, Frank
4800	Duffy, George
2636	Duffy, Michael
1900	Duffy, Percy Alexander
3502	Duffy, Richard John
4799	Duignan, Donald
2628	Dummigan, John
1710	Dun, Albert James
53529	Dun, Charles William S.
59858	Duncan, Allen Robertson
4750	Duncan, George Robert
3665	Dunker, James
3658	Dunkley, Milton Rewi
2588	Dunmill, William
1640	Dunn, Herbert Bertie
5471	Dunn, Richard
5071	Dunn, William
2816	Dunne, John Herbert
1896	Dunshea, Claude Emanuel
58557	Dupre, Clarence Thomas
	Duprez, Arthur Alexander
1641	Durkin, George Herbert
920	Durling, George
6502	Dwyer, Herbert Timothy
	Dyball Charles Henry
3137	Dyball, Cecil Gregory
59866	Eade, Archibald John
3086	Eads, Sydney
3055	Eakin, James Leslie
3538	Earl, Belford Wellington
	Earnshaw, Percy Alan
3404	Eastham, Edward William
3054	Eaton, William
2145	Eccleston, Napoleon Sylvester
4103	Eddy, Fred
1189	Edgley, Norman
3252	Edwards, Austin George
3516	Edwards, Charles Henry
53531	Edwards, Colin Munro
3056	Edwards, Edmund John
1646	Edwards, George
59737	Edwards, Herbert Hockin
3504	Edwards, James
207	Edwards, William
53454	Edwards, William Walter
3403	Eedy, Neil Arthur Matthew
4802	Egan, Edwin
1542	Egan, James Michael
1901	Eirth, Arthur Ernest
59735	Ekert, Herbert Henry
1135	Elbel, Henry Edward
3642	Elkins, Edward George
3141	Ellen, James
3759	Ellen, James
5367	Ellington, Victor Emil
1643	Ellingsworth, John Henry
2896	Elliott, Allen Charles
3142	Elliott, Cyril
4804	Ellis, Joseph Henry Lancaster
2400	Ellis, Raymond Stanley
3057	Ellis, Ronald Charles
3777	Ellis, Verner Murray
1642	Ellis, William
1644	Ellison, Edward
2020	Ellison, Harry
	Ellsmore, Rupert
2897	Elphick, Stephen Henry
3228	Elrick, James Robertson
3140	Elrington, William Field
2146	Elton, Arthur

NOMINAL ROLE

2648	Emerson, Albert Henry Thomas
3058	Emerson, George
3732	Ephraim, Edgar Parry
2279	Erkins, John Thomas
5366	Erskine, Reginald Orme
1900	Espie, Wesley
58562	Evans, Alfred
2401	Evans, David William
1645	Evans, John
1643	Evans, Walter Biggs
1110	Evans, William
122	Evans, William Thomas
754	Everingham, John
2946	Evitt, Thomas
2642	Ewer, Herbert
4803	Ewing, John Ernest
3797	Eyre, Garnet Wolseley
5368	Eyre, Reginald Henry
3059	Ezzy, Charles
3060	Fahey, Patrick William
1291	Fairweather, Andrew C. B.
3052	Fallon, James Benedict
	Farmer, Eric Maynard
1647	Farrall, Frederick Theodore
5076	Farrell, Coulson Montague
4951	Farrell, Daniel Dominic
1919	Farrell, Edward Albert
1921	Farrell, George Charles
3144	Farrell, Thomas Michael
1920	Farrell, Timothy Oxley
1550	Farrell, Wilfred
3295	Faust, George Frederick
1513	Fawcett, Albert Charles
2652	Featon, Malcolm Joseph
1903	Fenton, Donald Bruce
3145	Fenton, George
1904	Fenwick, Percival Roy
3061	Ferguson, Peter
1113	Fern, Victor Roy
3063	Ferns, Albert
3957	Ferris, Edward
2407	Field, David Ignatius
4939	Field, Harold E. (alias W.H. O'Shea)
1650	Fifield, Charles Arnold
2404	Finch, John
25659	Finegan, William Henry
3778	Fischer, Gordon William
2898	Fisher, Alfred
2408	Fisher, Thomas Francis Edward
3229	Fishwick, Reginald
2147	Fisk, Glynn Moulton
1649	Fitzgerald, Charles William
1648	Fitzgerald, Edward Lawrence
53727	Fitzgerald, Harold Desmond
3768	Fitzgerald, John Eli
2402	Fitzgerald, Michael Joseph
1690	Flanagan George B. C. (alias Fenton, G)
2150	Flanagan, Herbert Ellis
17184	Flanigan, Percy
5083	Flannery, William Thomas
2881	Flegeltaub, Travers
1998	Fleming, Edwin Bothwell
2405	Fleming, Frederick John Gidley
1353	Fleming, George Walter
1651	Fleming, Henry John
3491	Fleming, William Escott
3310	Fletcher, Frederick
1358	Fletcher, Harry
3143	Flood, Hilary Nicholas
1652	Florence, Robert John
4950	Flynn, Ernest
2153	Flynn, John Thomas
1653	Flynn, William Augustine
	Folkard, George D'arcy
3054	Foote, William
53456	Foran, Patrick Joseph
2273	Forbes, Arthur Henry
2899	Ford, Albert James
2900	Ford, Alfred William
539	Ford, Henry
3291	Ford, Norman Clarence
1992	Ford, Percy Oswald
1902	Ford, Rudolph Sidney
3055	Foreman, William
2646	Foret, Theodore Ambrose
2901	Fortune, Charles

3718	Foster, Albert George
2149	Foster, Clarence Herbert
2152	Foster, Frank Ernest
2002	Foster, Frederick Arthur
2650	Foster, James Steele
1115	Foster, John Hastings
2651	Foster, Stephen
53726	Fox, Rupert Edward
2902	Fox, Walter
3230	Foxley, Frederick Thomas North
3601	Francillon, Leslie Charles
3256	Francis, Alwyn Frank
2903	Francis, Frank
2403	Franklin, George Edward
3602	Franklin, John Frederick
2596	Franklin, Rowley
3803	Franks, Edgar Russell
	Fraser, Alexander Francis
2904	Fraser, Duncan Malcolm
5085	Fraser, Herbert John
2825	Fraser, James Simon
2647	Fraser, William
2765	Fraser, William Joseph
2148	Freebody, Edwin James T.
2151	Freebody, Lionel Edward
3510	Freeman, Francis John F.
3507	Freeman, George
1741	Freeman, George Harold
3051	Freeman, Norman Lister
2653	Friend, Alan Travers
2592	Frith, Harry
3066	Froggatt, Bertram
3570	Frost, Harold
5369	Frost, Robert
3520	Fryklund, Charles (alias C. Franklin)
2017	Fulford, Edward
53728	Fulford, William
	Fuller, Claude Elliot
1116	Fuller, Walter Clifford
2650	Furniss, Joseph Ross
2654	Gaffney, Andrew James
1655	Gaffney, Lawrence John
5370	Gainsford, James Alfred

5371	Gale, Philip William
59742	Gallagher, Charles Joseph
3728	Gallagher, Stanislaus Vaughan
2598	Gallagher, William Ernest
3067	Galloway, James McCullock
2661	Gamble, John Joseph
59741	Gardiner, John Wilson
	Gardiner, Reginald Scott
27	Gardiner, William
	Gardner, Archibald Ramsay
3513	Gardner, Norman Leslie
1658	Garie, George
3518	Garland, Arthur George
1553	Garland, Thomas
2870	Garrad, John James
225	Garrett, William
4952	Gask, William
3058	Gay, Bertie Lloyd
3405	Gayfer, Allan Bokenham
2906	Gaylard, Henry Charles
5372	Gear, Francis William
2409	Gearie, Alfred
4811	Geason, Percy
3521	Gebbie, Ernest
3808	Gell, George Lenton
2783	Geoghegan, Henry Herbert
5373	George, Frederick William A.
2599	Geppert, Fred Thomas
3762	Gerrard, Roy
5374	Gett, George
5375	Gibbes, Frederick Charles
2830	Gibbins, Frederick Charles
	Gibbins, Norman
	Giblett, William Norman
3796	Gibson, George William
2652	Gibson, Reginald Stanley
3503	Gibson, William Alexander
2157	Gilbert, Lyle Ernest
2418	Gilfillan, Robert
2915	Gill, George Thomasson
1656	Gill, Harold Percy
3407	Gillatt, Thomas Bertie
1657	Gillespie, Robert

2001 Gillet, Arthur James
2663 Gillett, Charles Leslie
Gilligan, Reginald Thomas
4808 Gilmour, John
4810 Gilmour, John
2664 Gilpin, Kenneth Elmer
2661 Ginns, Albert Ernest
2158 Girvan, Roderick Roland C.
4536 Gitsham, James
Gitsham, Thomas Louis
2155 Gleeson, John Joseph
2410 Gleeson, Thomas James
2415 Glennan, Morris
2871 Glossop, Edwin William
4129 Glossop, Walter Lionel W.
3727 Glover, Joseph Herbert
2658 Glover, Samuel Steel
3517 Glynn, Harry
3149 Golden, Phillip John
3604 Goldfinch, William Robert
1905 Goldrick, John
Goldrick, Roy Arthur
Goldstein, Percy Hirsch
2659 Goodchild, Frederick George
3235 Goodiff, William Cecil
2154 Goodman, Albert William
3408 Goodman, Richard Garfield
1909 Goodman, William Edwin
3146 Goodwin, James
3071 Goodwin, Walter
3147 Goodwin, William
2138 Gordon, Arthur John
2417 Gorman, James
3810 Gorman, Raymond Sydney
5214 Gorton, John Henry
2655 Gosden, Charles
1123 Gosling, Clarence
3811 Gough, Errol Frank
3492 Gough, John Myles
1928 Gould, Henry John
3239 Gould, Thomas Henry
Gow, Carl Beeston
2656 Graham, Albert
6373 Graham, Arthur James
3231 Graham, Ewart Gladstone
3519 Graham, George Edwin
3148 Graham, Paul Sidney
3298 Graham, Sydney Charles
2413 Granger, Edgar David
2414 Granger, Hubert Stanley
2427 Grant, Douglas Campbell
1910 Grant, John Osmond
58567 Grant, Robert
2907 Grantley, J. T. (alias J. T. Granville)
1090 Graves, Ambrose John
3801 Graves, Richard
53583 Gray, Albert Edward
53457 Gray, Albert Ernest
1906 Gray, Alexander William
1654 Gray, Alfred
2662 Gray, Cyril George
2660 Gray, Kenneth Harold
5094 Gray, William Fordyce
869 Graydon, Malcolm
1908 Greathead, Frederick
1907 Greathead, Henry Percy
53463 Green, Albert Oswald
2832 Green, George Edward
2663 Green, Henry James Cyril
Green, Herbert
4812 Greenaway, William Harold
2657 Greenwood, John
5095 Gregory, Christopher George
2908 Gregory, Joseph Reid
2412 Grene, James
2655 Greville, Harry Gale
2909 Gribble, Percival Patrick
2728 Gribbon, George
3814 Griffith, William George
2659 Griffiths, Ernest Sydney
4814 Griffiths, John Gordon
53586 Griffiths, Thomas Edward
585 Grimes, Bernard
3237 Grimes, Joseph Henry
1911 Grimston, Ernest George
3786 Gross, Joseph

243	Groutsch, Francis James
1502	Groves, Henry William
3234	Groves, Percy George Wallace
4953	Guillaume, Louis
2910	Gulson, Kelvedon
2911	Gumbrill, Ernest Alfred
2662	Gunderson, Carl Einer
3803	Gunn, Henry
2156	Gunning, Frank
3483	Gusaroff, Alexander
1930	Hackett, John Burkmans
3411	Hackman, William
3156	Haggar, Charles William
2548	Haggar, George Albert
2165	Haggar, John Walter James
2429	Haines, Frederick Gaman
2666	Haley, William
3151	Hall, Alfred William
4822	Hall, Arthur Thomas
1667	Hall, Leslie John
1918	Hall, Reginald Eric
1668	Hall, William Wingate
4820	Halpin, Lawrence
3409	Halstead, Walter
3605	Hamblin, William
53737	Hamilton, Alexander
1665	Hamilton, Ernest
3742	Hamilton, Herbert
1666	Hamilton, Hugh
3415	Hamilton, Percy Oliver
2425	Hammond, George William
4828	Hammond, James Hugh
3240	Hancock, Eric Floyd
3816	Hancock, Frederick
3504	Hancock, Wilfred Roy
1129	Hand, Ernie
53592	Hand, Henry
1917	Handley, Bertie Roy
3073	Handley, William James
4819	Hankinson, John Ramornie
2491	Hannah, John Leonard
3081	Hannan, Oliver George
342	Hansen, Arthur William
3006	Hansen, John (aka Andrews, Jack)
3150	Hansen, Joseph William
1669	Harden, William Henry
5212	Harding, William
3078	Hardy, Arthur Henry
53589	Hardy, Ben
569	Hardy, Charles
2835	Hardy, Joseph Henry
2912	Hardy, Robert Vincent
3507	Hare, David Edward
1292	Hare, Thomas
2844	Hargreave, Wilfred James
1344	Harper, Adrian Charles
2963	Harpley, Robert William
2431	Harpur, Patrick Bertrand
5397	Harris, Albert Edward
4944	Harris, Arthur Ethelbert
53466	Harris, Cecil
3797	Harris, Harold
3077	Harris, Herbert Henry
3819	Harris, James John
3815	Harris, Robert Henry
2673	Harris, Robin Burns
3076	Harris, Vivian Glen
3414	Harris, William Ernest
2668	Harrison, Francis James
3311	Harrison, Frederick George
53459	Harrison, Raymond
	Harrison, Walter Rupert
5100	Hart, Charles Joseph
	Hart, Christopher Crosby
3323	Hart, Frederick William
2674	Hart, William Barker
5237	Hartcher, Andrew Gregor
2505	Hartmann, Arthur Rubert
3080	Hartney, Joseph
4821	Harvey, Christopher Samuel
598	Harvey, James Donald
3498	Harvey, Joseph
5380	Harvey, Leslie Ambrose
2913	Harvey, Richard Ernest
2684	Harvey, William Raymond
3588	Hassett, James

3505	Hassett, Thomas Patrick
53741	Hathaway, George
3298	Hatherley, William Henry
1664	Hatton, John Thomas
1663	Hatton, William
1920	Hauber, Henry Ernest
2506	Haugh, James
3252	Havenstein, Charles Frederick
1931	Hawke, Martin Stanley
53747	Hawkes, Cyril Winston
	Hawkins, Alfred Roy
3607	Hawkins, Robert Samuel
888	Hawkins, William John
2163	Hayden, John Arthur
4824	Hayden, Norman Leslie
3315	Haydon, Herbert Ernest
2166	Hayes, Francis Thomas
2671	Hayes, John William Patrick
1204	Hayes, Leonard
53591	Hayes, Stanley George
3631	Hayes, William
3484	Haylock, Alfred Edgar
3248	Haylor, Walter Norman
2914	Haymer, Arthur Edward
2164	Hayse, Ernest Edward
3158	Hayston, Joseph Robinson
3152	Haywood, William
3016	Hazell, William
2278	Head, George William
2611	Healey, Wilfred Charles
3242	Hearnden, William Henry
1913	Hearne, Felix Clarence
1915	Hearne, Francis Thomas
1794	Hearne, Frederick
2427	Heath, Edward Alexander
3157	Heazlett, Rupert Clarence
1912	Hebblewhite, Thomas John
2351	Heber, Lawrence Paul
25	Heber, Vincent Ernest
2162	Hedger, Frederick William
2420	Heighington, Henry
9638	Helmore, Reginald William
3482	Henderson, Arthur Hilton
1662	Hennessy, John Francis
53588	Henrickson, Anton Wicktor
4933	Henry, Claude Lewis Charles
3499	Henry, Donald
25651	Henry, Neville
3779	Henry, William Hilton
2915	Herrmann, Alfred Conrad
53465	Hestbak, Einar
3243	Hester, William
3249	Hetherington, David
2160	Heyland, Thomas Reginald
2916	Heywood, Joseph William
2429	Hicks, William James Clifford
2671	Hicks, William Robert
5381	Hickson, Arthur Carlton
1342	Hickson, Earnest Henry
2917	Higgins, Edward John
231	Higgins, Samuel McCulloch
5104	Higginson, John
3129	Higgs, Isacchar Pearson
59749	Hill, Charles Godfrey
3119	Hill, Frederick Thomas
2918	Hill, George Ernest
3770	Hill, James William
59753	Hill, John Owen
3507	Hill, Walter
5110	Hill, William Richard
3241	Hilliard, George Franklin
1670	Hilton, Bertie Allen
2847	Hincks, Wallace Roy
3066	Hind, James
2761	Hinley, James
25530	Hinson, Rupert Sykes
516	Hinton, Herbert Henry
2667	Hinton, Oliver Richard
3083	Histon, Thomas Henry
1131	Hitchenor, William Albert
3606	Hitchens, Nicholas George
2668	Hoare, Bertram Deacon
3410	Hobbs, Frederick George
2004	Hockey, John Andrew
2609	Hocking, Francis Almund W.
1614	Hodder, Harry

3822	Hodges, Herbert	3154	Horan, Matthew Thomas
2676	Hodgkinson, Enoch Charles	3698	Hordern, Theodore Charles
3495	Hodgson, Noel	5238	Horgan, Daniel
2677	Hodgson, Oswald	1555	Horne, Eric James
3246	Hodkinson, Arthur	2681	Horne, Sydney
2673	Hoey, Edward Rockwood B.		Horniman, Lancelot Vicary
3510	Hogan, Edward	3090	Horsfield, Frederick
1660	Hogan, Edward Thomas	2670	Horsley, John
1929	Hogan, John Joseph	4181	Hottes, William
1136	Hogan, John Patrick	1718	Hourigan, Frank
3730	Hogan, Michael		Hourn, Midford Stanley
2848	Holden, Edward	3088	Housego, Charles
53460	Holden, Richard	2922	Howard, Francis
	Holder, Frank	2540	Howard, Harold Ernest
5106	Holford, Frank	2445	Howard, Henry George
	Holland, Austin Claude Selwyn	3745	Howard, Joseph Harry
5387	Holland, Cecil Frank	132	Howell, Frederick
4825	Holland, Frederick Thomas	1921	Howell, Kenneth James
2423	Holland, Reginald Hubert	3638	Howlett, John Claude
3245	Holley, Alfred Leslie	3062	Howlett, Reginald Augustus
2152	Holliday, Sydney Charles	2849	Hubbard, Francis Joseph
2424	Hollingsworth, Clyde	53535	Hubble, Albert Edward
5388	Hollis, Roy Edward	1965	Hudson, Francis William
5382	Hollis, Sydney Parker	4823	Hudson, Osmen John
1661	Holloway, Ernest James	3828	Hughes, Frederick Cox
2919	Holloway, Leslie James	2685	Hughes, Frederick Walter
53746	Holmes, George Alfred	1542	Hughes, George
1919	Holmes, Herbert Francis	777	Hughes George Edwin (alias Hughes, W.R.)
1994	Holmes, Mark Ulhorne		
2161	Holmes, Percival	5486	Hughes, Griffith Robert
3493	Holton, George Herbert	4826	Hughes, James Augustus
2430	Homann, Bertram Charles	3412	Hughes, Richard
2920	Hone, William John	2428	Hughes, Thomas George
3494	Honeyman, Thomas	3417	Hume, Charles
2921	Hoodless, Henry James	3413	Humphries, Henry James
3824	Hooper, Stewart	1364	Hunnikin, George William
59860	Hope, Thomas Arthur	25546	Hunt, Austin George
1659	Hopkins, Frank Mouat	1969	Hunt, Frank
3825	Hopkins, John	3737	Hunt, Henry
3830	Hopkins, Thomas	5385	Hunt, John Bede
4144	Hopkinson, Maticen	5389	Hunt, Joseph Samuel
1614	Hopper, Harry	1139	Hunt, Raymond
2426	Horan, John Joseph	5112	Hunt, Richard George

	Hunt, Sinclair Edward
	Hunter, David de Venny
3608	Huntington, Harold
3613	Hurley, Edward
3313	Hurlstone, Frederick George
2422	Hurn, James Edward
2669	Hush, Joseph John
2672	Hutchings, Frederick Charles
5386	Hutchings, Leslie John
1916	Hutchins, Sidney William
3501	Hutchinson, George Francis
2035	Hutton, William Arthur
3326	Hybenett, Albert Frederick
2675	Hyde, Thomas
2419	Hyde, William James
3829	Hyslop, Alexander
3539	Ibbott, Albert Edward
3095	Inall, William Frederick
2852	Ingham, Frederick
	Inglis, Archibald John
1931	Ings, Cecil William
3502	Inskip, Sylvester
2985	Ipkendanz, Eric Pirival
5115	Issanchon, Russell
1658	Ireland, George
2021	Irwin, Vincent Adrian
5478	Isedale, Harold Richard
1671	Iverach, James Samuel Peter
	Jackson, Allen Raymond
3832	Jackson, Joseph
1673	Jackson, Leslie Alfred
2723	Jackson, Leonard W. (alias R.W. Mayhew)
1346	Jackson, William Henry
3673	Jackson, William James
2167	Jacobs, Arthur Samuel
3421	Jacobs, Horace John
3160	Jacobs, Samuel John
99	Jaeger, Wilfred
897	Jagelman, Herman Oscar
53470	James, Albert Edward
2766	James, Francis
1548	James, George Leslie
4829	James, John Willey
3254	James, Thomas
2168	James, William
3076	James, William John
13956	James, William Joseph Stanley
2679	Jamieson, William Linton Ross
53750	Jansen, James
3419	Janson, Andrew
1672	Janus, John
3078	Jaques, Arthur Henry
3833	Jarman, James Richard
2677	Jarvis, Leslie Thomas
3834	Jeans, Octavius Thomas
2678	Jeffery, John Edward
2854	Jeffries, John
3747	Jeffs, George Alfred
1977	Jenkins, Benjamin Eric
1663	Jenkins, Joseph Thomas
4800	Jenkins, William Harold
1923	Jennings, Alfred Edward
1924	Jennings, Arthur Frederick
2435	Jennings, Leslie Raymond
3330	Jennings, Patrick George
4831	Jennings, Walter James
3835	Jensen, Alfred Ernest
2955	Jensen, Francis Campaign
1675	Jensen, Gunerius Oscar
2923	Jerman, Frederick William
2680	Jewell, James Henry
3420	Jex, Roy
3161	Jobson, Henry Ernest
3418	Johansen, John Edward
1953	Johns, Sidney
2619	Johns, William Thomas
5391	Johnson, Anton Elwood Lincoln
3836	Johnson, Charles
1928	Johnson, Edgar Bootes
5390	Johnson, Ernest
1922	Johnson, Francis Martin
2924	Johnson, Frank
3162	Johnson, Frank
3716	Johnson, John
53469	Johnson, Lawrence Leonard

2437	Johnston, Arthur McPherson
1218	Johnston, Benjamin D
2438	Johnston, David Joseph
3098	Johnston, Francis James
3253	Johnston, William Henry
3838	Johnston, William Sinclair
3097	Jones, Albert
2153	Jones, Donald
1661	Jones, George
2434	Jones, George Frederick
3840	Jones, Griffiths
1927	Jones, Harold William R.
59760	Jones, Herbert Cecil
2433	Jones, Hugh Henry
1930	Jones, Hugh Lewis
1925	Jones, Isidore Isaac
1926	Jones, James
2925	Jones, John Charles
3075	Jones, John Edward C.
3976	Jones, John Thomas
3329	Jones, Joseph
3841	Jones, Joseph
53468	Jones, Kenneth Austin
3839	Jones, Leonard Arthur
4710	Jones, Roger Arthur
3552	Jones, Ronald Arthur
2008	Jones, Thomas
4830	Jones, Waldo Emerson
3159	Jones, William
59759	Jones, William
1929	Jones, William Evan
4153	Jones, William Henry
2862	Jones, William Westley
2691	Jordison, Joseph Stanley
3253	Joy, James (alias J. Quinn)
5239	Judson, Frederick Joseph
2692	Juergens, Eric Charles
3485	Julliff, Lewis Robinson
3101	Kanaan, Edmund Michael
1682	Kasel, Joseph
1932	Kates, Arthur Stanley
2682	Kavanagh, Ambrose
2927	Kavanagh, William
2160	Kay, James Robert
4833	Kay, Ralph George Hosley
3154	Kay, Sydney Deakin
2928	Keane, Henry Robert
1941	Kearney, Edward David
1676	Kearns, Arthur Andrew
2265	Keating, Ernest Henry
3164	Keefe, Abraham
53472	Keeley, Charles Henry
2695	Keeling, George Henry
3842	Kell, Christopher
721	Kellett, Patrick
2930	Kelley, William James T.
2022	Kellond, John Henry
2171	Kelly, Charles Reginald
2685	Kelly, Frank
2929	Kelly, Herbert Albert
2859	Kelly, James Joseph
2684	Kelly, Philip
3143	Kelly, Robert William
1940	Kemp, Francis Robert
3764	Kempe, Hubert Holland
3256	Kendell, Arthur George
2876	Kennedy, John
5120	Kennedy, John Herbert Ross
3166	Kennedy, Joseph McNamara
3713	Kenney, Leonard James A.
3165	Kenny, Charles Peter
3512	Kenny, Francis Leslie
1942	Kenny, Phillip
1678	Kenny, Richard
252	Kent, Charles
2170	Keogh, Vears John
1953	Keown, Sidney John
4834	Kerin, Michael
3634	Kernot, Edgeworth William Pitt
	Kerr, James
5121	Kerr, Patrick James
	Kerr, Robert Baynton
3010	Kerslake, Godfrey Edward
58589	Keshan, Norman Daniel
58591	Kevan, John McClure
3163	Keyte, Robert

1664	Kibble, John Samuel
3715	Kilduff, John Edward
3515	Killen, Thomas Michael
1935	Kimball, Cyrus Richard
2440	Kinealy, William Thomas
2172	King, Alfred Mark
2701	King, George
2552	King, Jack
3846	King, James
1679	King, James Michael
2264	King, Joseph
	King, Renshaw Weyland Hugh
2767	King, William Albert
1680	King, William Joseph
3167	Kinsela, James Frederick
2931	Kinstler, Frank Louis
2932	Kippin, George Edmund
3780	Kirkland, Henry Alexander A.
2158	Kirton, Charles Frederick
2702	Kitney, George James
1681	Klein, Henry Charles
2937	Klemm, Frederick
2858	Klemm, Michael George
1665	Knight, Albert Lilley
1719	Knight, Wilfred Charles
1934	Knopp, John Andrew
1933	Knowlman, Eric
2681	Kolesnikov, William
1227	Kruss, August
2160	Kumsay, Cecil George L.
1696	Kurtz, Harold Thomas
2933	Lacey, Harold Neil
3171	Lach-Szyrma, William George
2712	Lakeman, Leslie Joseph
3423	Laman, Harold
3677	Lambert, Victor George
2934	Lang, George Bertram
1941	Langdon, Herbert John
1999	Lanham, Joseph
5392	Lannon, John
4505	Laraghy, Charles Royde
3348	Larbalestier, Percy William
2935	Larkham, Charles Henry
3856	Larkin, Robert Samuel
3109	Larmour, Robert Griffith J.
3509	Larsen, Harold Kenneth Leo
5400	Lasham, Alfred Theodore
53540	Lauder, James
4841	Laurence, Thomas Bernard
1580	Law, Horace Henry
3851	Lawler, John Peter
3346	Lawless, Albert
3168	Lawless, George Alexander
3516	Lawrence, Augustus Harvey
1938	Lawrence, Charles Alfred
	Lawrence, Charles William
3258	Lawrence, Francis John
3257	Lawrence, Thomas
3172	Lawrenson, James
2534	Laws, Sidney Alfred
3011	Lawson, Albert Bowman
2444	Lawson, Fred Lindsay
1685	Lawson, Reginald
2000	Lawton, James Arnold
1668	Layton, William James
2936	Le Cerf, Albert
2954	Leane, Albert Charles
53476	Leaney, Howard Henry
1683	Leaver, Thomas James
1937	Ledger, Francis Cornelius
1684	Ledwell, William Charles
1689	Lee, Charles
3141	Lee, Charles Lester
5393	Lee, Francis Herbert
	Lee, Leslie William George
5394	Lee, Thomas Randolph
5695	Lees, Charles
3173	Lees, Joseph
1573	Legg, Horace John
741	Leigh, Robert
4840	Leister, Leslie
1940	Lemon, Harry Marcus
4775	Leneard, E.W.(alias E.W. Clifton)
3085	Lennard, Augustus Thomas
4228	Lennard, Henry
4229	Lennard, Samuel Robert

1936	Lennon, Thomas
2686	Leonard, Michael James
3771	Levey, Leonard William
2173	Levi, Andrew John
25562	Levy, Coleman
3110	Lewin, Joseph Frederick
1168	Lewis, Andrew
5896	Lewis, Francis Thomas
2179	Lewis, George Henry
5129	Lewis, Harry
56102	Lewis, John Henry
2937	Lewis, Raymond Henry
1686	Lidden, William
53473	Liddiard, William Jams
	Liggins, William Frederick
149	Lincoln, William Daniel
2634	Linderup, Thomas Conrad R.
3422	Lindquist, Gustaf Albert
3676	Lindsey, George Stanley
2177	Linnegar, Charles Aloysius
3089	Lithgow, Peter Fleming
552	Livermore, Edgar Leslie
5401	Livesley, William
508	Livingstone, Sydney Philip
87	Llewellyn, Jenkin James
2938	Lloyd, Francis James
3170	Loader, James
1939	Locke, Hilary Xavier Joseph
2446	Lodge, Arthur Ernest
2720	Logan, John Alexander
1388	Long, George Francis
2486	Long, Richard Anthony
2175	Longhurst, Vincent James M.
1687	Looby, Lawrence Herbert
1688	Loughlin, John
3661	Loughrey, Gordon Alfred
3169	Lovejoy, Charles George
59768	Lovell, George
59771	Loveridge, Ernest William
25563	Lovett, William Sidney
1023	Lowe, Arthur Albert Kelly
5397	Luby, Arthur Alfred Harold
2939	Luby, Reginald

3106	Luke, William Herbert
4837	Luland, Frederick Henry
2055	Lulham, Herbert Otto
1942	Lund, Svend Holger
2705	Lund, William James
2174	Luton, Walter Patrick
3349	Luxford, Elexus
3347	Luxford, Percy
3088	Lymbery, Francis H. W.
1760	Lynam, Benjamin Leo
3108	Lynch, Arthur Leslie
2445	Lynch, James
5398	Lyons, Cornelius Matthew
3964	Lyons, Frederick Norman
2861	Lyons, George
2541	Lyons, James Patrick
2096	Lyons, Thomas Vincent
5399	Lyttle, Roy Frederick
	MacAuley, Clarence John
3187	MacCallum, Hector Archie
2847	MacDiarmid, Sydney D.
3186	MacDonald, John
53486	MacDougall, Bruce F. S.
3272	MacIntyre, Charles Thomas
1709	Mack, Richard Lynn
3519	Mackay, Alexander Leslie Gordon
3545	Mackay, Frank James
3536	Mackay, John
	Mackay, Norman John
2695	Mackellar, D'arcy Canning
3093	Mackenzie, Forbes Burnside
2465	Mackenzie, Roderick Munro
2691	Mackenzie, William Guilar
	Mackenzie, William Kenneth C.
3666	Mackison, James
1171	Mackney, Edward Albert
3857	Mackney, George Edmund
3442	MacNamara, Victor
3510	MacNeill, James
1953	Madden, John
2940	Madden, Martin Ernest
16446	Madden, William George
3610	Maddison, Archibald Robert

3627	Magnus, James Henry	3871	Martin, Arthur Edward
2473	Maguire, Phillip James	3385	Martin, Cecil George
4481	Maguire, Thomas	2517	Martin, Daniel Thursby
4934	Maher, Edward Peter Patrick	1658	Martin, Owen Patrick
3885	Maher, Joseph	3704	Martin, Roy Henry William
5402	Maher, Joseph	3267	Martin, Thomas Watt
270	Maher, Linus Edward	3787	Martin, Walter
3424	Maher, Reginald Joseph	3375	Martin, William James
3520	Maher, Thomas Martin	1990	Mason, Arthur Frederick
	Mahoney, Clarence Russell	3863	Mason, Francis
3750	Mahony, Leonard James	3864	Mason, John William
4954	Mainger, Nicholas	2872	Mason, William Arnold
3174	Mainwaring, Sidney	2718	Masterton, Alexander
53837	Maley, Harold	3614	Matheson, Donald
3175	Malone, Hector	3432	Matheson, William
2941	Malone, Patrick	2942	Mathieson, George Stephen
6732	Manderville, John	2722	Matterson, Charles Robert
3486	Mangan, Thomas		Matthews, Albert Reginald
1696	Manning, Harold Thomas	6102	Matthews, Alexander
3858	Mansfield, Charles	3436	Matthews, Frank
1697	Mansfield, John	153	Matthews, Frank Walter
5403	Mansour, John	3609	Matthews, Harold James C.
1554	Mantle, William John	2192	Matthews, Hugh
2690	Manwaring, Percy		Matthews, John Hamilton
2689	Manwaring, Walter Haynes	2186	Matthews, John Owen
1172	Manwarring, Francis Daniel	3865	Matthews, William John
5410	Marler, Edgar John	3135	Maude, Frederick William
3556	Maroney, William James		Maudsley, Henry Fitzgerald
2459	Marriott, Henry Charles	2449	Maxwell, Alfred Ernest
53541	Marriott, Wilfred De Garris	2202	May, Claude
2644	Marsden, Thomas Samuel	2873	Mayer, Henry
2738	Marsh, Charles	3448	Maylor, Ernest Joseph
3183	Marsh, Frank	1394	Maynard, James
59783	Marsh, Reginald John	4843	Maynard, John Lucius
3012	Marshall, Albert John	3357	Mayor, Lynton Cooper
	Marshall, Bertie	2035	McAlary, Hugh Joseph
2458	Marshall, Frank	1695	McAlister, Frederick William
3425	Marshall, George William	2462	McAlister, Hugh Wallace
1690	Marshall, James Percival	2471	McArthur, Henry Lindsay
	Marshall, John James	1952	McBrien, Arthur William C.
345	Marshall, Samuel William	5418	McCabe, Christopher
3861	Marshall, Sidney	3866	McCabe, Joseph George
3382	Marshall, William	1702	McCabe, William Jenkins

3270	McCallum, Arthur Bruce
3271	McCallum, John Archibald
1195	McCann, Arthur Reginald
4941	McCann, Phillip Joseph
2699	McCarroll, Frederick
3574	McCarthy, Ernest William
3097	McCarthy, Francis
3126	McCarthy, Francis Benedict
	McCarthy, James Stanley E.
1955	McCarthy, Samuel Harley
2878	McCartney, Harry
1185	McCauley, Andrew Joseph
651	McCluskey, Peter
1558	McCluskie, Hugh
931	McColl, Eion
3437	McColl, James
	McConaghy, David Mcfie
3447	McCormack, Ernest Albert
3184	McCormick, William Hugh T.
5413	McCosker, William Joseph
3708	McCoy, Thomas William
3445	McCracken, William
1694	McCrory, Thomas
4855	McCulloch, Thomas
2468	McCullough, Alexander
1947	McDonald, Alexander Scott
1953	McDonald, Cecil William
2467	McDonald, Charles
3269	McDonald, Denis
2466	McDonald, Duncan David
2195	McDonald, Harold John
1946	McDonald, Henry Beresford
2280	McDonald, Hubert John
3440	McDonald, John Alexander
2200	McDonald, Leslie
2197	McDonald, Roderick Peter
2199	McDonald, Roderick Sylvester
1188	McDonald, Walter
3439	McDonald, Walter Campbell
3123	McDonald, William
2850	McDonald, William Arthur
2700	McDonald, William Austin
2697	McDowell, William Jackson
3185	McEachern, John Gilbert
3568	McEwan, Leslie Thomas
959	Macfadyen, Horace Gillespie
2728	McGee, James
	McGee, William James John
3871	McGill, Arthur Cecil
2469	McGill, James Laurence
3124	McGoldrick, Charles Dominic
833	McGovern, John
3967	McGowan, John
3487	McGrath, Charles
3714	McGrath, John Bingham
59786	McGregor, Albert George
2464	McGregor, Percy
3190	McGrory, James Bernard
3873	McGuarr, Richard James
2944	McGuigan, Hugh Owen
838	McIlwain, Andrew Eyles
1189	McIlwain, Joseph Oscar
5414	McInnes, Alfred
2198	McInnes, Arnold
3188	McInnes, John Alexander
3189	McInnes, John Lachland
2218	McInnes, Richard
3874	McInnes, William John
3672	McIntyre, Dugald
3441	McIntyre, Frederick
3186	McKay, William John
1251	McKellar, Daniel
2497	McKenna, Edward Francis
1401	McKenna, Patrick
4857	McKenzie, Alexander
2701	McKenzie, Alexander
3114	McKenzie, Hector
58609	McKenzie, Nathaniel James
3577	McKenzie, Robert
1995	McKeogh, James Joseph
3099	McKeown, William Arthur
1693	McKernan, Samuel
1676	McKervey, Thomas
3191	McKie, John James
3192	McKinnon, Archibald
2881	McKinnon, Kenneth

2945	McKnight, James Taylor
5146	McLachlan, John Allan
2703	McLachlan, Samuel
3523	McLaren, Henry Thomas
2946	McLaren, William Robert
1943	McLaughlin, William G.
3443	McLean, Archibald
58612	McLean, James Duncan
3438	McLean, Sidney
4856	McLeod, Archie Buchanan
1707	McLeod, Charles Roderick
3273	McLeod, Hector John
1692	McLeod, Paul Henry
3446	McLure, Harry James
5416	McMahon, Albert William
2698	McMahon, Sylvester Luke
2472	McMahon, Thomas
2196	McMahon, Timothy E.
3576	McManus, Bernard Joseph
218	McManus, James Bernard
1691	McManus, John
780	McManus, Michael Joseph
2463	McManus, Thomas Francis
3118	McMaster, George Stewart
1708	McMeekin, George
58614	McMillan, Alfred Charles
2740	McMillan, William
5417	McMullen, Edward John
4180	McMullen, Stanley Wiser
1704	McNeich, William Henry
4858	McPhee, Alexander Norman
3098	McPhee, John Alexander
3782	McPherson, Erle Ewan
3355	McPherson, John
58611	McPherson, John
2947	McPherson, John Walter
2470	McPherson, Leonard Donald
3567	McPherson, Leslie Allan
1706	McPherson, Robert
2201	McRae, George Malcolm P.
2848	McRae, George William
1146	McRitchie, William
3117	McWhinney, Laurence H.
3157	McWilliams, Leslie James
2181	Mead, Samuel Henry
3251	Meade, Stewart Michael
3369	Meagher, James Donald
5405	Meale, Stanley Victor
3429	Mealey, William Frederick
2688	Meedan, Thomas James
1948	Meehan, Eric Llewellyn
1705	Mehegan, John William
2948	Meldrum, Thomas Alfred
4846	Mellor, Frank Renshaw
4969	Melrose, Donald Arthur Aloysis
3651	Mendall, Harry
	Mendelsohn, Berrol Lazar
2525	Menser, Leslie Maurice
3103	Mercer, John James
2768	Meredith, Alfred Charles
1951	Meriless, William Thomas
3176	Merkel, Norman Phillip
4966	Merrell, Thomas
2864	Merritt, Hurtle Augusta
823	Metcalfe, Arthur Charles
3877	Metcalfe, Robert George
2693	Mewburn, Earl Roland
3264	Meyer, Ernest Sydney R.
3138	Meyer, Frederick Herman G.
909	Michie, George Leslie
3134	Micklewright, Robert
53542	Middleton, Charles Bertram
	Midgley, Richard James D.
5406	Miles, Leonard
2949	Milham, Hubert James
3161	Millar, Robert McHardy
3880	Millard, Charles Westbrook
6731	Millard, Stanley
1703	Miller, Arthur Edward
3689	Miller, Arthur Elsner
4847	Miller, Cecil Letton
3128	Miller, Frank Edgar
	Miller, Harold
5409	Miller, Harold Edward
2694	Miller, John Christian
3435	Miller, Morrey Hill

53548	Miller, Vincent Lewis
1736	Milligan, Francis Frank
3431	Milliken, David Thomas
1599	Millingen, Cedric
5220	Mills, Andrew
2450	Mines, Joseph
2950	Minter, William Vincent
3180	Miosge, Arthur Ernest Herbert
5411	Mitchelhill, Douglas Shaw
3378	Mitchell, Albert Edward
3427	Mitchell, Alec Raymond
3563	Mitchell, George Pacificus
	Mitchell, John Alexander S.
2769	Mitchell, Joseph William
3177	Mitchell, Leslie Hugh Lindsay
4848	Mitchell, Patrick
3428	Mitchell, Richard William
1698	Mitchell, Robert Oliphant
3179	Mitchell, Walter Charles
3434	Mitson, Henry Percival
2953	Mitten, William John Richard
53546	Mobbs, Allan George
1699	Mobey, John
59779	Moffatt, Alan Keith
3360	Moloney, John
2733	Moloney, Thomas Edward
3266	Monaghan, Thomas Patrick
2183	Monck, Charles James M.
2184	Monck, Peter Eugene
2185	Monck, Stephen Bertram
3511	Monnelly, John Thomas
1537	Monteith, William James
4854	Montgomery, Robert
3559	Montrose, Charles Otho
5412	Mooney, James
2715	Mooney, Vincent
2641	Moore, Edward Alexander
2739	Moore, Eric Osborne
1700	Moore, Ernest John
5221	Moore, Harry Hugh
5407	Moore, John William
3590	Moore, Phillip Ernest William
2457	Moore, Robert Thompson
3612	Moore, William John
1945	Moore, William Martin
3105	Morgan, Edward Joseph
2719	Morgan, John
3599	Morgan, Pritchard Llewellyn
2651	Morgan, Reginald Steve
	Morgan, William Black S.
1949	Moriarty, Merion Morton
2860	Morrall, Leslie Alexander
3882	Morrall, Sydney
5222	Morris, Benjamin Lewis
2712	Morris, Cyril Daniel Victor
27490	Morris, Frank
1674	Morris, Frederick
2448	Morris, John
3371	Morris, Norman Sydney
	Morris, Reginald Sydney
3353	Morris, Richard James
2447	Morris, Robert Owen
	Morris, William George
1739	Morrison, Alexander
2692	Morrison, James
2954	Morrison, James
2191	Morrison, John Charles
3691	Morrison, Joseph
2454	Morrow, George Frederick
4851	Mortlock, Harry
2189	Mortlock, Henry
2182	Mortlock, William Henry
4852	Mountford, Oswald Denis
2460	Moxon, Alfred Ernest
3678	Moy, George
3611	Moy, William
5140	Moylan, Joseph Patrick
3426	Muir, Arthur George
2605	Muir, William Beresford
2955	Mullins, William
3182	Mulvihill, Lawrence Rupert
2245	Mundy, George Albert
	Munro, Stuart
3578	Munson, William
2956	Murdie, Hughie
2957	Murdie, Robert

2721	Murdock, Albert William	2960	Nelligan, Thomas Amos
1950	Murphy, Alphonsus Edward	4860	Nelson, Alexander Edward
2958	Murphy, Edward Paul	2961	Nelson, Arthur
3875	Murphy, Edward Peter	4746	Nelson, Francis Harry
2716	Murphy, Harold Richard	3107	Nelson, John McDonald
59780	Murphy, James	5421	Nelson, John
3884	Murphy, Stephen	946	Nelson, William Edgar
5408	Murphy, William	59797	Nettleship, Kenneth Walter
1701	Murphy, William James	2474	Neuss, William
3178	Murray, David		Neville, Dalton Thomas W.
3883	Murray, Dick Clifford	1709	Newman, Edric
2537	Murray, Harold	2962	Newman, Martin
2187	Murray, James	1197	Newman, Murdock McIntosh
	Murray, James Edward		Nicholls, Charles Walter
3139	Murray, John Francis	2704	Nicholls, William Gourley
2846	Murray, Stanley Clive	2204	Nichols, Henry
2188	Murry, James Joseph	2203	Nicholson, Arthur Douglass
2180	Murry, Odillo William	2903	Nicholson, Wallace Edwin
4853	Mussett, George Henry	59796	Nickisson, William Stafford
2190	Mustard, William Henry	2476	Nield, Vincent
2725	Mutton, John William	4861	Nixey, William George F.
2959	Myers, Harry	3193	Noakes, Abraham Theodore
2455	Myers, Thomas Leslie	1955	Noble, Alfred Lawrence
3451	Nagle, Thomas	2475	Noble, Joshua Oswald Earl
3141	Naldrett, George William A.	4226	Nolan, John
59792	Narbeth, Edward	2509	Nolan, Justin
3639	Nash, Daniel	2963	Nolan, Patrick Joseph
	Nash, George Edward	3194	Nolan, Thomas Francis
2707	Nash, James	2423	Noldart, Albert Stanley
2543	Nash, Roy James	2965	Norris, Charles
2712	Naylor, Joseph	2478	Norris, Henry
3588	Neal, George Henry	3449	Norton, Edwin
1194	Neal, John Powell	2964	Norton, Thomas Joseph
5150	Neal, Victor Henry James	1956	Nott, Henry William
59793	Neale, Walter	1954	Nowland, Frederick Ernest
59794	Neasmith, Keith	2706	Nutt, Horace Robert
2477	Neave, William Arthur	2770	Oakes, Arthur
53781	Neil, Walter	1711	Oakman Herbert
2849	Neild, Edwin	1710	Oakman, Thomas
53547	Neill, Edward Arthur	1724	Oastler, Walter
1634	Neill, John Harold	3890	O'Brien, Charles Percy
5419	Neill, William Emmett	3453	O'Brien, Frank Tasman
4859	Nell, Harold	3195	O'Brien, James Francis

2745	O'Brien, James Patrick
3686	O'Brien, Thomas
4865	O'Brien, Thomas Francis
3454	O'Brien, Thomas James
2207	O'Connor, Cyril Eugene
4940	O'Connor, John Oliver
3198	O'Connor, Michael Joseph
3144	O'Dea, Denis John
2481	O'Dea, Peter
2480	O'Donnell, Michael
3254	Offley, Lawrence William
2208	O'Grady, John Joseph
4863	O'Grady, Reginald Standish
2665	Ohlsson, Maurice Albert Carl
2479	O'Keeffe, Ralph Selwyn
4864	Oldfield, John James
2181	Olive, Arthur Horace
3199	Oliver, Alexander John
2966	Oliver, James Ingles
845	Oliver, Wilfred Roland
4956	O'Mara, Thomas
4866	O'Neill, John Joseph
2206	O'Neill, Joseph John
2551	O'Neill, Patrick William
2850	Orchard, George
1596	O'Regan, Clifford Lindsay
3636	O'Regan, Robert Edward
2205	O'Reilly, Donald Henry
2482	O'Reilly, Edward Henry
3452	O'Rourke, Thomas
2414	Osborne, Frederick Robert
3393	Osborne, Walter Charles
3200	O'Shea, Geoffrey
3196	Osmand, Sydney Alexander
4957	O'Sullivan, Vincent Dominic
165	Outridge, Alfred
3197	Overend, David Henry
1639	Owens, Jack
3143	Owens, Michael Frank
4876	Packer, Charles Albert
3279	Packer, Frederick Ryland
7193	Painter, Percy
2717	Painter, Richard John
2752	Pakes, Charles Edwin
5422	Pakes, George
2420	Palmer, Arthur
3668	Palmer, Ernest
	Palmer, Herbert Leopold
3456	Pannell, Charles Albert
2747	Panton, Alexander Waugh
871	Pardey, Leslie James
2752	Pares, Charles Edwin
2023	Parker, Andrew William
3109	Parker, Harold Leslie
1403	Parker, Hilton Joseph
2212	Parker, Hugh Archibald D.
2968	Parker, Melville Gordon
2081	Parker, William Charles
1962	Parker, William John
2484	Parkes, George Albert
2487	Parkes, Thomas
933	Parkinson, Edward Vivian L.
870	Parryman, John Edward
1742	Parry-Okeden, Uvedale Edward
4869	Parsonage, Anson Ronald
3457	Parsons, William Howard
2969	Partridge, Charles
2758	Partridge, William Matthew
1404	Pascoe, J.H.(alias J. Greenwood)
5428	Passmore, Arthur
9791	Patchell, Clarence Avandale
1712	Patek, Charles
3281	Paterson, James Sutherland
	Patman, Norman William C.
1211	Paton, David
1716	Pattenden, William George
2549	Patterson, Abbie
4870	Patterson, Albert William
3314	Patterson, Robert Malcolm
3353	Pattison, Cyril William
2216	Pattrick, Edgar Eric
4874	Paul, Albert Raymond
4875	Paul, George Hindmarsh
3113	Payne, Charles
2748	Payne, Edgar Ashton
4873	Peacey, Henry John

4871	Peacey, Percy Wallace
3186	Peachey, Cecil Richard
	Peake, Robert Algernon B.
1347	Pearce, Charles
1715	Pearce, Glenworth Francis
3200	Pearce, Oliver Austin
2268	Peckham, William James
3204	Pegram, Albert George
2214	Pegram, Arthur Joseph
2281	Pegram, Henry Thomas
2778	Pegram, William Thomas
2213	Peisley, John Richard Bede
53788	Pender, James Samuel
2668	Penfold, Neil
	Penly, William Charles M.
1957	Pennay, John William
2217	Penney, William Benson P.
2415	Penny, Alec Gordon
3205	Perceval, Thomas
3255	Perfect, Hubert
3150	Perkins, James Alfred
3114	Perkins, William Henry
3896	Perrim, George James
2198	Perriman, Frederick John
2545	Perry, Frank
3201	Perry, Frederick Joseph
3152	Perryman, George
3604	Peters, George Luke Antony
2544	Peters, Jack
4958	Peterson, Arthur Thomas
3283	Peterson, Robert Edward
1713	Pettiford, Miles Robert Edward
2771	Pettit, Harry
2751	Petitt, Albert Edward
167	Petts, William
1959	Phillips, Alfred Leslie
2714	Phillips, Clarence Henry
3458	Phillips, Harry Rudolph
	Phillips, Horby Walter
3206	Phillips, Leslie James
3203	Phillips, Reginald William
3202	Phillips, William John
4757	Pickering, Edward Colburn
5481	Pickering, Frederick North
2839	Pickering, Harrie
5482	Pickersgill, Reginald
	Pickup, Frank Ainsworth
	Piddington, William Thomas
5427	Pike, Philip George
2972	Pilley, Charles Robert
2625	Pillow, William
	Pinkstone, Norman Ewart F.
	Pinkstone, Sidney Albert
3166	Pogson, Frank Sidney
2490	Poidevin, Henry Leslie
4946	Polley, George Alfred
4906	Pont, Robert James
268	Poole, Edwin William
1204	Poole, Frederick Alexander
2518	Pope, Josiah Thomas
1964	Porteils, John Ernest
4064	Porteils, William
	Porteous, George William
3968	Porter, Arthur Henry
3117	Porter, William Charles
2716	Postlethwaite, William
3899	Potter, John Henry
4872	Potter, Paul Nevil
1730	Power, Alfred James
2210	Power, Ernest
3773	Power, George Alexander
2711	Pratt, Gilbert Harding
2718	Pratt, Sydney Roderick
2489	Price, Austin Herbert
3149	Price, Donald Hornsby
2713	Price, Henry
2973	Priddle, Albert James
3277	Pride, William Arthur
4945	Priest, Harold Adrian
2974	Priestly, Sydney Bilton
2486	Prigge, Charles Henry
2209	Primmer, Ernest Montie
2853	Pringle, George
2841	Pringle, George McDonald
2854	Pringle, John Leslie
3110	Priston, Reginald Joseph

3455	Pritchard, Nathaniel	2844	Read, Charles
5424	Proberts, Charles	4885	Reay, John Gilbert
1723	Procter, Edmund James K.	3160	Redford, George Rees
3688	Prowse, William Henry	1720	Redhead, Mark
2215	Pryce, William Louis	1963	Redman, Horace Frederick
1714	Puckeridge, James	4947	Redmond, George
3900	Puckeridge, Joseph	2725	Reed, Thomas
1958	Puckett, Sydney Walter	2283	Rees, Francis John
2211	Pugh, Ellis Lisle	2495	Rees, John
3901	Purdon, Reginald Arthur	3794	Reeve, Cyril
3902	Purdon, Sydney Victor	5430	Reeves, Joseph Robert
920	Purling, George	4883	Reeves, Leslie John
	Pye, John Bruce	25533	Reeves, William John
6588	Quartermaine, Frank	2728	Regan, Thomas
2491	Quigg, George Harold	1911	Reginald William Cracknell
1962	Quigg, Wallace Miller	1172	Reid, Alexander Waddington
1960	Quigg, Walter Cecil	2726	Reid, Andrew Laurie
2218	Quigley, Leslie Joseph	2220	Reid, James Henry
25616	Quilkey, Hugh Cornelius	1964	Reid, William Laidlaw
2721	Quinlivan, John Patrick	2765	Reidy, Henry Stephen
2720	Quinlivan, William Joseph	5225	Renouf, Albert
59806	Quinn, Bernard Harold	1722	Resso, Robert William
3153	Quinn, James Barnabus	1727	Reviere, John William
2018	Quinn, Stanley James	3907	Rex, Edgar Hartzenberg B.
3667	Quinn, Thomas Joseph	4892	Reyner, John Rawling
3154	Quirk, Thomas Alfred	3725	Reynolds, Charles
1719	Rake, Percy Harold	4880	Reynolds, Charles
2896	Ralph, Victor	2722	Reynolds, Clarence Edward
1728	Ramsey, John William	1966	Reynolds, Frank Bishop
4878	Randall, Albert	1977	Reynolds, George
2225	Randall, Ronald James	4888	Reynolds, John Thomas
4881	Randell, Esric Leslie James	2975	Reynolds, Reuben James
2222	Rankin, Hugh Alexander	2493	Reynolds, William Morton
1718	Rankin, Leo Vincent	3419	Richards, Alfred
2492	Rankin, Norman Christian	3908	Richards, Branxton
1618	Rankine, William Harold	1721	Richards, William James
3917	Rapmund, Alfred	1996	Richardson, Clarence
4882	Ratcliffe, John	3209	Richardson, Eric Oliver
3905	Raymon, Robert William	3909	Richardson, John
3906	Rea, Bertram William	2372	Richardson, John Harold
3459	Rea, Mervyn	1686	Richardson, Joseph
3611	Rea, William John	3157	Richardson, Sydney Arthur
2224	Read, Cecil	882	Richardson, William

2769	Ricketson, Joseph
4960	Riddle, Alexander
3603	Ridgley, William Stanley
3284	Ridings, Herbert Charles
1213	Ridley, Edward
59809	Rietdyk, Walter
3249	Rigby, Arthur
1723	Rigg, James Morrison
1764	Riley, Frank Neil
3213	Riley, James
1965	Riley, James Richard C.
3669	Riley, Thomas Patrick
2223	Rixon, George William
5168	Rixon, John Osbourne M
2219	Roach, Walter
1688	Roach, William
4887	Robbins, Albert Edgar
3158	Robens, Ernest Neill
2976	Roberson, Alfred
3617	Roberts, Arthur Hamilton
2916	Roberts, Aubrey Kingston
3733	Roberts, Edward August
59814	Roberts, George Adolphus
3699	Roberts, Harry James
2763	Roberts, Hugh
83	Roberts, John
3911	Roberts, Thomas
3119	Roberts, Wesley John
4936	Robertson, Alexander
1168	Robertson, Alexander C.
2226	Robertson, Henry (alias Harry)
3207	Robertson, Lindsay McKinnon
3619	Robertson, Malcolm Clyne
2000	Robertson, Wallace Reid
2977	Robertson, William
5444	Robinson, Allan Howard
1427	Robinson, Arthur William
1974	Robinson, Christopher
3159	Robinson, Edward
3616	Robinson, Frederick
4889	Robinson, George
3124	Robinson, Herbert
2959	Robinson, James Spencer
4890	Robinson, John
	Robinson, Locksley Stuart F.
	Robinson, Norman Alexander
1724	Robinson, Percy
	Robinson, William Ernest
1975	Robottom, Robert Bertram (alias R.B. Robinson)
1725	Rodd, Joseph Jackson
3158	Rodda, Arthur Edward
1967	Roddan, Alexander Reginald
2724	Rodgers, John
1647	Roffey, George Street
3679	Rogers, Raymond Sylvester
2987	Rogers, Walter
3912	Rogerson, Charles Francis
2723	Rolfe, William George
3210	Rolles, Alan Leslie
3212	Rolles, Frederick William Louis
53796	Ronalds, William Alfred
2989	Rootes, Henry Leonard
3285	Rosborough, Archibald
1087	Rose, William Henry Christian
3748	Ross, Alexander Samuel
1646	Ross, Cecil James
4891	Ross, Donald Lachlan
2201	Ross, James
4961	Roth, Harold Wallace
4767	Routley, John James
3710	Rowan, Sydney
3415	Rowe, Richard Courtney
3610	Rowledge, Albert
1726	Rowley, Thomas Mountford
3913	Rue, Albert Edward
58629	Rumble, Rupert Elliott
53500	Rumney, Harry Laurenson
2978	Rumph, Arthur Cecil
3211	Ruse, Samuel Boley
3156	Russell, Benjamin Hillier
59815	Russell, James
2727	Russell, Joseph Michael
1221	Russell, William Walter
4877	Rutherford, Walter Roy
3121	Ryall, William Patrick

2221	Ryan, Cyril James
3460	Ryan, Edward Mitchell
59812	Ryan, Henry Archibald
1717	Ryan, John
4935	Ryan, John
4884	Ryan, Michael Newman
2979	Ryan, Michael Thomas
1997	Ryan, Patrick
3916	Ryan, Patrick
25534	Ryan, Patrick Andrew
3289	Ryder, Amos John
3804	Ryder, Michael Edward
1743	Sainsbury, Alfred Ernest
4897	Sale, Henry
2018	Salisbury, Joseph Crompton
89	Salisbury, William Murray
3469	Sammons, Edward Alexander
5483	Sampson, Harry Stephen
	Sampson, Victor Horatio Buller
1987	Sams, Ernest George
5431	Sanders, Richard Edgecombe
3006	Sargent, W.E.(alias L. W. Bennett)
53505	Sault, Albert Henry
25567	Saunders, Eric Davenport
3179	Saunders, Stewart James
2980	Saville, John
3463	Saxby, Bert
3464	Saxby, Charles
4267	Saxby, John Henry
2787	Sayer, Thomas
2526	Scale, Frank
1741	Scallon, Thomas
2527	Schafer, Albert Raymond
2528	Schafer, Frederick James
3174	Schindler, Ernest George
2981	Schofield, Albert Percival
3462	Schubert, William Gladstone
994	Schwind, Arthur
1738	Schwind, Bert
1739	Schwind, Francis
3136	Scotson, John
1254	Scott, James Harper
5432	Scott, John Russell
2772	Scott, Walter
	Scott-Olsen, Eric
6581	Scown, Edmund
1692	Scully, Joseph Francis
53558	Seabrook, Alfred William J.
3215	Seager, Arthur Spencer
2924	Seaman, George Alexander
5446	Seater, William Albert
1735	Seckold, Percival Frederick
2550	Seears, Arthur Henry Harold
1420	Sefton, Harry
1746	Seiffert, Frederick Ernest
3178	Seinor, William Ernest
3918	Selberg, Ernest
4899	Sellen, Donald Alexander
5177	Sellers, Joseph
1970	Semlitzky, Cecil Glentworth
3703	Semmler, Gustave Edwin
5434	Sewell, Joseph Augustine
3921	Seymour, Albert John
3920	Seymour, Aubrey Issac
3440	Seymour, Claud Cambourne
3784	Seymour, Leonard Edward
	Seymour, Thomas
3437	Shanahan, John
1744	Shannon, Edward Michael P.
5442	Shannon, Matthew
4907	Sharam, Horace Owen
1729	Sharkey, Edward Francis
5435	Sharp, Charles Arthur
2735	Sharpe, Bernard Aubrey
2242	Sharpe, Charles William
2743	Sharpe, Thomas John
	Shaw, Charles Edgar
2919	Shaw, Frederick
3774	Shaw, Frederick Archibald G.
1224	Shaw, Isadore Ross
54	Shaw, Parkes
2739	Shea, Harold William George T.J.
53498	Shean, Frederick William
3643	Shearen, Ernest Robert Dudley
3790	Shearing, Frederick Allen
4898	Shedden, David

2923	Sheedy, Claver Joseph
5448	Sheedy, Robert William
1226	Sheehan, Thomas Phillip
2742	Sheils, James Clive
2833	Sheldon, Rupert Fairfax
2738	Sheldrick, Alfred James
3173	Shelly, Patrick
179	Shephard, Leslie Clarence
2516	Shepherd, Leslie
3680	Shepherd, Wallace
	Sheppard, Lionel Duncan
1972	Sheridan, Fred Parker
3922	Sheridan, John Francis
	Sheridan, John Patrick
53806	Sherman, Alfred Nicholas
3294	Sherwin, Harold Austin
3662	Shields, John Edward
2779	Shiell, George Watt
3305	Shirley, George
4902	Shirley, William
53816	Shirt, Arthur John
3791	Shoesmith, Arthur James
2780	Short, Richard
6558	Short, Robert
2227	Siddins, Norman Macquarie
1632	Siddons, Bruce Lionel
6561	Sidey, Alfred James
4962	Sidgreaves, Robert
2789	Siely, Frank Ephraim
3469	Simmonds, David John
2982	Simmonds, Edward Ernest
3465	Simmonds, James Henry
1730	Simmons, Donald
3126	Simmons, Michael
4865	Simons, James Lovel
2443	Simons, John Sidney
3217	Simpson, John William
1655	Simpson, Norman Medwyn
1175	Sims, Henry
5182	Sinclair, Charles Robert
2831	Sinclair, George Watt
1731	Sinclair, James Edward
2983	Sinclair, Roland Osborne
2909	Sinclair, Thomas Cyril
5436	Skeates, James
3235	Skelly, James
3291	Skinner, Gordon Patrick A.
2830	Skulander, Peter Neilson
	Slater, Henry Ernest
2736	Sleeman, Frank
2240	Sloan, William Walter
4900	Sloss, David
2910	Small, Harold
2277	Smart, Herbert Leslie S.
2663	Smeeth, Albert George
2012	Smith, Abraham Gorham
5443	Smith, Albert
1968	Smith, Alexander
2784	Smith, Alexander Robert
3185	Smith, Alfred Edward
	Smith, Alfred Edward
2244	Smith, Archibald Ross
2282	Smith, Arthur Emanuel
	Smith, Arthur Henry
2786	Smith, Arthur Josiah
3374	Smith, Charles
5183	Smith, Charles Thomas
3135	Smith, Colin Lindsay
2740	Smith, D'Arcy Benedict
2500	Smith, David Joseph
2505	Smith, Edward Henry
4901	Smith, Eric William
3181	Smith, Ernest
3221	Smith, Ernest Athol
3170	Smith, Frederick
4903	Smith, George Henry
3327	Smith, George William
5437	Smith, George William Norris
5447	Smith, Harold Edward
1303	Smith, Harold Reid
1696	Smith, Henry
3290	Smith, Henry
3628	Smith, Henry James
2781	Smith, Herbert Charles
2913	Smith, Herbert James
1969	Smith, John Clifford

6060	Smith, John Henry
2237	Smith, John William
5438	Smith, Joseph William
5439	Smith, Laurence
3134	Smith, Leslie Cameron
180	Smith, Leslie John Shaw
1732	Smith, Leslie Morris
2675	Smith, Percy Carlton
	Smith, Reginald Arthur
2236	Smith, Reginald Norman
2241	Smith, Reuben
1808	Smith, Robert Henry
3186	Smith, Robert Stanley
3222	Smith, Roy Henry Joseph
2778	Smith, Sidney Absell
2230	Smith, Thomas
3183	Smith, Thomas Gardiner
3620	Smith, Thomas Ray
2607	Smith, Walter Ravenscroft
2984	Smith, William
2832	Smith, William George
3169	Smith, William Roy
4906	Smith William Thomas Bob
1660	Smyth, George Australia
2277	Smyth, Owen James
3172	Snoddy, William Thomas
2985	Snowden, Charles Edward
3218	Snowden, Thomas Joseph
5440	Solomon, Philip
	Somerville, James Frederick
3180	South, Harold Thomas
1971	South, Walter James
3800	Southam, Edwin Henry
17233	Southwell, William James
5186	Spain, Michael
1650	Spalding, Andrew
782	Spalding, John
3175	Sparkes, Valentine Camps
556	Spicer, William Henry
2741	Spindler, Albert Ernest
2986	Spink, Raymond
2987	Spink, Wilfred Stanley
3927	Spinks, Gordon
3691	Squires, George Searle
3924	St Smith, Herbert Newey
2504	Stadtmiller, Phillip James
3438	Staff, James
4905	Stafford, Cyril
	Stafford, Louis Norman
1993	Stafford, William Percival
4896	Stagg, John Henry
2239	Stalker, Cecil Ernest Clive
2238	Stalker, Raymond Leslie
2243	Stalker, William Edward John
5228	Standen, Robert Henry
1736	Stanfield, Edward George
1585	Stanwell, Frederick
5441	Stapleton, John
53805	Starling, William
2730	Starr, Herbert Albert
1291	Starr, William Gordon
1733	Stearman, Arthur
1734	Stearman, George
1661	Stearman, George Thomas
3293	Stearn, William Louis
3171	Steavens, John Edward
2506	Steel-Bennett, John Joseph
3220	Steele, Robert Crawford
2772	Stephen, William John
3214	Stephens, William Charles
1742	Stephenson, Arthur
2014	Stephenson, John Edward
1740	Stevens, Gus
1651	Stevenson, John
686	Stevenson, Robert Cross
4895	Stevenson, William Henry
1232	Stevenson, William James
2228	Stewart, Henry Wallace
2219	Stewart, Jack
3634	Stewart, James Millard
3466	Stewart, John Henderson
2535	Stewart, Ronald Esmond
2732	Sticker, Paul Oscar
2737	Stillwell, Oliver
1663	Stirling, Richard
3129	Stokeld, George

No.	Name
2233	Stokes, Charles Henry
2232	Stone, Alfred Ernest
3468	Stone, Roger Humphrey
53804	Stone, William Stanley
5190	Storey, Frank
2503	Stormont, Fulke Greville Le Poer
3467	Stott, James
3618	Strain, James
1570	Street, Edmund Harrington
2836	Stringer, Arthur Frederick
3436	Stringfellow, George Henry
2235	Stroud, Thomas
1177	Stubbins, Walter
	Stutchbury, Eric William
3256	Styles, Charles Henry
2773	Styles, Mortimer
2920	Sullivan, Andrew
2731	Sullivan, Eugene
1737	Sullivan, George
3013	Sullivan, John Thomas
2005	Sullivan, Thomas John
5230	Sutcliffe, Joseph George
3931	Sutherland, George C.C.
3932	Sutherland, John Joseph W.
3806	Sway, Harold
3754	Sweetman, Leon William J.
3219	Swift, Clarence Eugene
2733	Swinbourne, Clarence H
184	Sykes, Henry Alexander
3216	Sykes, John Cyrus
2776	Sykes, Walter William
1745	Syran, Karl
1974	Taber, Percy Ernest
1975	Taber, Reuben
2798	Tait, Francis
5451	Tanko, Stanley Vincent
	Tarn, Joseph Teesdale
1748	Tarrant, George Herbert
3304	Taylor, Arthur
1747	Taylor, Charles William Sydney
3471	Taylor, David Henry
1999	Taylor, Edward
3933	Taylor, Edward Henry
2910	Taylor, Eugene Godfrey
1977	Taylor, Francis William
5449	Taylor, Frank
2797	Taylor, Frederick Andrew
10100	Taylor, George
17170	Taylor, George
3192	Taylor, George Edward
3223	Taylor, Hamilton Isaac
3946	Taylor, Harry
3470	Taylor, Herbert James
3141	Taylor, James
2510	Taylor, Jason
1185	Taylor, John
2747	Taylor, John Joseph
1973	Taylor, Sydney John
2746	Taylor, Victor Cecil
1705	Teague, William Francis
	Teasdale, William James
59828	Terry, Frank
2799	Terry, Hemp George
2452	Terry, William Edward
3188	Tharme, Herbert Sampson
3190	Thatcher, Sidney Rupert
1902	Theo Agassiz (alias M. Agassiz)
2744	Thew, Arthur James
58691	Thomas, Alan Jocelyn
2988	Thomas, Charles Victor
2794	Thomas, Robert
2989	Thomas, Thomas
2247	Thompson, Alexander Robert D.
2990	Thompson, Alexander Vivian
2825	Thompson, Alfred
4908	Thompson, Charles
1998	Thompson, Charles Raymond
3935	Thompson, Leonard
2248	Thompson, Peter Denis
3138	Thompson, William Edward
2745	Thoms, Walter
5231	Thomsen, Hans Peter
3140	Thomson, Francis Charles
1698	Thomson, George
786	Thomson, John Whitmill
3302	Thomson, Thomas

1247	Thornthwaite, Edward John
2462	Thornthwaite, Ernest Hubert
1984	Thoroughgood, Oswald Harry
2748	Thorpe, Arthur
3224	Thwaites, Leslie Frederick
53822	Tidball, George
3629	Tidyman, Christopher
3632	Tidyman, William Colledge
5450	Tighe, Alban James
4909	Tighe, Reginald George
2001	Timson, Walter
2919	Tollis, Horace Charles
	Tonking, Reginald Llewellyn
1978	Toone, William John
2508	Torpy, Patrick Joseph
3939	Touhill, Joseph
2245	Tozer, Alfred Sydney Albert
3446	Travers, James Arthur
3225	Travers, Oliver Henry
2218	Traynor, Timothy
53565	Trench, Robert Le Poer
1665	Trew, William Charles
2991	Trinder, Henry Harold
3734	Tucker, Alfred Cyril
4948	Tufnell, George
1664	Tuite, Nicholas Arthur
2749	Tunney, Charles Bede
5452	Turner, Albert Harold
177	Turner, Arthur Brown
3531	Turner, Bernard Colin
1667	Turner, Frank
1976	Turner, Richard Ernest
2246	Turner, Simon Henry
3943	Twining, James George
2774	Tyler, Joseph James
3014	Tyrrell, Patrick
5453	Ulherr, Carl Fredrick Frank
3305	Underhill, Charles Clyde
2509	Underwood, Charles Edward
380	Upton, Walter
2694	Urquhart, William
3515	Vane, Thomas
2469	Vass, David
1461	Veale, Vosper
2249	Venables, Ernest Cecil Walter
3621	Verrills, John
1749	Vincent, Harold Leslie
	Vine-Hall, Austin Pickard
	Vogan, Arthur Henry
1838	Vogt, Peter Christian
1731	Wade, Hubert
2755	Waite, Clarke Henry
3148	Waldon, Harold Victor
2776	Wales, Harold Ambrose
3476	Walford, Charles Edward
1193	Walker, Arthur Hedley
2269	Walker, Arthur Washington
3312	Walker, Frank Owen
5455	Walker, George
3474	Walker, John Edward
3307	Walker, Mathew
1751	Walker, Walter James
4927	Wall, Herbert Joseph
2756	Wallace, Arthur Richard
1270	Wallace, Harold
3752	Waller, William
1979	Walsh, Frederick Martin
4918	Walsh, Harold Austin Madigan
3952	Walsh, Patrick Joseph
3196	Walsh, Thomas
2803	Walshe, Frederick Alston
3310	Walter, Arthur Barton
	Walter, Reginald
4943	Walters, Samuel
3479	Walton, Francis Cyril
1750	Walton, Robert Francis
3257	Warburton, James Thomas
6106	Warburton, Richard Dunbar
2513	Warburton, William Arthur
3944	Ward, Arthur Ernest
2251	Ward, James Ernest
2992	Wark, William John Appleton
3230	Warne, Walter John
2993	Waters, Leslie
2994	Wates, Cecil James
5463	Watkins, Oliver

2220	Watkins, Sydney
3241	Watling, Bertie
3231	Watling, Reginald
3654	Watson, Clarence Sinclair
2995	Watson, Prosper Victor
1530	Watson, Richard
6750	Watt, Douglas Robert
3655	Watt, Hugh Roy
3646	Watt, Norman Pitt
1758	Watt, William
5776	Watters, Aubrey
	Watters, Thomas Clinton
3948	Watts, Albert Victor
4920	Waygood, Percy John
2996	Wealands, Arthur
3242	Wearne, Richard
2486	Weaver, Reginald Walter
2463	Webb, Henry William
2005	Webb, Jock
2257	Webb, John
2994	Webb, Victor Eugene
1752	Webb, William Henry
3744	Webster, Francis
2997	Webster, Harold George
1646	Webster, Hedley Vincent
1983	Webster, James
2807	Wedd, Frank
3311	Weddup, George
3233	Weekes, John William Leslie
59839	Weekes, Raymond Walter
3513	Weekes, William John
2804	Weigand, Albert William
2494	Weinrabe, Lewis Byron
3650	Weir, John Francis
2705	Weitzel,. Frank
1981	Wensor, Edwin John
3477	West, Arthur George
2515	West, Edward Henry
2019	West, Frank
3203	West, Frank Leslie
3150	West, Frederick Harold
59836	West, John Boucher
2700	West, William Dickson
4922	West, William John
4921	Westerway, Henry
2378	Westlake, Harold Bass
2758	Westphal, Karel Hendrick J.
1728	Whalan, David William
3239	Whalan, Edwin Simpson
2281	Whan, Alexander
2253	Whealey, Joseph
3657	Wheeler, Arthur Daniel
3656	Wheeler, Thomas Edward
1753	Wheeler, William James
58659	Wheelhouse, Frank Campbell
2998	Whelan, Thomas
3237	Whipp, Edward Hercules
3226	Whipp, William Arthur Stanley
4930	Whit, Rupert Percival
804	Whitaker, Fred
4923	Whitbread, Herbert Manning
4915	Whitby, Robert Allan John
1497	White, Alfred
2035	White, Bertram Arthur
5456	White, Clarence
2999	White, Frederick Samuel
5233	White, John Maitland
3753	White, William
5458	White, William
3651	Whitehouse, Charles Henry B.
1258	Whitehouse, William
2754	Whitelaw, Arthur Robert
2256	Whiteman, Vernon John
3641	Whitfield, George Henry
3471	Whiting, David Ralph
3472	Whiting, James Ernest
4963	Whiting, William Herbert
3308	Whitnall, Alexander Charles
3151	Whitney, Marshall
2514	Whittaker, Thomas
3001	Whitting, Frank
3152	Whittingham, Frederic John
4926	Wicks, Alfred Kenneth
2009	Wiggins, Frederick James
5459	Wigglesworth, Horace
1754	Wild, Roy Arnold

33552	Wiles, Dexter Freemantle
	Wilkin, Reginald Howes
1259	Wilkins, Alfred
2702	Wilkinson, William Robert
59832	Willcocks, Norman Henry
2512	Williams, Alfred Ernest
5464	Williams, Alfred Norman
3236	Williams, Amos Edward
59835	Williams, Benjamin Parry
2753	Williams, Benjamin Victor
6000	Williams, Charles Henry
2751	Williams, Charles Joseph
287	Williams, Edgar Richard
1755	Williams, Edward
2529	Williams, Frederick John
3293	Williams, Harry Charles
3002	Williams, Henry
3199	Williams, Henry Francis
3795	Williams, James Patrick
1261	Williams, John
3719	Williams, Joseph
1759	Williams, Lloyd E C
2752	Williams, Oscar Sylvester
4924	Williams, Percy
2750	Williams, Reginald Alexander
3003	Williams, Thomas
5460	Williams, Thomas
4802	Williams, William Frank
1988	Williamson, Francis Roy
3478	Willis, Alfred
1987	Willis, Cecil
3229	Willis, Francis Michael
3622	Willis, George Johnston
3807	Willis, Lionel Charles
3235	Willis, Ubert Victor
3655	Willmette, Harry
5468	Willmore, Richard
2003	Wilmot, James Giles
3475	Wilson, Alfred Benjamin
58667	Wilson, Arthur John
1677	Wilson, Clement
3004	Wilson, Clyde
	Wilson, Cyril Ashley
2802	Wilson, David
3159	Wilson, Frank
4919	Wilson, Frank
3197	Wilson, Frank Alexandria
3005	Wilson, Frederick
1986	Wilson, George Harold
2775	Wilson, George Henry
1985	Wilson, George William
	Wilson, Harry Lionel
4925	Wilson, Henry O'Brien
2014	Wilson, James
5469	Wilson, James
575	Wilson, Mark
	Wilson, Norman Lewis
3473	Wilson, Oliver William
1637	Wilson, Richard Edward
2068	Wilson, Richard Henry Linfield
3238	Wilson, Sydney Leonard
3232	Wilson, Thomas Harry
3309	Wilson, Thomas William
3228	Wilson, William Ramsay
3234	Wilton, Alfred Thomas
2252	Winner, Arthur James
1982	Winter, Alexander
3204	Winter, Archibald Thomas
3244	Wiseman, Stephen Edward
1980	Withers, Clement Percival
3243	Wolfe, Thomas Edward
2917	Wolrige, Edgar Claude R.
3147	Womsley, Vivian
5470	Wood, Arthur
3681	Wood, Charles Francis
3006	Wood, George
2039	Wood, George
5103	Wood, H.H. (alias C. Hewson)
2806	Wood, James
	Wood, Louis Cecil
3240	Wood, Sidney Arthur
1757	Woodcroft, Thomas
2250	Woodger, John Robert
2546	Woodhart, Edward Grey
	Woodhouse, Charles James
2254	Woodhouse, Thomas Henry S.

4928	Woods, Clarence Merton
1756	Woods, Edward George
4916	Woods, Leslie Robert
	Woods, Percy William
5462	Woods, Walter
4914	Woodward, Edgar Hubert
	Woolley, Charles Russell
4929	Wootten, Frederick Charles
	Wren, William James
1273	Wright, Bert
3625	Wright, Claude Kendall
1804	Wright, Duncan Claude
3783	Wright, Ernest Roy
3841	Wright, Herbert Isaac
2980	Wright, James Kerr
3227	Wyatt, Ransome Tovey
3206	Wye, Henry James
	Wyllie, Hugh Alexander
3205	Wyllie, James Arthur
	Wyllie, Kenneth Robert
1277	Yardy, Arthur David
1989	Yates, Sydney
3517	Yeark, Alfred John
2258	Yelds, Stanley Darwin
3480	Yelland, Henry Oswald
2759	Young, George Alton
2706	Young, Harold
2812	Young, James
	Zander, Charles Leonard

Bibliography

Australian War Memorial (AWM)

5th Australian Division War Diary, AWM4, 1/50.

14th Infantry Brigade War Diary, AWM4, 23/14.

15th Infantry Brigade War Diary, AWM4, 23/15.

55th Infantry Battalion War Diary, AWM4, 23/72.

55th Infantry Battalion Nominal Roll and Next of Kin, AWM 9, 87/28.

55th Infantry Battalion, Statements made by Prisoners of War, AWM30, B14.6, B14.7, B14.8.

Allen, H.D., 55th Australian Infantry Battalion: History 12 Feb - 30 Sep 1916, AWM224, MSS183.

Bean, C.E.W. Notebooks, AWM 38 /3DRL 606/178/2.

Law, Francis, 'Recollections of the Battle of Fromelles, France, July 19th 1916 by No 246 Sgt F. Law, Platoon Sergeant, No 2 Platoon, A Coy, 31st Btn, 8th Inf Bgde, 5th Div. AIF', 3DRL606/243A/1.

Sheppeard, Alfred Edward, 'Extracts from an account written at Ferry Post 27th May'.

Stegemann, William C., The Snowy River Marchers, their war effort and afterwards, MS50821.

Thomson, Alistair, 'Oral interview with Mr F. Farrall by Alistair Thomson dated 7 July 1983', AWM S01311.

Personal files held by the AWM

Barker, James, PR86/367.

Bean, Charles Edwin Woodrow, AWM38.

Berryman, Gilbert, PR88/210.

Bishop, Raymond Charles, 1DRL/0123.

Brown, Frank Cecil, 1DRL/0160.

Buckingham, Arthur E., PR04782.

Campbell, Rupert J., 3DRL/5087(B).

Chapman, Percy Wellesley, 1DRL/0198.

Chapman, Percy Wellesley, PR86/389.

Cosgrave, William Patrick, 1DRL/0215.

Farrall, Frederick, PR84/327.

Farrall, Frederick, S01311.

Gardiner, Reginald Scott, 1DRL/0304.

Gibbins, Norman, PR02053.

Giblett, William Norman, PR83/095.

Gulson, Kelvedon, 3DRL/7294.

Harpley, Robert W., 3DRL/3663.

Hinson, Rupert Sykes, PR86/277.

Horniman, Lancelot Vicary, 1DRL/0357.

Horniman, Robert Geoffrey, 1DRL/0358.

Hunt, Sinclair, 2DRL/0277.

Lynch, Arthur Leslie, PR84/077.

MacCallum, Hector Archie, PR89/022.

Marshall, John James, 1DRL/0481.

Moriarty, Merion Morton, PR86/277.

Murray, James E., PR83/166.

Nicholson, W.E., PR89/107.

Smith, Darcy Benedict, PR86/264.

Stewart, James C., 3DRL/1459.

Street, Edmund Harrington, PR85/179.

Sykes, John Cyrus, S03732.

Vogan, Arthur Henry, PR82/007.

Willis, Ubert Victor, PR04786.

Winter, Archibald Thomas, PR89/163.

National Library

Hunt, Atlee, 'Unidentified newspaper clipping, 1914', Papers of Atlee Hunt, NLA MS 1100.

McCallum, John Archibald, MS 9872.

Mitchell Library

Gill, George Thomasson, ML MSS 2765.

Harris, Herbert Henry, ML MSS 2772.

Walker, John, ML MSS 7328.

Private Collection

Hardy, Charles B., Diary.

O'Grady, John Joseph.

Official Histories

C.E.W. Bean, *Official History of Australia in the War of 1914–1918*, Vol. III, *The AIF in France, 1916*, Angus & Robertson, Sydney, 1941.

C.E.W. Bean, *Official History of Australia in the war of 1914–1918*, Vol. IV, *The AIF in France 1917*, Angus & Robertson, Sydney, 1941.

C.E.W. Bean, *Official History of Australia in the War of 1914–1918*, Vol. VI, *The AIF in France May 1918 – the Armistice*, Angus & Robertson, Sydney 1942.

Books

Bean, C.E.W., *Letters from France*, Cassell, London, 1917.

——, *Anzac to Amiens*, Australian War Memorial, Canberra, 1946.

Bishop, Walter Herbert (Bert), *The Hell, The Humour, The Heartbreak: A Private's View of World War 1*, Kangaroo Press, Kenthurst, 1991.

——, *Dear All; Letters from World War 1*, compiled and edited by Pamela Goesch, Brynwood House, Sydney, 2010.

Cobb, Paul, *Fromelles 1916*, Tempus Publishing, Stroud, UK, 2007.

Coombes, D., *The Lionheart: A Life of Lieutenant-General Sir Talbot Hobbs*, Australian Military History Publications, Loftus, NSW, 2007.

Corfield, Robin S., *Don't forget me, cobber: the Battle of Fromelles, 19/20 July 1916: an inquiry*, Brown, Prior, Anderson, Burwood, 2000.

Cutlack, F.M., *The Australians: their final campaign, 1918: an account of the concluding operations of the Australian divisions in France*, Sampson, Low, Marston, London, 1919.

Ellis, Alexander D., *The Story of the Fifth Australian Division*, Hodder and Stoughton, London, 1920.

Farrall, Lois, *The File on Fred: a biography of Fred Farrall*, High Leigh Publishing, Carrum, 1992.

Fohlen, Yves, *He came from Australia: Lieutenant Percival Ralph, MM, MID (53 Battalion - Australian Imperial Force), Australian Military History* Publications, Loftus, NSW, 1998.

Hunt, Sinclair, 'Overcoming terror and winter hardship on the Somme 1917', *We fought the battles: in letters and diaries written home, World War One soldiers relate their own personal experiences of conflict, frustrations, anxieties and tragedies all perpetrated by war*, compiled and edited by James M. Woolley, Eastwood, 1998.

Kelly, Darryl, 'Corporal Ernest Corey, MM: One of a kind', *Just Soldiers: Stories of Ordinary Australians Doing Extraordinary Things in Times of War*, ANZAC Day Commemoration Committee Inc., Queensland, 2004.

Kennedy, John J., *The Whale Oil Guards*, James Duffy and Co, Dublin, 1919.

Lindsay, Patrick, *Fromelles: the story of Australia's darkest day the search for our fallen heroes of World War One,* Hardie Grant Books, Prahan, 2007.

Maddrell, Roslyn, *Letters from the front: Boer War to WWII through letters sent by servicemen to their families in Braidwood, self-published,* Braidwood, NSW, 2004.

McAndrew, Alex, *Milton Ulladulla in the Wars*, self-published, Epping, NSW, 1994.

McMullin, Ross, *Pompey Elliot*, Scribe, Melbourne, 2002.

Monash, John, *The Australian Victories in France in 1918*, Angus & Robertson, Melbourne, 1936.

Mongan, Cheryl and Reid, Richard, *We have not forgotten: Yass & districts war 1914-1918*, Milltown Research and Publications, Yass, 1996.

Perry, R., *Monash: the outsider who won a war: a biography of Australia's greatest military commander*, Random House, Milsons Point, NSW, 2004.

Pye, Rosemary, *To all the dear ones at home: the letters and narratives of Sinclair Hunt 1915-1918*, compiled by Rosemary Pye, Barbara Brady and Wendy Henningham, JM Executive Printing Service, Sydney, 2012.

St Claire, Ross, *Our gift to the Empire: 54th Australian Infantry Battalion, 1916-1919*, self-published, The Junction, 2006.

Stegemann, William C., *A Monaro mosaic: patterns of Monaro life: events and people*, Linbarroo Press, Barneyview, 1985.

Williams, Harold Roy, *Comrades of the Great Adventure: an Australian soldier's story of 1915 – 18*, Angus & Robertson, Sydney, 1935.

——, *The Gallant Company: An Australian Soldier's Story of 1915-18*, Angus & Robertson, Sydney, 1933.

Wray, Christopher, *Sir James Whiteside McCay: a turbulent life*, Oxford University Press, South Melbourne, 2002.

Wren, Eric, *Randwick to Hargicourt: History of the 3rd Battalion, AIF*, Ronald G. McDonald, Sydney, 1935.

Articles

Bazley, Arthur William, 'Military Medal and Three Bars', *Stand-To*, January–March 1951.

Bean, C.E.W., 'Doignies', *The Mercury*, 20 June 1917.

——, 'The Leadership of Norman Gibbins', *The Link*, 1 September 1926.

Bishop, Walter Herbert (Bert), 'From France', *The Ulladulla and Milton Times*, 30 September 1916.

Dalley, W. B., 'The Moascar March: A Tragedy of the Desert. Full Story of One of the Most Pitiless Episodes in the History of the AIF', *Smiths Weekly*, 2 December 1922.

Fitzsimons, Peter, 'One fight that even 'Nutsy' Bolt couldn't win', *Sydney Morning Herald*, 24 April 2010.

Gray, Cyril G., 'Huns Await Australians: Charge across 'No Man's Land': Waist Deep in Water', *Evening News*, n.d. (poss 1916).

Hetherington, Les, 'The Kangaroos March: Wagga Wagga to Sydney, December 1915 – January 1916', *Journal of the Australian War Memorial*, Issue 26, April 1995.

Maynard, James, '55th Battalion 'Kangaroos' and Others', *Reveille*, 1 May 1935.

——, '55th Bn. at Polygon Wood – a few reflections', *Reveille*, 1 September 1935.

——, 'Cookers in the Front Line. 55th Battalion's Apprenticeship at Fleurbaix', *Reveille*, 1 February 1936.

——, 'Five Xmases with the AIF. Liverpool Camp, Lemnos, Dernancourt, Le Turne, Weymouth', *Reveille*, 1 December 1935.

Pedersen, Peter, 'Reflections on a Battlefield', *Wartime*, Issue 44, 2008.

Slater, Henry Ernest, 'Lieut-Col P.W. Woods. DSO (and Bar), M.C.', *Reveille*, 1 February 1933.

——, 'Polygon Wood Memories', *Reveille*, 1 October 1935.

Smith, William, 'Leaves from a Sapper's Diary: The Battle of Fromelles', *Reveille*, 1 July 1936.

Unknown author, 'The World's Great War. Exciting Raids. Game the ANZACs like. Bewildered Germans', *The Sydney Morning Herald*, 30 November 1916.

Woods, Percy, 'Corporal Corey: Won M.M. (3 Bars)', quoted in *Reveille*, 31 March 1931.

Websites

Jamesion, Sue, *WW1 – For this Sydney family it was 'on for young and old'*. http://www.awm.gov.au/blog/2009/03/18/wwi-for-this-sydney-family-it-was-on-for-young-and-old/ accessed 19 June 2012.

Taplin, Harry, 'Neville, Dalton Thomas Walker (1896–1969)', *Australian Dictionary of Biography*, Australian National University, at: http://adb.anu.edu.au/biography/neville-dalton-thomas-walker-7822/text13577, accessed 26 December 2011.

Endnotes

Front section

1. Bert Bishop, quoted in Alex McAndrew, *Milton Ulladulla in the Wars*, self-published, Epping, NSW, 1994.
2. The Lewis gun (or Lewis Automatic Machine Gun) is a light machine-gun distinguished by its wide tubular cooling shroud around the barrel and top-mounted drum-pan magazine. It was more portable than a heavy machine-gun and could be carried and used by a single soldier.

Chapter 1

1. Andrew Fisher, then leader of the opposition Labor Party, in a speech during the 1914 election campaign. Labor won the election and Fisher became Prime Minister in September 1914. Papers of Atlee Hunt, NLA, MS 1100.
2. Also called Tall al Kebir. The name means 'great mound' in Arabic.
3. Godfrey George Howy Irving.
4. David McFie McConaghy. DOW 9 April 1918.
5. McConaghy had been awarded the CMG (Companion in the Most Distinguished Order of Saint Michael and Saint George).
6. Robert Orlando Cowey.
7. Austin Claude Selwyn Holland.
8. Percy William Woods.
9. Norman Gibbins. KIA 20 July 1916.
10. John James Marshall. KIA 27 April 1918.
11. Eric William Stutchbury.
12. John Hamilton Matthews.
13. Sidney Albert Pinkstone. His brother, Norman, also served as an officer in the 55th Battalion.
14. Herbert Leopold Palmer. KIA 11 March 1917.
15. Lionel Duncan Sheppard.
16. Harry Lionel Wilson. Described by Fred Farrallas as 'a very brave man, but a real militarist'. His nicknames were 'Von' or 'Von Hindenburg'. Lois Farrall, *The File on Fred: a biography of Fred Farrall*, High Leigh Publishing, Carrum, 1992.
17. James Stanley Edwin McCarthy.
18. William Norman Giblett.
19. Fredrick James Cotterell. KIA 2 September 1918.
20. Norman Ewart Franklin Pinkstone.
21. Percy Wellesley Chapman. KIA 12 March 1917.

22. Eric Wren, *Randwick to Hargicourt: History of the 3rd Battalion, AIF*, Ronald G. McDonald, Sydney, 1935, p. 129.
23. Also referred to as the 'Fresh Water Canal' or the 'Ismailia Canal'.
24. A regional centre situated midway between both ends of the Suez Canal.
25. 2553 Herbert Daniel Allen.
26. Hugh Alexander Wyllie.
27. H.D. Allen, 55th Australian Infantry Battalion: History 12 Feb - 30 Sep 1916, AWM224, MSS183.
28. Ibid.
29. 3204 Archibald Thomas Winter.
30. Wren, *Randwick to Hargicourt*, p. 116.
31. Ibid., p. 117.
32. 6000 (also 723 and 2923) Charles Henry Williams. Williams (real name Williamson) rejoined the AIF at the end of 1916 and was attached to the 46th Battalion. He won the Military Medal and was KIA in May 1918.
33. 2619 William Patrick Cosgrave. KIA 17 April 1918.
34. Allen, *55th Australian Infantry Battalion History*.
35. 1677 Frederick Theodore Farrall.
36. Farrall, *The File on Fred*, pp. 51–52.
37. James Whiteside McCay.
38. The battalion was not yet at full strength.
39. Also called Kassassin.
40. Sinclair Edward Hunt. KIA 1 September 1918.
41. A 'dixie' is a small aluminium cooking and eating pan issued to individual soldiers.
42. 2915 George Thomasson Gill.
43. 2963 Robert William Harpley.
44. Actual capacity was two pints.
45. Arthur Henry Vogan. DOW 4 February 1917.
46. 55th Infantry Battalion War Diary, AWM4, 23/72.
47. Edward, Prince of Wales, son of King George V of the United Kingdom (UK).
48. W.B. Dalley, 'The Moascar March: A Tragedy of the Desert. Full Story of One of the Most Pitiless Episodes in the History of the AIF', *Smiths Weekly*, 2 December 1922.
49. Arthur Edmund Shepherd.
50. 3077 Herbert Henry Harris.
51. 3270 Rupert Joseph Campbell.
52. Percy Wellesley Chapman, AWM 1DRL/0198.
53. Alexander D. Ellis, *The Story of the Fifth Australian Division*, Hodder and Stoughton, London, 1920, p. 46.

54. 3762 Walter Herbert Bishop.
55. Walter Herbert (Bert) Bishop, *The Hell, The Humour, The Heartbreak: A Private's View of World War 1*, Kangaroo Press, Kenthurst, 1991, p. 43.
56. A military practice whereby all men stand ready in the trench in event of a surprise attack. A 'stand-to' can be ordered to occur at any time but it occurs at sundown and sunrise as a matter of routine.
57. Ellis, *The Story of the Fifth Australian Division*, p. 46.
58. 'Stand-down' is the point at which the alert state of 'stand-to' ceases.
59. 'Crimed' is the charging of an individual with a military offence.

Chapter 2

1. 'Burgoo' is oatmeal gruel or thick porridge eaten by sailors.
2. 3108 Frank Brown.
3. Francis Edwin Armstrong.
4. 'Very lights' (also called 'Very flares') were signal flares fired from a pistol-shaped flare-gun. Very lights used combinations of green, red, and 'white star' flares to communicate with forward troops. White-coloured Very flares were also used to illuminate a small area at night.
5. 'Fritz' was a nickname for the Germans used by Commonwealth troops.
6. The parapet is the side of the trench directly facing the enemy line.
7. 'Jim Magee' is 2728 James McGee.
8. The Stokes mortar was a light trench mortar issued to Commonwealth armies. It was a simple weapon, consisting of a smooth bore metal tube fixed to a base plate with a lightweight bipod mount.
9. 2497 Edward Francis McKenna. KIA 13 July 1916.
10. German trench mortars, also nicknamed 'Minnies'.

Chapter 3

1. The Central Powers consisted of the German Empire, the Austro-Hungarian Empire, the Ottoman Empire, and the Kingdom of Bulgaria. The term 'Central Powers' is derived from the location of these countries; all four were situated between the Russian Empire in the east and France and the UK in the west.
2. Herbert Charles Onslow Plumer.
3. Charles Carmichael Monro.
4. Richard Cyril Byrne Haking.
5. The battle was the initial British component of the combined Anglo-French offensive known as the Second Battle of Artois. In support of French operations, the First Army made a lunge for Aubers Ridge. This same area had been targeted in the Battle of Neuve Chapelle two months earlier. The northern pincer of the British attack was against precisely the same sector as that attacked by the Australians on 19 July 1916. The British troops were cut down by German machine-gun fire, suffering 11,000 casualties in one day of fighting. No significant progress was made and the following day the offensive was abandoned.

6. C.E.W. Bean, *Official History of Australia in the War of 1914–1918*, Vol. III, *The AIF in France, 1916*, Angus & Robertson, Sydney, 1941, pp. 336–37.
7. In French Rivière des Layes. In the Fromelles area the Laies River is little more than a water-filled ditch.
8. Lieutenant Eric Stutchbury assisted Holland in this role.
9. 3230 Merton James Desmond Austin.
10. 569 Charles Briggs Hardy.
11. Bishop, *The Hell, the Humour, The Heartbreak*, p. 61.
12. Ibid.
13. 3134 Robert Micklewright.
14. 169 John Bain.
15. 3760 Sydney Bell. KIA 20 July 1916.
16. 3050 John William Doyle.
17. 4747 George Alfred Blunt. DOW 19 July 1916.
18. Roy Arthur Goldrick.
19. 2662 Cyril George Gray.
20. The parados forms the side of the trench furthest away from the enemy line; that is, the back of the trench.
21. 2423 Albert Stanley Noldart. KIA 20 July 1916.
22. 3657 Horace Samuel Blackburn. KIA 20 July 1916.
23. A formation or position is in enfilade if weapons' fire can be directed along its longest axis. A line of advancing troops is enfiladed if fired on from the side (flank) such that bullets travel the length of the line. The benefits of enfilading an enemy formation are: by firing along the long axis it is easier to hit individual troops; and projectiles that miss the intended target are more likely to hit a different target within the formation.
24. Bishop, *The Hell, The Humour, The Heartbreak*, pp. 62–63.
25. 3348 Percy William Larbalestier in a letter to Gibbins' sister dated 27 January 1917. A batman is a soldier assigned to a commissioned officer as a personal servant. A batman's duties often included: acting as a 'runner' to convey orders from the officer to subordinates, maintaining the officer's uniform and personal equipment as a valet, acting as the officer's bodyguard in combat and other miscellaneous tasks.
26. Kenneth Robert Wyllie. Robert Cowey, in a letter to Charles Bean, wrote: 'Captain Wyllie has for long been admitted to be the most successful company commander in the Battalion. His men worship him.'
27. Bishop, *The Hell, The Humour, The Heartbreak*, p. 63.
28. 4229 Samuel Robert Lennard. KIA 19 July 1916.
29. The other two brothers were 3085 Augustus Thomas Lennard and 4228 Henry Lennard. Augustus was badly wounded at Fromelles, leading to his discharge in early 1917. Henry was KIA on 21 October 1917.

30. Claude Elliot Fuller.
31. 2722 Charles Robert Matterson.
32. In some reports this advanced line is also referred to as the 'German Second Line' or even the 'German Third Line' as another minor gutter lay between the German front line and the advanced line.
33. 2601 John William Coates. During the fighting at Fromelles, Coates was knocked unconscious by a German grenade explosion at around 10.00 pm that also wounded him in the head, eyes and right arm. On regaining consciousness the next day he found himself in Lille Hospital, a prisoner.
34. VC Corner, a supply dump, was located at the crossroads of the D171 road and Rue Delvas in the village of Petillon.
35. Bishop, *The Hell, the Humour, the Heartbreak*, pp. 63–64.
36. Walter Edmund Hutchinson Cass.
37. 2668 Neil Penfold. KIA 20 July 1916.
38. 3436 George Henry Stringfellow.
39. Second Lieutenant Cecil Agassiz was the battalion's Lewis Gun Officer.
40. 'Hun' was a nickname for the Germans used by Commonwealth troops.
41. Bean, *Official History*, Vol. III, p. 406.
42. 3181 Ernest Smith.
43. Cowey is referring to George Cummins.
44. 1701 Henry Anson.
45. Stanley Colless. KIA 1 September 1918.
46. 3007 Vinton Battam Baker. KIA 20 July 1916.
47. 3029 Frederick Lloyd Carpenter.
48. George Darcy Folkard.
49. Bishop,'From France' in *The Ulladulla and Milton Times,* 30 September 1916.
50. 3219 Luther Chadwick.
51. 3160 James Alfred Perkins.
52. APM (Assistant Provost Marshal) was the AIF's military police function.
53. Wyllie received a gunshot wound to the chest but was able to rejoin the battalion in May 1917.
54. Eric Maynard Farmer.
55. To prevent accidental explosions, Mills grenades were transported in boxes of 12 with the detonators in a separate tin located in the middle of the box. In order to 'detonate' the grenade, the base plug of the grenade had to be unscrewed, the detonator inserted, and the base plug screwed back. This essential but tedious chore was undertaken as part of preparations for front-line service. In his report on the conduct of the battle, Lieutenant Colonel McConaghy highlighted the fact that many men involved in bomb fights had no knowledge of how to use the Mills grenade correctly, some even throwing the bomb without first pulling the

safety pin. This admission is extraordinary. The Mills grenade had been issued to the British Army since February 1915 and the fact that, 18 months later, soldiers in the 55th were still untrained in their use is a poor reflection on McConaghy as the battalion's CO.

56. Berrol Lazer Mendelsohn. KIA 20 July 1916.
57. Peter Pedersen, 'Reflections on a Battlefield', *Wartime,* No. 44, pp. 24–29.
58. 2873 Henry Mayer. Second Lieutenant Agassiz described Mayer as 'a quiet boy, but a fighter'. KIA 20 July 1916.
59. 3235 James Skelly.
60. 1833 Francis Thomas Stringer. He received a DCM for his actions.
61. Probably 2nd Lieutenant Charles Leonard Noble Zander. During the fighting at Fromelles Zander suffered a gunshot wound to the eye from which he never fully recovered. He spent the remainder of the war in training establishments in the UK.
62. Denoon's wounds required lengthy hospitalisation, followed by his return to Australia. It is unlikely that he ever recovered from the physical or mental trauma of his wounding. He died in 1923 aged 41. He was awarded the Military Cross for his actions.
63. 3240 Eric Floyd Hancock. KIA 20 July 1916.
64. 946 William Hurley.
65. Bean, *Official History*, Vol. III, p. 420.
66. 246 Francis Law.
67. Norman Alexander Robinson.
68. 186 Leonard Adams Davis.
69. 3061 Peter Ferguson. KIA 20 July 1916.
70. The Vickers gun was the standard heavy machine-gun for Commonwealth forces during the Great War. The tripod-mounted Vickers was loaded from a 250-round fabric belt and weighed less than 20 kilograms. It had a rate of fire of 450 rounds per minute. While it enjoyed a reputation for reliability, its heavy weight made it unwieldy as an offensive infantry weapon, although it was ideal for defensive operations.
71. Bean, *Official History*, Vol. III, p. 432.
72. 2728 George Gribbon.
73. 3009 Herbert Thomas Bolt. 'Nutsy' was a well-known player for the Newton Rugby League Club. He had represented NSW twice on a tour of Queensland in 1913; in the second match he was sent off for brawling. KIA 20 July 1916.
74. 3098 Francis James Johnston.
75. 4885 John Gilbert Reay. DOW in a German hospital 26 July 1916.
76. Bean, *Official History*, Vol. III, p. 434.
77. 3237 Joseph Henry Grimes.
78. Bean, *Official History*, Vol. III, p. 434.
79. Colless was awarded the Distinguished Conduct Medal for his actions.
80. 3761 Raymond Charles Bishop. KIA 20 July 1916.

81. Possibly 1168 Alexander Christopher Robertson. KIA 20 July 1916.
82. Bishop, 'From France'.
83. 4811 Percy Geason. KIA 20 July 1916.
84. 3291 Norman Clarence Ford.
85. William Ernest Robinson.
86. 2035 Bertram Arthur White.
87. 2609 Francis Albert Wearne Hocking. He was awarded the Distinguished Conduct Medal for his actions. DOW 4 February 1917.
88. Ross McMullin, *Pompey Elliot*, Scribe, Melbourne, 2002, p. 216.
89. William Smith, 'Leaves from a Sapper's Diary: The Battle of Fromelles', *Reveille,* 1 July 1936, pp. 4, 5, 14.
90. Bishop, 'From France'.
91. Second Lieutenant George Rolla Barlow was one of these. In March 1917 he was sent to the UK, ostensibly with laryngitis or mild bronchitis. However, his service record states that he was suffering from neurasthenia (shell shock). He returned to Australia in October 1917.
92. 4742 James Lochee Barker.
93. Robin S. Corfield, *Don't forget me, cobber: the Battle of Fromelles, 19/20 July 1916: an inquiry*, Brown, Prior, Anderson, Burwood, 2000, p. 123.
94. 3271 John Archibald McCallum.
95. Every man for himself.

Chapter 4

1. Ellis, *The Story of the Fifth Australian Division,* pp. 117–18.
2. Clarence John Hobkirk, appointed commander of the 14th Brigade with effect 22 July 1916.
3. This was Lieutenant Percy Chapman.
4. Harris means 2383 Robert Edward Peterson.
5. 5333 Robert Victor Buckenham. KIA 4 August 1916.
6. 5447 Harold Edward Smith. KIA 9 August 1916.
7. Larbalestier was formerly Captain Norman Gibbins' batman, but was performing this task for Lieutenant Chapman following Gibbins' death at Fromelles.
8. A small, usually shabby, café selling snacks, wine, beer and coffee.
9. Bishop, *The Hell, The Humour, The Heartbreak*, p. 79.
10. Ibid., pp.74–75.
11. A duckboard is a board or boards laid to form a floor or path over wet or muddy ground.
12. Bishop, *The Hell, The Humour, The Heartbreak*, p. 79.
13. In '*The Hell, The Humour, The Heartbreak*', Bishop gives the name of his platoon sergeant as 'Harlock'. In fact, 'Harlock' is 2445 Henry George Howard.

14. 'Reg' may have been 3777 Reginald Gladstone Conlon. Conlon was in C Company (the same company as Bishop) and was shot in the left hand on 28 August 1916. Conlon returned to the battalion on 17 January 1917 and was wounded again on 31 January 1917. He was transferred to England and was invalided to Australia in July 1917.
15. 5397 Arthur Alfred Harold Luby. Accidental (injuries) 1 September 1916.
16. 1670 Bertie Allen Hilton.
17. 5341 Harry Banwell. KIA 7 September 1916.
18. 3759 Francis Wilson Beaven. KIA 7 September 1916.
19. 3952 Patrick Joseph Walsh. KIA 7 September 1916.
20. Two 6-inch howitzers and five 18-pounder batteries (each battery consisting of six guns) were allocated in direct support of the raid.
21. Thomas Louis Gitsham.
22. 5338 Arthur Edwin JohnBuckingham.
23. The CO, Lieutenant Colonel David McConaghy, was on leave.
24. Unknown author, 'The World's Great War. Exciting Raids', *The Sydney Morning Herald,* 30 November 1916, p. 3.
25. Short trenches dug in advance of the main trench line to allow patrols, raiders and sentries to safely leave the front line and move into no man's land.
26. 3905 Robert William Raymon. KIA 30 September 1916.
27. 2812 James [Jim] Young gave his age at enlistment as 18 but his family attests that he was 16.5 years old at the time of his death. After the raid Captain Eric Stutchbury and 2854 John Jeffries (who was Young's platoon sergeant) brought the badlywounded Young to the aid station; Young had a bullet wound in his right arm and had also been shot in the groin. Young DOW 1 October 1916. Jeffries informed the Red Cross that he had always taken Young with him on patrol as he was 'a game lad'. According to Sue Jamesion (see website in bibliography), in World War I, for this Sydney family it was 'on for young and old'. Perhaps the youngest member of the battalion was 2723 Leonard Walter Jackson who joined the AIF on 6 August 1915 aged 13 years 11 months and 10 days (although he claimed to be 18). Len's two brothers were in the AIF and his father, 3832 Joseph Jackson, joined the 55th with his youngest son. On enlistment, Joseph stated his age as 44 years and 11 months but he was actually 52, making him perhaps the oldest man in the unit. Father and son served with the 55th at Fleurbaix and Fromelles. On the death of the eldest Jackson boy, 1766 Harry Melville Jackson, in August 1916, Joseph admitted he was over age and that Leonard was serving without his consent. Joseph and Leonard were discharged and returned to Australia in March 1917. Another known under-age soldier was 99 Wilfred Jaeger, who joined the AIF on 10 January 1916 aged 15 years and seven months. He was taken on strength of the 55th on 15 May 1916 and served until February 1918 when his under-age status was identified.
28. 194 Louis Cecil Wood. He was awarded the Military Medal for his actions.
29. 1131 William Albert Hitchenor. He was awarded the Military Medal for his actions.
30. 2923 Claver Joseph Sheedy.
31. 3010 Edwin Nicholson Brookes.

32. 3771 Edward Malcolm Butler.
33. 5406 Leonard Miles. KIA 2 April 1917.
34. 4758 Alfred Browne. DOW 19 March 1918.
35. 5348 James Thomas Cox.
36. Eric Stutchbury was awarded the Military Cross for his actions.
37. Probably 2201 James Ross.
38. 5477 John Francis Dawes.
39. Sidney Pinkstone was Mentioned in Despatches for his organisation of the raid.
40. Alexander John Godley, GCB, KCMG.
41. 2626 John Templeton Davies. KIA 30 September 1916.
42. 3028 Roy Cameron. DOW 1 November 1916.
43. Farrall had become ill in Egypt and had been left behind when the remainder of the battalion sailed for France. On recovery, he went to England, where he joined the 2nd Reinforcements, 55th Battalion. He travelled to France with this group.
44. Farrall, *The File on Fred*, pp. 65–67.
45. The fatalities were: 3043 Thomas Crooks, DOW 13 October 1916; 1740 Gus Stevens, DOW 13 October 1916; 3138 Frederick Herman George Meyer, DOW 14 October 1916; 2499 Robert Cornick, DOW 15 1916; 5368 Reginald Henry Eyre, DOW 15 October 1916; and 1660 Edward Thomas Hogan, DOW 15 October 1916.
46. 1740 Gus Stevens.

Chapter 5

1. Henry Seymour Rawlinson.
2. The 1st, 2nd and 4th divisions were engaged in low-intensity defensive operations around Ypres.
3. Order from Haig to the Second Army dated 9 October 1916.
4. Bishop, *The Hell, The Humour, The Heartbreak*, p. 80.
5. Ibid., p. 81.
6. Buire-sur-l'Ancre.
7. Ellis, *The Story of the Fifth Australian Division*, p. 138.
8. Bishop, *The Hell, The Humour, The Heartbreak*, p. 81.
9. Ibid., p. 82.
10. Walter Herbert (Bert) Bishop, *Dear All; Letters from World War 1*, compiled and edited by Pamela Goesch, Brynwood House, Sydney, 2010, p. 33.
11. Nowadays Montauban-de-Picardie.
12. 'Bing' Hall is 1667 Leslie John Hall. DOW 2 September 1918.
13. The 14th Brigade relieved the British 21st Brigade (30th Division).
14. 'Passed over our wet blankets' means that they handed their blankets to the quartermaster staff.

15. Sinclair Hunt, 'Overcoming terror and winter hardship on the Somme 1917', *We fought the battles: in letters and diaries written home, World War One soldiers relate their own personal experiences of conflict, frustrations, anxieties and tragedies all perpetrated by war*, compiled and edited by James M. Woolley, Eastwood, 1998, p. 61.
16. Farrall, *The File on Fred*, p. 74.
17. Ellis, *The Story of the Fifth Australian Division*, p. 141.
18. 1626 Charles Stanley Bingley. DOW 2 November 1916.
19. 1693 Samuel McKernan.
20. 2469 David Vass was badly wounded by shrapnel in the eyes, face and leg and contracted tetanus. DOW 3 November 1916.
21. Dalton 'Dally' Thomas Walker Neville.
22. 3144 Denis John O'Dea. He was awarded a Military Medal for this action.
23. 2886 Alexander Connelly. KIA 23 October 1916.
24. 5428 Arthur Passmore. KIA 23 October 1916.
25. 5339 Patrick Barry. KIA 22 October 1916.
26. 838 Andrew Eyles McIlwain and 1189 Joseph Oscar McIlwain, both KIA 22 October 1916.
27. Farrall, *The File on Fred*, p. 81.
28. Ibid.
29. Ibid.
30. Bishop, *The Hell, The Humour, The Heartbreak*, p. 86.
31. Trench foot is caused by prolonged exposure of the feet to damp, unsanitary and cold conditions. The feet were affected by poor vascular supply, occasioned by the wet, cold conditions in the trenches. Trench foot begins with numbness of the feet. The next stage is marked by pain, especially in the toes, accompanied by considerable swelling of the feet. If medical intervention is not obtained before this stage develops, the feet often break out in open sores and become gangrenous, frequently necessitating amputation of the foot or leg.
32. 3604 George Luke Anthony Peters.
33. Fred Farrall in 'Oral interview with Mr F. Farrall by Alistair Thomson dated 7 July 1983', AWM S01311.
34. 1969 Frank Hunt. KIA 1 November 1916.
35. Farrall, *The File on Fred*, pp. 80–81.
36. Ribemont-sur-Ancre.
37. Also known as 'elephant shelters', these structures were made of pre-shaped, curved corrugated iron sheet which was also used to reinforce dugouts.
38. John J. Kennedy, *The Whale Oil Guards*, James Duffy and Co, Dublin, 1919, p. 89.
39. Farrall, *The File on Fred*, p. 89.
40. 4794 Philip De Jongh. DOW 8 December 1916.
41. Farrall, *The File on Fred*, pp. 90–91.

42. 2230 Eric Broomfield Biden.
43. Hunt, 'Overcoming terror and winter hardship on the Somme 1917', p. 63.
44. 5488 Oscar Waldermar Frank Detmers. KIA 5 December 1916.
45. 5388 Roy Edward Hollis.
46. 1889 Leslie Raswell Elton Coleman rejoined the unit after recovering from his wound in May 1917. Coleman was wounded in action on 23 September 1917 at Glencorse Wood (fractured skull) and succumbed to his injuries on 29 September 1917.
47. 3393 Walter Charles Osbourne.
48. 1754 Roy Arnold Wild.
49. 2607 Walter Ravencroft Smith.
50. Private Farrall made it to Montauban Camp, from where he was evacuated with trench foot. He did not rejoin the battalion until 27 August 1917.
51. 2021 Vincent Adrian Irwin.
52. Joseph John Talbot Hobbs.
53. 2208 John Joseph O'Grady.
54. 2605 William Beresford Muir.
55. 2684 William Raymond Harvey. KIA 4 July 1918.
56. The reinforcements included the men who had participated in the 'Men from Snowy River' recruitment march.
57. By this time Lieutenant Colonel McConaghy was suffering from rheumatism and exhaustion. He was invalided to England and spent the majority of 1917 in training appointments. McConaghy returned to France in January 1918 and took command of the 54th Battalion. On 9 April 1918 he was mortally wounded by shellfire just outside Villers-Bretonneux.
58. Woollen leg wrappings worn to stop water and mud from getting into boots and pants.
59. 1517 Edward Alexander Bourke.
60. 2751 Albert Edward Petitt. KIA 3 February 1917.
61. 2215 William Louis Pryce. KIA 3 February 1917.
62. 1904 Percival RoyFenwick. KIA 3 February 1917.
63. 'Tommy Atkins' or 'Tommy' was used by the AIF as a generic name for a British soldier.
64. 2720 William Joseph Quinlivan. KIA 13 February 1917.
65. 2721 John Patrick Quinlivan. KIA 26 September 1917.
66. 1570 Edmund Harrington Street.
67. The 'Kangaroo' recruitment march left Wagga Wagga on 1 December 1915 with 88 volunteers. They marched through the Riverina to Campbelltown and then travelled from Campbelltown into Sydney by train, arriving on 7 January 1916 with somewhere between 210 and 230 recruits.

68. Sandbags were wrapped around the legs between the pants and the boots and secured using laces. They performed an equivalent function to puttees. In extremely muddy conditions, puttees tended to fall down around the wearer's ankles, making them uncomfortable and dysfunctional to wear.
69. David De Venny Hunter. KIA 28 September 1917.
70. 2367 Alexander Rankin Cameron.
71. 1701Henry Anson was Mentioned in Despatches for his patrolling activities during this period.
72. 1921 George Charles Farrell. KIA 9 March 1917.
73. 1550 Wilfred Farrell.
74. 1920 Timothy Oxley Farrell. KIA 22 May 1918.
75. 1908 Frederick Greathead. KIA 8 March 1917.
76. 1907 Henry Percy Greathead. KIA 2 April 1917. Another brother, a member of the 20th Battalion, was killed on 9 October 1917 near Ypres.
77. 2770 Arthur Oakes. KIA 9 March 1917.
78. 2528 Frederick James Schafer. KIA 9 March 1917.
79. 5434 Joseph Augustine Sewell. KIA 9 March 1917.
80. Probably 2713 Henry Price who had joined the unit in February 1917 having just turned 19.
81. Projectiles from the *Granatenwerfer*, the lightest of the German trench mortars used during World War I.
82. Captain Harry Wilson, the Officer Commanding C Company, was on a course and Captain Herbert Palmer had been given temporary command of the company.
83. Chapman's body was discovered on 16 March 1917.
84. Eddie Street mentions that there were eight dead Germans around Chapman.
85. Bean refers to Chapman as 'one of the gentlest of men' (Bean Notebook 243A).

Chapter 6

1. C.E.W. Bean, *Official History of Australia in the war of 1914–1918*, Vol. IV, *The AIF in France 1917*, Angus & Robertson, Sydney, 1941, pp. 210–11.
2. Ellis, *The Story of the Fifth Australian Division*, p. 191.
3. Bean, *Official History*, Vol. IV, p. 224.
4. 2257 John Webb. KIA 2 April 1917.
5. Artillery formation was a dispersed tactical formation used by infantry to minimise casualties from enemy artillery. A platoon moving in artillery formation had its four sections advancing with the men in column, one or two men wide, with the four sections in the shape of a diamond and the platoon commander in the middle.
6. To further add to the confusion surrounding the actual time of H-Hour, the supporting artillery had been informed that H-Hour was 5.15 am.
7. Arthur Alexander Duprez.

8. William Black Stewart Morgan.
9. Charles Russell Woolley. DOW 4 April 1917.
10. 2924 George Alexander Seaman. KIA 2 April 1917.
11. 2786 ApsleyBradford. KIA 2 April 1917.
12. 3285 Archibald Rosborough. He was awarded the Military Medal for his actions.
13. 2633 William Dobson. He was awarded the Military Medal for his actions. KIA 26 September 1917.
14. Wyllie was awarded the Military Cross for his actions.
15. Fire and movement is an infantry tactic useful in both attack and withdrawal. In essence, one group of soldiers suppresses the enemy with fire while another group advances. After a short distance, the advancing soldiers halt and begin firing on the enemy, allowing the group previously involved in suppressing the enemy to advance to a new position. This cycle continues until the enemy is engaged in close-quarter battle.
16. 3078 Arthur Henry Hardy. He was awarded the Military Medal for his actions. He was subsequently wounded in the consolidation of Doignies.
17. C.E.W. Bean, 'Doignies', *The Mercury*, 20 June 1917, p. 6.
18. 3568 Leslie Thomas McEwan.
19. Bean, 'Doignies', p. 6.
20. Stutchbury received a Bar to his Military Cross for his actions.
21. Major Woods received a Distinguished Service Order for bringing 'the interior economy of the Battalion to a high standard' as well as his instrumental part in the planning and conduct of the operation.
22. Bean, *Official History*, Vol. IV, p. 230.
23. Ellis, *The Story of the Fifth Australian Division*, p. 192.
24. 2022 John Henry Kellond. KIA 2 April 1917.
25. 5354 Frank Alfred Catterall.
26. 2917 Edward Claude Rhodes Wolrige.
27. 2913 John Wilfred Buckeridge.
28. David Coombes, *The Lionheart: A Life of Lieutenant-General Sir Talbot Hobbs*, Australian Military History Publications, Sydney, 2007, p. 167.
29. Official records state that approximately 12 German prisoners were captured by the 55th and 56th battalions during the operation. It seems that most of these were taken prisoner by Private O'Grady, who related: 'I completed my run to contact [Captain] Stutchbury collecting about 8 German soldiers who on seeing me there, threw all their guns away including their steel hats, they accompanied me to D Coy.'
30. Elliot to Bean, correspondence dated 2 August 1929 in AWM38 /3DRL, 606/260/1.
31. Ibid.

Chapter 7

1. Nowadays Bécordel-Bécourt.
2. Probably 3041 Frederick James Edward Douglas.
3. Nearly six months' salary.
4. The 55th's nucleus at Bullecourt comprised a senior major; two company commanders; one subaltern per company; two company sergeant majors; one sergeant, one corporal, one lance corporal and three privates per company; two Lewis gunners per section; 12 signallers; 12 runners; five snipers and five scouts.
5. Ellis, *The Story of the Fifth Australian Division*, p. 201.
6. Ibid., p. 202.
7. 'Bosker' means outstandingly good or wonderful.
8. This was 5036 William George Bennett. He was found guilty of self-inflicted wounding and sentenced to 28 days' forfeiture of pay. He deserted from the unit as it moved into Flanders in September 1917. Arrested in March 1918, he was sentenced to five years' in prison. Bennett returned to Australia in 1919.
9. Kennedy, *Whale Oil Guards*, p. 105.
10. Medication used for treating minor headaches and other ailments.
11. Archibald Ramsay Gardner. KIA 9 May 1917.
12. 179 Leslie Clarence Shephard. KIA 9 May 1917.
13. 2919 Leslie James Holloway. KIA 9 May 1917.
14. 2431 Patrick Bertrand Harpur. DOW 12 May 1917.
15. 2618 Arthur Burch. KIA 14 May 1917.
16. 4871 Percy Wallace Peacey.
17. 243 James Groutsch.
18. 2247 Alexander Robert David Thompson.
19. 2143 Ernest Albert Corey.
20. 1673 Leslie Alfred Jackson.
21. 2133 John Joseph Buckley. DOW 3 September 1918. Eddie Street wrote of him: 'Jack Buckley M.M. was shot through the heart as he went forward in that calm determined manner of his. He was a kindly, lion-hearted Lewis gunner whom I thought to have a charmed life.'
22. Darryl Kelly, 'Corporal Ernest Corey, MM: One of a Kind' in *Just Soldiers: Stories of Ordinary Australians Doing Extraordinary Things in Times of War*, ANZAC Day Commemoration Committee Inc., Queensland, 2004.
23. 1123 Clarence Gosling. He was awarded 21 days' forfeiture of pay. He rejoined the unit and was wounded in action in May 1918.
24. 3107 John McDonald Nelson. KIA 20 May 1917.
25. 2001 Private Arthur James Gillett remained at his post for five hours before being evacuated. He was awarded a Military Medal for his actions.

26. James Edward Murray.
27. A popular means used in Australia for a group of people to express disapproval of the actions of a person in authority. It took the form of the group counting '1-2-3-4-5-6-7-8-9' then shouting 'out'.
28. Bean, *Official History*, Vol. IV, p. 683.

Chapter 8

1. Ypres in French; Ieper in Flemish. The French spelling, Ypres, was used by the British and Commonwealth forces during the Great War.
2. A creeping barrage was designed to place a constantly shifting or 'creeping' curtain of artillery fire just ahead of advancing infantry.
3. A stop-butt is an earth mound on a shooting range in which the projectiles safely terminate.
4. Bean, refers to Jetty Trench by its German name – Flandern 1.
5. These guns and howitzers came from the 7th and 8th Field Artillery Brigades (FAB) and had been detached from the 3rd Division (their parent formation) in support of the attack on Polygon Wood.
6. Ninety-four 6-inch howitzers, thirty-two 8-inch howitzers, thirty 9.2-inch howitzers, thirty-six 60-pounders, four 12-inch howitzers, one 15-inch howitzer and eight 6-inch guns.
7. Bishop, *The Hell, The Humour, The Heartbreak*, pp. 107–08.
8. Ibid., p. 108.
9. Ibid.
10. Ellis, *The Story of the Fifth Australian Division*, p. 239.
11. Bishop, *The Hell, The Humour, The Heartbreak*, p. 108.
12. Ibid., pp. 108–09.
13. Bachtold was awarded the Distinguished Service Order for his work in preparing the assembly area for the brigade and later for his efforts during the consolidation phase.
14. 2848 George William McRae. KIA 26 September 1917.
15. Bishop, *The Hell, The Humour, The Heartbreak*, p. 111.
16. 1917 Alfred John Everleigh Carmichael. KIA 26 September 1917.
17. 2523 Varney Eric Stanford Davis. DOW 28 September 1917.
18. Bishop, *The Hell, The Humour, The Heartbreak*, p. 112.
19. Bean, *Official History*, Vol. IV, p. 825.
20. Ellis, *The Story of the Fifth Australian Division*, p. 244.
21. 2142 William Henry Cook. KIA 26 September 1917.
22. Bishop, *The Hell, The Humour, The Heartbreak*, p. 112.
23. Ibid., p. 113.
24. 55th Battalion, *Narrative of the operations of the 55th Battalion AIF against the Enemy defences in the vicinity of Polygone Wood and Jetty Trench, on 26th September 1917 and subsequent days.*

25. 'Squirt' is a slang term for a pistol.
26. 'Na Poo' or 'Narpoo' is slang from the French 'il n'y a plus' meaning 'no good'.
27. 1204 Frederick Alexander Poole. KIA 26 September 1917.
28. Street attributes Poole's actions to a desire to avenge his dead brother, 268 Edwin William Poole, a member of the 55th Battalion who had been killed in action at Doignies on 2 April 1917.
29. 132 Frederick Walter Howell.
30. 2012 Abraham Gorham Smith.
31. 2684 Philip Kelly. KIA 30 September 1918.
32. 3226 William Arthur Stanley Whipp. KIA 26 September 1917.
33. 4851 Harold Mortlock.
34. 3134 Ronald Guy Dowling.
35. So confident was Major General Hobbs that the VC recommendation had gone through that he personally congratulated Slater on receiving the award.
36. C.E.W. Bean, Notebooks, AWM 38 /3DRL 606/178/2, p. 5.
37. Possibly 3119 William Breen. KIA 26 September 1917.
38. Bishop, *The Hell, The Humour, The Heartbreak*, p. 114.
39. 4787 Norman Dark. KIA 26 September 1917.
40. 25655 John Frederick Babbage. He had retained the rank he held when serving in the 5th Divisional Artillery Column. KIA 26 September 1917.
41. Cotterell was awarded the Military Cross for his actions at Polygon Wood.
42. William Frederick Clark. KIA 17 April 1918.
43. The Regimental Sergeant Major (RSM) of the 55th Battalion, 2816 John Herbert Dunne, was awarded the Distinguished Conduct Medal for his actions.
44. 'Bars' are awarded to a bravery medal in recognition of the performance of further acts of gallantry meriting the award.
45. For his inspiring leadership during the consolidation and for his part in establishing the night defensive perimeter, Clark was awarded the Military Cross.
46. 2019 Frank West was awarded a Military Medal for his activities.
47. Bishop, *The Hell, The Humour, The Heartbreak*, pp. 115–16.
48. Ibid., p. 116.

Chapter 9

1. 3383 Stanley Crane. DOW 11 October 1917.
2. 4820 Lawrence Halpin. KIA 11 October 1917.
3. 3243 Thomas Edward Wolfe. KIA 11 October 1917.
4. 871 Leslie James Pardey. DOW 12 October 1917.
5. 1614 Thomas John Bald. DOW 11 October 1917.

6. Bishop, *The Hell, The Humour, The Heartbreak*, pp. 116–17.
7. Ibid., p. 117.
8. 'Drum fire' is a heavy artillery barrage fired not in salvo but by each gun in succession, the resulting sound resembling rapid drumbeats.
9. Bishop, *The Hell, The Humour, The Heartbreak*, pp. 118–19.
10. Gongs and bells were positioned along the front line so sentries could raise the alarm in the event of a gas attack.
11. 3455 Nathaniel Pritchard. KIA 19 October 1917.
12. 3182 Lawrence Rupert Mulvihill. KIA 19 October 1917.
13. 2718 Alexander Masterton. KIA 19 October 1917.
14. 4865 James Lovel Simons. KIA 19 October 1917.
15. 2691 William Guilar Mackenzie. KIA 19 October 1917.

Chapter 10

1. Bishop, *The Hell, The Humour, The Heartbreak*, p. 122.
2. Ibid., pp. 122, 127.
3. 1729 Edward Frances Sharkey. KIA 26 November 1917.
4. 3267 Thomas Watt Martin.
5. 4853 George Henry Mussett.
6. 'Potato Masher' is the slang used for a German hand grenade.
7. Bishop, *The Hell, The Humour, The Heartbreak*, p. 127.
8. Ibid., p. 128.
9. Ibid.
10. Ibid., p. 129.
11. Ibid.
12. Ibid., pp. 133–34.
13. Ibid., p. 136.
14. Ibid., pp. 135–36.
15. A 'trench cooker', also known as a 'Tommy Cooker', was a compact, portable, solidified alcohol-fuelled stove.
16. Anson was awarded the Distinguished Service Medal for his patrolling work at Wytschaete and Villers-Bretonneux.
17. Ellis, *The Story of the Fifth Australian Division*, p. 274.
18. The raiding party consisted of Lieutenant Stanley Colless, 2816 Warrant Officer Class 2 John Herbert Dunne, 1942 Corporal Svend Holger Lund (German speaker), 2012 Private Abraham Gorham Smith, 1364 Lance Corporal George William Hunnikin, 5452 Lance Corporal Albert Harold Turner, 5358 Private William Donald Drew, 2742 Private James Clive Sheils, 5409 Private Harold Edward Miller, 3204 Corporal Archibald Thomas Winter,

1630 Private Leslie William Curran, 2217 Private William Benson Percy Penney (KIA 17 April 1918), 3570 Private Harold Frost, 2001 Private Walter Timson, 3386 Private Joseph Carter, 3394 Private Sydney John Clinch, 2942 Private George Stephen Mathieson, 4745 Private John William Bell (DOW 15 May 1918), 1658 Private George Ireland (DOW 4 July 1918), 2663 Lance Corporal Henry Charles Cyril Green, 2429 Private William James Clifford Hicks, 2241 Private Reuben Smith (KIA 3 March 1918), 2260 Private Charles Victor Arnett and 2123 Private Alexander Samuel Arnett (KIA 27 April 1918).

19. Bourke was awarded the Distinguished Conduct Medal for undertaking reconnaissance activities prior to the raid.
20. Reginald Scott Gardiner.
21. 2260 Charles Victor Arnett.
22. 1630 Leslie William Curran.
23. 5452 Albert Harold Turner.
24. 2241 Reuben Smith. KIA 3 March 1918.
25. 2429 William James Clifford Hicks.
26. Bishop, *The Hell, The Humour, The Heartbreak*, p. 137.
27. Ibid., p. 136.
28. 3163 Robert Keyte. DOW 17 March 1918.
29. 823 Arthur Metcalfe. KIA 16 March 1918.
30. 1967 Alexander Reginald Roddan. DOW 21 March 1918.
31. 1949 Merion Morton Moriarty. KIA 12 March 1918.
32. 3167 James Frederick Kinsela. KIA 19 March 1918.
33. 25530 Rupert Sykes Hinson. He had retained the rank he held when serving in the artillery. KIA 14 March 1918.
34. John Samuel Byrne.
35. Miller received a 'Mention in Corps Orders' for his patrolling work during this period.
36. 3026 Thomas Andrew Buckingham.

Chapter 11

1. James Campbell Stewart.
2. Bishop, *The Hell, The Humour, The Heartbreak*, p. 145.
3. Ibid., pp. 146 –47.
4. Coombes, *The Lionheart*, p. 226.
5. Since 1937, Hill 104 has been the site of the Australian National Memorial recognising the service of the Australian forces in France during the Great War.
6. 1621 Archie Lorne Box. KIA 6 April 1918.
7. One of the killed was 2659 Ernest Sydney Griffiths. 2009 Fred Wiggins, a great friend of both Private Box and Private Griffiths, buried them where they fell.

8. Infested with body lice. 'Chat' was a slang word for a louse.
9. 1708 George McMeekin. KIA 18 April 1918
10. 2632 James Donovan. KIA 14 April 1918.
11. 5215 William Harding. KIA 14 April 1918.
12. 3102 William John Affleck.
13. 3015 James William Burns. DOW 18 April 1918.
14. 2272 RoyWilliam Christie.
15. Arthur Joseph Collins. KIA 16 April 1918.
16. These were: 2619 William Patrick Cosgrave; 845 Wilfred Roland Oliver, KIA 17 April 1918; 2403 George Edward Franklin, KIA 16 April 1918; and 870 John Edward Parryman, KIA 16 April 1918. Killed in the shell-burst that mortally wounded Collins were 1920 Henry Ernest Hauber (Collins' batman), KIA 16 April 1918; 1427 Arthur William Robinson, KIA 16 April 1918; 3755 David Edward Arnold, KIA 16 April 1918; and 5207 Charles William Day, KIA 16 April 1918.
17. 3029 Frederick Lloyd Carpenter.
18. 5418 Christopher McCabe won the Military Medal for his many acts of courage while on patrol during this period.
19. Only 2% of mustard gas casualties died, mainly from secondary infections.
20. 2969 Charles Partridge. KIA 27 April 1918.
21. 1554 George Hughes. KIA 27 April 1918.
22. 'Flare king' is the nickname given to a soldier, typically German, who fires rockets or flares from the front line or an advanced post.
23. Several bravery awards were presented after the raid. Lieutenant Neville was awarded a Military Cross for his leadership. 2184 Private Peter Eugene Monck received the Belgian Croix de Guerre for killing one German in a hand-to-hand struggle and capturing another. 651 Private Peter 'Scotty' McCluskey was awarded a Military Medal when he saved one of his fellow attackers by clubbing a bayonet-wielding German to death with the butt of his rifle.
24. 3474 John Edward Walker.
25. 2170 Joseph Francis Coughlan. KIA 3 July 1918.
26. 2765 Henry Stephen Reidy. DOW 29 June 1918.
27. 2228 Henry Wallace Stewart. KIA 29 June 1918.

Chapter 12

1. One platoon from the 56th also moved into the front line to replace the platoon from C Company tasked with participating in the raid.
2. George Shirley Donaldson.
3. William Thomas Piddington. KIA 4 July 1918.
4. Louis Norman Stafford. KIA 4 July 1918.
5. Norman Lewis Wilson.

6. Norman John Mackay. He was awarded a Military Cross for his bravery under shellfire and compassion in dealing with the wounded over the period he was with the battalion.
7. Bishop, *The Hell, The Humour, The Heartbreak*, pp.181–82.
8. Bean repeated Woods' claim that it was German fire. There is a strong likelihood that Bean was simply using Woods' report as the basis for his own interpretation of events.
9. For example, Corporal George Gill recorded: 'It seems that our own trench mortars caught them owing to some fault in the bedplate' and Brigadier General Stewart wrote in his diary: 'The 55th Coy Comdr who carried out the raid [Wyllie] is very cut up at his casualties and blames our trench mortars.'
10. 1612 Walter Otto Herbert Allen. KIA 4 July 1918.
11. 1665 Ernest Hamilton. KIA 4 July 1918.
12. 2674 William Barker Hart. KIA 4 July 1918.
13. 2415 Alec Gordon Penny. He was awarded the Distinguished Conduct Medal for his actions.
14. 3219 Clarence Eugene Swift.
15. Chadwick was awarded a Military Cross for his leadership during the raid.
16. 2491 John Leonard Hannah. He was awarded the Distinguished Conduct Medal for his actions during the raid. Hannah was slightly wounded while returning to the Australian lines after the raid.
17. The raiders carried trench mortar bombs for this purpose, as hand grenades were not deemed powerful enough by the men to guarantee the destruction of a dugout and any German sheltering in it.
18. 2167 Arthur Samuel Jacobs. He was awarded the Military Medal for his actions.
19. 3129 Isacchar Pearson Higgs. He was awarded the Military Medal for his actions.
20. 5182 Charles Robert Sinclair. KIA 4 July 1918.
21. 3351 Frederick William Doherty. He was Mentioned in Corps Orders for his actions.
22. 3254 Thomas James. He was awarded the Military Medal for his actions.
23. The sergeant referred to by Corporal Gill is 3121 Sergeant William Patrick Ryall, who was awarded a Military Medal for his actions.
24. 2646 Theodore Ambrose Foret. He was awarded the Military Medal for his actions. KIA 1 September 1918.
25. William Ernest Campbell.
26. Bishop, *The Hell, The Humour, The Heartbreak*, p. 180.
27. Lancelot Vicary Horniman. KIA 1 September 1918.
28. Bishop, *The Hell, The Humour, The Heartbreak*, p. 181.
29. Captain Norman Mackay wrote in his RMO's report for June 1918: 'Influenza is very prevalent with high temperature and severe course. This affects general health to some considerable extent.' However, the battalion's rate of hospital admittance for non-battle casualties remained at historical averages, and there are no reports of deaths in the unit caused by influenza.

30. 5330 Ernest Arthur Burnaby. KIA 6 July 1918.
31. Bishop, *The Hell, The Humour, The Heartbreak*, p. 181.
32. Ibid., p. 182. The full name for the village of Mericourt is Méricourt-l'Abbé.
33. The full name of the village is Frémont, Vaux-en-Amiénois.

Chapter 13

1. Bishop, *The Hell, The Humour, The Heartbreak*, pp. 193–94.
2. The full name of the village is Bussy-lès-Daours.
3. Bishop, *The Hell, The Humour, The Heartbreak*, pp. 194–95.
4. In 1974, Warfusée merged with nearby Lamotte en Santerre to give the village its contemporary name of Lamotte-Warfusée.
5. Bishop, *The Hell, The Humour, The Heartbreak*, p. 195.
6. Ibid., pp. 195–96.
7. Archibald John Inglis. KIA 2 September 1918.
8. Bishop, *The Hell, The Humour, The Heartbreak*, p. 200.
9. This division had recently been placed under command of the Australian Corps.
10. Bishop, *The Hell, The Humour, The Heartbreak*, pp. 201–02.
11. Julian Frank de Meyrick.
12. 1704 William Henry McNeich. He was awarded the Military Medal for his actions. KIA 22 August 1918.
13. 2557 Victor Bernard Anchor. KIA 17 August 1918.
14. 2854A John Leslie Pringle. He was awarded the Military Medal for his actions.
15. 5121 Patrick James Kerr. KIA 17 August 1918.
16. 1286A Walter Caldwell. KIA 17 August 1918.
17. 2692 James Morrison.
18. 1667 Leslie John Hall. DOW 2 September 1918.
19. Bishop, *The Hell, The Humour, The Heartbreak*, pp. 207–08.
20. 2535 Ronald Esmond Stewart.
21. 1676 Arthur Andrew Kearns.
22. Kerr's body was never recovered. His wounds were so severe he was likely to have died soon after his capture.
23. 2157 Lyle Ernest Gilbert. KIA 21 August 1918.
24. 2774 Joseph James Tyler. KIA 21 August 1918.
25. 2450 Joseph Mines. DOW 21 August 1918.
26. Bishop, *The Hell, The Humour, The Heartbreak*, p. 209.
27. 3346 James Carson Alcorn. DOW 22 August 1918.
28. 2644 Thomas Samuel Marsden. KIA 22 August 1918.

29. 1924 Arthur Frederick Jennings. KIA 22 August 1918.
30. Clarence Russell Mahoney.
31. 3164 Abraham Keefe. KIA 22 August 1918.
32. 1956 Henry William Nott. KIA 22 August 1918.
33. 2763 Hugh Roberts. KIA 22 August 1918.
34. This attack became known as the Battle of Chuignes.
35. Bishop, *The Hell, The Humour, The Heartbreak*, p. 213.
36. Ibid., p. 214.

Chapter 14

1. The Canal de la Somme is a roughly 200-kilometre-long shipping canal that joins the English Channel at Saint Valery sur Somme to the Canal de Saint-Quentin at St Simon.
2. Nowadays Omiécourt-Lès-Cléry.
3. C.E.W. Bean, *Official History of Australia in the War of 1914–1918*, Vol. VI, *The AIF in France May 1918 – the Armistice*, Angus & Robertson, Sydney 1942, p. 822.
4. Bert Bishop also witnessed the death of these men: 'As we rushed to the bridge a shell exploded before us. Three casualties, the bridge was still there.' Bishop, *The Hell, The Humour, The Heartbreak*, p. 215.
5. This area is known as the Vallée de Berlinval.
6. John Monash, *The Australian Victories in France in 1918,* Angus & Robertson, Melbourne, 1936, p. 166.
7. Bishop, *The Hell, The Humour, The Heartbreak*, pp. 216–17.
8. 2372 Arthur Amos Bellchambers. KIA 1 September 1918.
9. In contemporary sources, Anvil Wood is also called Quinconce Wood after the farm of Le Quinconce situated on the edge of the forest.
10. Probably Johannes Trench.
11. Sidney Pinkstone won the Military Cross for this initiative and for his role in reorganising the battalion in preparation for the assault on 2 September.
12. Bishop, *The Hell, The Humour, The Heartbreak*, pp. 218–19.
13. William Joseph Robert Cheeseman.
14. Renshaw Weyland Hugh King. He survived the injuries he suffered at Péronne.
15. Midford Stanley Hourn. He was awarded the Military Cross for his actions.
16. Allan Raymond Jackson. KIA 30 September 1918.
17. 3619 Malcolm Clyne Robertson. KIA 1 September 1918.
18. 2269 Arthur Washington Walker. KIA 1 September 1918.
19. Bean mentions that there were only enough uninjured troops for the 53rd to form 'one thin line'. *Official History*, Vol. VI, p. 853.
20. 5392 John Lannon. KIA 1 September 1918.

21. 3860 Bertie Marshall. In 1919, Marshall was commissioned to the rank of lieutenant.
22. 1086 William Bourke. He retained the rank he had previously held as a member of the battalion's Transport Section. DOW 2 September 1918.
23. 2205 Donald Henry O'Reilly. Private Street described O'Reilly as 'our cheerful and smiling runner'. KIA 1 September 1918.
24. Street is referring to 2724 Walter Rogers.
25. Ellis, *The Story of the Fifth Division*, p. 351.
26. 2280 Hubert John McDonald.
27. William C. Stegemann, The Snowy River Marchers, their war effort and afterwards, MS50821, Australian War Memorial.
28. 1632 Bruce Lionel Siddons. KIA 1 September 1918.
29. 3803 Edgar Franks. KIA 2 September 1918.
30. Bishop, *The Hell, The Humour, The Heartbreak*, p. 222.
31. Ibid., p. 223.
32. Ibid., p. 224.
33. Ibid.
34. Ibid.
35. The 5th Division report on operations states that fire from the ramparts 'did not seem to be affected by our bombardment, which was still continuing upon it'.
36. 2965 Alexander Ignatius O'Connor. He was recommended for a Victoria Cross for his actions but was awarded the Distinguished Conduct Medal.
37. 2973 Albert James Priddle. DOW 12 September 1918.
38. 1653 William Augustine Flynn. DOW 2 September 1918.
39. 2391 Patrick Joseph Corbett.
40. 1646 Hedley Vincent Webster.
41. 1188 Private Walter McDonald was recommended for the Distinguished Conduct Medal but was awarded the Military Medal.
42. Private Street was also a witness to the amputation of Harlock's arm, subsequently writing a graphic account of the 'surgery' and a macabre incident which followed: 'A very famous stretcher-bearer [Corey] and his mate [Bishop] came across a sergeant of his coy lying with a severely wounded arm. The arm hung by a shred of flesh. Taking a razor from his pocket (it was kept there for such occasions) the bearer severed the piece of flesh and cast the arm to one side. The flow of blood was stopped as far as possible and the stump bandaged after being liberally soaked in iodine. The bearers carried the patient back to the 1st Aid Post. The Sgt requested the famous bearer to do him a favour. "Go back and get my arm," he pleaded, "on the third finger there is a ring my wife gave to me. Save it for me and I'll send you the address of the hospital where they settle me." The bearer assented and went to find the arm. It was still lying where he had thrown it to one side. But there was difficulty in getting the ring off. The bearer brought it along to a support platoon where I happened to be at the time. He carried it like a leg of mutton under his arm, blood still oozing out. "Catch hold

of this end while I pull the ring off" he requested of me, but I really did not fancy it, neither did the others. Nothing daunted he took out the razor and really severed the finger at the appropriate joint and jubilantly slipped the ring off, exclaiming, "I learned the trade killing ration sheep on a station."' Private Corey was awarded a Military Medal for his actions at Péronne; in part his citation reads:'... all were greatly impressed with the manner in which he cut off a sergeant's shattered arm with a razor, dressed the wound and carried him on his shoulder to safety.'

43. Bishop, *The Hell, The Humour, The Heartbreak*, pp. 225–26. Private Street wrote that Sidney Pinkstone 'suffered very severely in civilian life as a result of his awful experiences in France'.
44. 1756 Edward George Woods.
45. 2138 Martin Hilary Clement Cant.
46. 3736 William Neil Christie.
47. Horby Walter Phillips.
48. Ellsmore was awarded the Military Cross for his actions.
49. John Joseph Scanlan.
50. Bishop, *The Hell, The Humour, The Heartbreak*, p. 230.
51. 2698 Sylvester McMahon. He was recommended for the Distinguished Conduct Medal for his actions but was awarded the Military Medal.
52. 3093 Forbes Mackenzie. He was awarded the Distinguished Conduct Medal for his actions.
53. 2704 William Gourley Nicholls.
54. 1690 James Percival Marshall.Born in Gundagai, NSW, hence the nickname 'Gundy'. Maynard, in 55th Battalion *'Kangaroos' and Others,* described Marshall in the following terms: 'There wasn't a great deal of 'Gundy' in regard to size, but he had the heart of a giant.' He returned to Australia a broken man — he died in 1924 of alcoholism.
55. 5067 James Leopold Diversi.
56. 1738 Bert Schwind.
57. 2225 Ronald James Randall. DOW 3 September 1918.
58. Arthur William Bazley, 'Military Medal and Three Bars', *Stand-To*, January–March 1951.
59. Bean, *Official History*, Vol. VI, p. 869.
60. Lieutenant Colonel Woods was awarded a Bar to his Distinguished Service Order for his role in consolidating the line.
61. Unknown officer quoted in Ellis, *The Story of the Fifth Division,* p. 354.
62. 3056 Edmund John Edwards. KIA 2 September 1918.
63. Bishop, *The Hell, The Humour, The Heartbreak*, pp. 230–31.
64. Ibid., pp. 232–33.
65. Alfred Roy Hawkins.
66. C.E.W. Bean, *Anzac to Amiens*, Australian War Memorial, Canberra, 1946, p. 483.
67. Coombes, *The Lionheart*, p. 283.

Chapter 15

1. Formally designated the Le Catelet–Nauroy system.
2. Nowadays Mesnil-Bruntel.
3. William Norman Giblett.
4. Bishop, *The Hell, The Humour, The Heartbreak*, p. 243.
5. Ibid., p. 246.
6. Ibid., pp. 244–45.
7. Reginald Sydney Morris. He was awarded the Military Cross for his actions.
8. Bishop, *The Hell, The Humour, The Heartbreak*, pp. 246–47.
9. It was later discovered the location for Headquarters 15th Brigade provided by Brigadier General Stewart at the conference was incorrect.
10. Given that the 53rd Battalion arrived in the same trench just after the 55th, it is conceivable that the 53rd had followed the 55th's column to the assembly area instead of making its way independently to the start line as ordered.
11. 3105 George William Ham Bush. KIA 30 September 1918.
12. 5331 Albert Leopold Bedson. KIA 30 September 1918.
13. 2848 Edward Holden. KIA 30 September 1918.
14. Reginald Valentine Hill. He was awarded the Distinguished Service Order for his actions.
15. 2731 Eugene Sullivan. KIA 30 September 1918.
16. Jackson had just been commissioned. His brother, 1623 Leslie Jackson, helped bury him.
17. 1695 William McAlister. He was awarded the Military Medal for his actions.
18. Goldrick's citation for the Military Cross has him shouting a more polite 'hurrah'!
19. Bishop, *The Hell, The Humour, The Heartbreak*, pp. 249–50.
20. 3030 George McIntyre Campbell. He was awarded the Military Medal for his actions.
21. Sydnie Lionel Cohen.
22. 1887 Kenneth Arthur Clark. He was awarded the Distinguished Conduct Medal for his actions.
23. Marshall was awarded the Distinguished Conduct Medal for his actions.
24. 3347 Arthur Auckland. He was awarded the Military Medal for his actions.
25. 1681 Henry Charles Klein. He was awarded the Military Medal for his actions.
26. 2974 Michael Henry Cantwell. KIA 30 September 1918.
27. Bishop, *The Hell, The Humour, The Heartbreak*, pp. 250–51.
28. John Bruce Pye.
29. 3452 Thomas O'Rourke. KIA 30 September 1918. He was 35 at the time of his death. Private Street had another anecdote to tell involving O'Rourke after his death: 'The dead diggers were brought back to HQs and there loaded in limbers to be transported to the rear and there buried. As the double limber moved off there was a loud explosion in the pile of

dead in the rear limber … On investigation it was found that a Mills Bomb had exploded. The Armourer-Sgt whom I had seen in the kneeling position had, before kneeling, pulled the pin out of the bomb. In death his grip on the bomb had been tight. The jolting of the limber, and of being dumped into it had caused the bomb to slip out of the dead man's hand. The handle was thus released causing the little demon to burst among the corpses.'

30. 2546 Edward Gray Woodhart. He was awarded the Military Medal for his actions.
31. Armstrong (formerly 3004A) was awarded the Military Cross for his actions.
32. Giblett was awarded a Bar to his Military Cross for his actions.
33. 3014 Harold Francis Barker. KIA 30 September 1918.
34. 3187 Hector Archie MacCallum. KIA 30 September 1918
35. 5357 William Donohue. KIA 30 September 1918
36. 1717 (Edward) John (Frances also Francis) Ryan (enlisted as John Ryan). He was badly traumatised by his wartime experiences and associated alcoholism. After his return to Australia, Ryan drifted around rural NSW and Victoria. He ended up in Melbourne, where he continued to flit between jobs. Destitute, he died of pneumonia in 1941.
37. Stewart was awarded the Military Medal for his actions.
38. Buckingham was awarded the Distinguished Conduct Medal for his actions.
39. 2907 James Thomas Granville (aka James Thomas Grantley). KIA 30 September 1918.
40. 2849 Edwin Neild. He was awarded the Military Medal for his actions. Neild's combat experiences affected him for the rest of his life: on 21 July 1957 his wife, Charlotte, wrote to Base Records in Melbourne 'to ask you to be so good as to forward me the citation for which my late husband … was awarded the Military Medal. He gave me a very slight account of what he did; and was afterwards a sick man.'
41. 1635 Albert Edward Cadd.
42. 2151 Lionel Edward Freebody.
43. 2258 Stanley Darwin Yelds.
44. 3188 John Alec McInnes. He was awarded the Military Medal for his actions.
45. 1168 Andrew Lewis. He was awarded the Military Medal for his actions.
46. Bishop, *The Hell, The Humour, The Heartbreak*, p. 247.
47. 2181Arthur Horace Olive.
48. 1940 Henry Marcus Lemon.
49. 2846 Stanley Clive Murray.
50. After Corey's wounding, 2743 Thomas John Sharpe, a stretcher-bearer, crawled out of the trench and made his way to Captain Goldrick; the Germans spotted the movement and fired towards him. Reaching Goldrick, Sharpe re-dressed the officer's wounded legs then carried him back to the trench. Sharpe was awarded the Military Medal for his actions.
51. Bazley, 'Military Medal and Three Bars'.
52. 3822 Herbert Hodges. He was awarded the Military Medal for his actions.
53. William Dillow (formerly 2625). KIA 1 October 1918.

54. Once again, a work party was procured from the American 119th Infantry Regiment to take rations and ammunition to the front line; this party augmented the organic resupply arrangements of the 53rd and 55th battalions.
55. Bishop, *The Hell, The Humour, The Heartbreak*, p. 252.
56. Peters was awarded a Military Medal for his actions.
57. Bishop, *The Hell, The Humour, The Heartbreak*, p. 255.
58. Ellis, *The Story of the Fifth Australian Division*, p. 380.

Chapter 16

1. Bishop, *The Hell, The Humour, The Heartbreak*, pp. 256–57.
2. Henry Slater wrote of Woods: 'Few Diggers had a more remarkable or brilliant career, than he. His meteoric risk from Sergeant on Gallipoli to Lieut-Colonel, commanding a front-line battalion in France, is an outstanding example of what can be achieved under active service conditions by a man gifted with pluck, military "sense", organizing ability, plenty of common sense and a personality which attracts men to him, and wins their loyalty and admiration.' Henry Ernest Slater, 'Lieut-Col P.W. Woods. DSO (and Bar), M.C.', *Reveille*, 1 February 1933.
3. Bishop, *Dear All; Letters from World War 1*, p. 64.
4. Bishop, *The Hell, The Humour, The Heartbreak*, p. 260.
5. The 56th had previously absorbed the 54th Battalion as part of the AIF's restructuring in October 1918.

Index

INDEX